Biophysics of Brain Imaging

This book takes a multiscale overview and theoretical framework for all existing brain imaging modalities. It also presents technological development for data fusion and underlying biophysical models. The book also provides problems and exercises useful for undergraduate and graduate courses.

Biophysics of Brain Imaging is broken up into 3 parts. Part 1: *Modeling Observations and Microscopic Phenomena* provides the physical and physiological foundation of all brain imaging modalities. Part 2 (Edited): *Measuring and Modeling Emergent Mesoscopic Phenomena* examines recording techniques and biophysical models used today to represent the mesoscale of the brain, the voxel. Part 3: *Hypothesis Testing and Applications* discusses methods to analyze brain imaging data and their applications to evaluate signatures in a variety of diseases, e.g., epilepsy, AD, and migraine.

This book is intended for undergraduate and graduate level students in Computational Neuroscience, Medical Imaging, Electrophysiology, and Brain Data Analysis and Modeling courses.

Figure slides are available to adopting professors.

Jorge Riera obtained a B.S. in Physics at the University of Havana in 1988. During 1995–1998, he was "Junior Associate" of the International Centre for Theoretical Physics, Trieste (Italy), where he completed the required credits for a master's degree in Biophysics. In 1999, he received the Ph.D. degree in Physics from the University of Havana. Part of his Ph.D. thesis was completed at the Cuban Neuroscience Center (CNC) and the Pitié-Salpêtrière Hospital in Paris. Dr. Riera's postdoctoral term was first at the RIKEN Brain Science Institute (Japan) and, after, at NICHe, Tohoku University (Japan). In 2004, he was appointed Assistant Professor at the Tohoku University School of Engineering and promoted to Associate Professor of the Tohoku University School of Medicine in 2006. From 2006 to 2011, his research was funded by three Japanese agencies: Japan Society for the Promotion of Science, Telecommunications Advancement Organization of Japan, and Japan Science and Technology. In 2011, he joined Florida International University (FIU), first as Visiting Professor and later as Associate Professor in the Department of Biomedical Engineering. For the past twelve years he has directed the Neuronal Mass Dynamics (NMD) lab. At FIU, he has also been appointed by the Honor College, the Herbert

Wertheim College of Medicine, and the STEM Transformation Institute. He is currently the Chair of the Biomedical Engineering Department. In 2024, he was elected fellow of the American Institute for Medical and Biological Engineering (AIMBE) for his outstanding contributions in advancing the technology of simultaneous functional magnetic resonance imaging and electroencephalography, from mouse to man.

Biophysics of Brain Imaging

Modeling Observations and Microscopic Phenomena

Jorge Riera

CRC Press
Taylor & Francis Group
Boca Raton London New York

CRC Press is an imprint of the
Taylor & Francis Group, an **informa** business

Designed cover image: Shutterstock ©

First edition published 2026
by CRC Press
2385 NW Executive Center Drive, Suite 320, Boca Raton FL 33431

and by CRC Press
4 Park Square, Milton Park, Abingdon, Oxon, OX14 4RN

CRC Press is an imprint of Taylor & Francis Group, LLC

ISBN: 978-0-367-90149-3 (hbk)
ISBN: 978-1-041-02285-5 (pbk)
ISBN: 978-0-367-90150-9 (ebk)

DOI: 10.1201/9780367901509

Typeset in Times
by Newgen Publishing UK

Contents

Preface

In 1978, a middle school teacher called by everybody "Angelito" gave me a book about basic principles in Physics. There might have been something in me that called his attention. He was not even close to imagining the impact that book would have in my life. That year, he prepared me for contests in Physics at different levels. I was able to compete and advance from the school level to the municipality, and finally to the province. A year later, another teacher at the school, who students nicknamed "Maconcha", prepared me for several contests in Mathematics. It was in that period that a little flame inside me for science and numbers began to grow. In the first year of high school, my teacher Niurka Paz fueled my passion for Biological Sciences. In these initial years of my life, one thing was very clear; I was committed to studying Physics at the university and specialize in Biophysics during my graduate education. What I didn't expect was to encounter a very difficult path with barriers and challenges. Being a First-Generation College Student (FGCS) was the first challenge. I would not be able to successfully navigate the first years of the university without the support and encouraging words from my grandmother Fidelia Rodriguez. During the first year at the University of Havana, I frequently visited the Department of Psychology. It was in that context that I found myself fascinated by Alexander Romanovich Luria's book "The Working Brain: An Introduction to Neurophysiology". This book prompted me to join the INIFUNCE (Institute for Fundamental Brain Research) in Havana, under the supervision of Dr. Jose Luis Hernandez Caceres. In this period, I learned the whole-cell voltage clamp technique and obtained extracellular electrical recording from epileptic rats under the penicillin model. After completing my undergraduate studies in Physics (1988), I joined the Cuban Neuroscience Center (CNC), where I learned Neural Network Theory and EEG Data Analysis under the supervision of Dr. Pedro Valdes Sosa. During this period, I received a scholarship from the International Center for Theoretical Physics (ICTP) to complete my MS degree in Biophysics by attending a series of the Antonio Borsellino College on Biophysics (1990–1992). In 1996, I received a CNRS scholarship to join LENA (Pitié- Salpêtrière Hospital) under the supervision of Dr. Line Garnero. At LENA, I learned about MEG and focused on the electrophysiological forward/inverse problems, which together with my work at the CNC was the final Dissertation topic for my PhD degree (1999). I had the opportunity to join different laboratories during my postdoctoral period, including The Institute of Statistical Mathematics (Tokyo, Japan) under the supervision of Dr. Tohru Ozaki, the RIKEN Brain Science Institute (Wako-Shi, Japan) under the supervision of Dr. Andreas Ioannides and finally, Tohoku University (Sendai, Japan) under the supervision of Dr. Ryuta Kawashima. I became Assistant Professor at Tohoku University School of Engineering (NICHe) in 2004 and was promoted to Associate Professor at the Tohoku University School of Medicine (IDAC) in 2006. From 2006, I created one of the most sophisticated laboratories for brain imaging in the world, i.e., The Neuronal Mass Dynamics Lab, to study the fundamental mechanisms underlying different imaging modalities and the imprints brain diseases leave on them. After a devastating earthquake and tsunami that hit Sendai in March 2011, I moved my laboratory to Florida International University (Miami, USA). This series of books summarizes my experience over these long forty-five years.

Introduction

Brain imaging refers to a set of non-invasive techniques that enable us to visualize the structure and function of the brain. We can classify brain imaging modalities into structural and functional. Structural imaging provides static pictures of brain anatomy describing tissue organization and its chemical composition. Functional imaging measures dynamic processes that reflect brain activity in the forms of neuronal activation with its related changes in blood perfusion and metabolic cost. Understanding the biophysical foundation of brain imaging is key to interpreting the data from these techniques and applying them in both research and clinical settings. The most common brain imaging modalities are associated to three basic physical principles: **1)** the resonance properties of atomic nuclei, **2)** the decaying properties of atomic nucleons, and **3)** the propagative properties of electromagnetic fields inside brain tissues. The magnetic resonance imaging (MRI) modalities, with its structural and functional subtypes, results from the physical principle (1). On the other hand, physical principle (2) underlies two types of imaging modalities: positron emission tomography (PET) and single photon emission tomography (SPECT). Propagation of electromagnetic fields (physical principle 3) inside the brain depends on the photon energy (or wavelength). Computer tomography (CT) uses X-rays, sometimes referred as to Röntgen radiation, with energy ranging between 100 eV to 100 keV (or equivalently 10 nanometer to 10 picometer wavelengths). On the other hand, functional near infrared spectroscopy (fNIRS) used photons with lower energies 1.77 eV – 2.75 eV (i.e., 700 nm – 1000 nm). Neuronal activation causes extracellular electric currents, with typical frequencies below 1000 Hz. As a result of the electric permittivity and conductivity for brain tissues, the electric and magnetic fields are decoupled for frequencies below 1 KHz. In this case, two large-scale electrophysiological techniques can be used to obtain tomographic images of brain neuronal activity: the electroencephalogram (EEG) and magnetoencephalogram (MEG). In Part 1 of this book series, we will present the "macroscopic" physical foundations of these imaging modalities.

Brain imaging relies on the biophysical principles that underlie a variety of brain cells, from neurons and astrocytes to vasculature-related cells. Understanding principles underlying functions and organizations of these brain cells is crucial for the interpretation of these imaging modalities. As a result of ion gradients across their membrane, neurons have an intrinsic potential difference between their cytosol and the extracellular space at rest. This property is a key component for their functioning and for the existence of life in general. During neuronal activation, voltage, and ligand -gated channels in the membrane of neurons open, causing temporary changes in the neuron membrane potentials, and consequently, extracellular electric currents. To reestablish the ionic gradients across their membranes, neurons need to use ATP to fuel active transport mechanisms. The production of ATP requires quick increases in oxygen supply and long-term refilling of glucose reservoirs in the form of glycogen. To that end, neurons send direct and indirect signals to the surrounding brain microvasculature to temporarily change regional blood perfusion. Indirect vasoactive signals are produced by neighboring astrocytes, which under a highly coordinated cellular machinery: **a)** quantify the overall metabolic need, **b)** activate glycogen utilization, and **c)** predict/modify excitatory/inhibitory balance. Changes in blood perfusion results from the integration of vasoactive signals by a complex multicellular system, involving endothelial and smooth muscle cells. In Part 1 of this book series, we will also present biophysical mechanisms at the cellular "microscopic" level by means of which brain cells work and interact.

In summary, the biophysical foundations of brain imaging are rooted in the principles of physics and physiology. Understanding how neuronal activity generates measurable signals, and how these signals are captured and interpreted by imaging modalities, is essential for using these tools effectively in neuroscience and medicine. Each technique offers unique insights, and often multiple methods are combined to gain a fuller picture of brain function and structure. Part 1 of this book series is mostly theoretically oriented with a focus on the macroscopic (observations) and microscopic level (cell physiology). Readers are encouraged to look at another similar book (Toga and Mazziotta, 2002) for interesting discussions and applications of brain imaging modalities. We encourage readers to evaluate a recent platform available for large-scale computer simulations and analyses of different brain imaging modalities (Sim4Life). Part 2 of this series will connect the observational (macroscopic) and cellular (microscopic) level by evaluating the most significant emergent phenomena at the mesoscopic level, i.e., the voxel. Part 3 will present statistical methods and their applications to study brain disorders.

REFERENCES

Sim4Life (ZMT, Zurich Med Tech, https://zmt.swiss/)

Toga AW, Mazziotta JC. *Brain Mapping: The Methods* (2nd Edition). Academy Press, California, 2002.

1 NMR-Based Techniques

1.1 OVERVIEW

Several brain imaging modalities are based on a particular type of nuclear magnetic resonance (NMR) phenomenon for molecules in solutions, consisting in reading certain physical parameters while nuclei in a tissue sample return to a thermodynamic equilibrium after being perturbed with a non-ionizing radiofrequency (RF) stimulus at an appropriate frequency. There are five major measurable NMR parameters, not all discussed in this book: **1)** chemical shift/exchange, **2)** spin-spin splitting, **3)** linewidths, **4)** the nuclear Overhauser effect, and **5)** relaxations. Perturbations of nuclei in NMR are performed by interactions of the RF field with the magnetic moment of the nuclei, which absorb the RF energy at a resonance frequency, called the Larmor frequency. This resonance frequency depends on: **1)** the type of nucleus (e.g., ^{1}H, ^{13}C), **2)** the intensity of the magnetic field, and **3)** the chemical environment around the nucleus. The NMR phenomenon can be explained from either a classic or a quantum mechanics viewpoint. With few exceptions, this chapter focuses the discussion on the NMR characteristics of the hydrogen atom (^{1}H), which is by far the most abundant in biological tissues. Depending on the sequence used to perturb and read energy radiation from the nuclei, NMR can generate different types of brain imaging modalities (i.e., anatomical/diffusional, functional, and spectroscopic). Research about the NMR phenomenon has generated several Nobel prizes. In 1944, Isidor I Rabi (1898–1988) received the prize in Physics for the measurements in 1938 of the nuclear magnetic resonance in molecular beams. The technique was further developed for liquids/solids by Felix Bloch (1905–1983) and Edward M Purcell (1912–1997) in 1946. For this contribution, they received the Nobel prize in Physics in 1952. Richard Ernst introduced the Fourier transform NMR spectroscopy, for which he received the Nobel prize in Chemistry in 1991. Paul Lauterbur (1929–2007) and Peter Mansfield (1933–2017) received the Nobel prize in Physiology or Medicine in 2003 for their technology to perform magnetic resonance imaging (MRI). For detailed descriptions of the mathematics and physical principles underlying the NMR phenomenon, readers are referred to the following books (Abragam, 1961; Ernst et al., 1987; Cavanagh et al., 2007; Brown et al., 2014).

DOI: 10.1201/9780367901509-1

1.2 NMR PRINCIPLES – THE NUCLEAR MAGNETIC MOMENTUM

Nucleons, i.e., protons and neutrons, possess two types of angular momenta: **(i)** the orbital and **(ii)** the spin. The orbital angular momentum $\vec{L} = \vec{r} \times \vec{p}$ is caused by the orbital movement of the nucleons inside the nucleus, which is defined as the vector product between the linear momentum $\vec{p} = -i\hbar\nabla$ and position $\vec{r}$ operators. On the other hand, the spin is an intrinsic property of matter, as it is mass and charge. The spin angular momentum is defined as $\vec{S}$. For simplicity, we assume henceforth the nucleon angular momentum points in the z-direction. The magnitude of the orbital $L = |\vec{L}| = \sqrt{l(l+1)}\hbar$ and spin $S = |\vec{S}| = \sqrt{s(s+1)}\hbar$ angular momenta are quantized, with quantum numbers $l = 0, 1, \cdots$ and $s = 1/2$ (i.e., nucleons are Fermions), respectively. The magnitude $\hbar = h/2\pi$ is the reduced Planck constant ($h = 6.62607015 \times 10^{-34}\, J \cdot s$). Similarly, the z-projections of the nuclear orbital $L_z = m_l\hbar$ and spin $S_z = m_s\hbar$ angular momentum are also quantized, with angular momentum quantum numbers $m_l = -l, -l+1, \cdots, l-1, l$ and $m_s = \pm 1/2$, respectively. Therefore, for nucleons $S = \sqrt{3}/2$ and $S_z = \pm\hbar/2$, respectively. The total angular momentum of the nucleon $\vec{J} = \vec{L} + \vec{S}$ is conserved. The magnitude $J = |\vec{J}| = \sqrt{j(j+1)}\hbar$ and z-projection $J_z = m_j\hbar$ of the total angular momentum of a nucleon are also quantized, with $|l - s| \le j \le |l + s|$ and $-j \le m_j \le j$. For nucleons, $j = n/2$, with $n = 1, 2, \cdots$.

It is convenient to express the magnetic momentum of a nucleon as a function of the nuclear magneton $\mu_N = e\hbar/2m_p = 3.15 \times 10^{-8}\, eV/T$, with $m_p = 1.672622 \times 10^{-24}\, g$ the mass of the proton and e the charge of the electron. By the Wigner-Eckart theorem, the orbital and spin magnetic moments of the nucleon are defined as $\vec{\mu}_L = \mu_N g_L \vec{L}/\hbar$ and $\vec{\mu}_S = \mu_N g_S \vec{S}/\hbar$, respectively. The total magnetic moment of nucleons is defined by the equation (1.1).

$$\vec{\mu} = \vec{\mu}_L + \vec{\mu}_S = \frac{\mu_N}{\hbar}\left(g_L\vec{L} + g_S\vec{S}\right) \tag{1.1}$$

Under the Schmidt limits, $g_L = 1$ and $g_S = 5.586$ for protons. Trivially, $g_L = 0$ for neutrons. However, the neutron is composed by 1 *u-quark* ($+2e/3$) and 2 *d-quarks* ($-1e/3$) distributed asymmetrically. Hence, although it has zero net charge, its spin magnetic moment is not zero, i.e., $g_S = -3.826$. By definition, $\vec{\mu}$ and $\vec{J}$ are not parallel. However, the angle formed by these two vectors $\vec{\mu} \cdot \vec{J}$ is conserved. Therefore, it is demonstrated that $\mu_z = g_J \mu_N m_j$, with the Landé factor g_J defined by the equation (1.2).

$$g_J = \frac{g_L\left(l(l+1) + j(j+1) - s(s+1)\right) + g_S\left(s(s+1) + j(j+1) - l(l+1)\right)}{2j(j+1)} \tag{1.2}$$

For nucleons ($s = 1/2$), the following simplification can be made:

$$g_J = g_L \pm \frac{g_S - g_L}{2l+1} \tag{1.3}$$

The total angular momentum $\vec{I}$ for a nucleus ${}^A_Z X$ with Z protons and N neutrons is also a quantized physical magnitude, i.e., $I = |\vec{I}| = \sqrt{i(i+1)}\hbar$ and $I_z = m_i\hbar$, with $-i \le m_i \le i$

for a given total angular momentum quantum number i of the nucleus. Under this notation, $A = Z + N$ represents the total number of nucleons (i.e., the mass number). Z is called the atomic number. Note that A or Z are sometimes omitted in the notation for any element X. Unfortunately, calculating i is not a trivial task. The Pauli exclusion principle suggests that all nucleons move freely within nucleus in the ground state. Hence, a spherically symmetric central potential (e.g., the Saxon-Woods potential) could be used to approximately represent the movement of nucleons within the nucleus. To reproduce the observed magic numbers for nucleus, spin-orbit interactions must be included in the central potential framework. The extreme limit of the nuclear shell model must be used to calculate i for particular nuclei. The model proposes energy levels formation according to two principles: **1)** quantum energy state of potential wells and **2)** splitting due to spin-orbit coupling. The first three energy levels suffice to describe the most useful nuclei in NMR. From high to low energy, the order of the first three energy levels is: 1st level ($l = 0$, 2 states of $i = 1/2$, $p = +1$), 2nd level ($l = 1$, 4 states of $i = 3/2$ and 2 states of $i = 1/2$, $p = -1$), and 3rd level ($l = 2$, 6 states of $i = 5/2$, 2 states of $i = 1/2$, and 4 states of $i = 3/2$, $p = +1$). The parity, defined as $p = (-1)^l$, is another important property of each energy level (or shell). The parity represents an inversion of sign for all spatial coordinates. The Pauli exclusion principle operates independently for protons and neutrons; hence, they are independently distributed in these split energy levels from low- to high-energy states. This model can be used as a first approximation to estimate total nuclear angular momentum. Let us apply it to calculate the total angular momentum of the stable ^{13}C nucleus, for which $Z = 6$ and $N = 7$. Protons contribute with a zero net angular momentum due to pairing in their nuclear shells. Two neutrons will be paired in the 1st level. In the 2nd level, the two paired neutrons will be allocated in the lowest energy level ($i = 3/2$) and the unpaired one will be in the highest energy level ($i = 1/2$). Therefore, the total nuclear angular momentum quantum number for ^{13}C is $i = 1/2$, with $p = -1$ for a final magnitude $I = \sqrt{3}\hbar/2$ and z-projection $I_z = \pm\hbar/2$. The same strategy can be used to determine the total angular momenta for other important stable nuclei (^{1}H, ^{17}O, ^{31}P, and ^{15}N) in NMR (Table 1.1). Regrettably, the nuclear shell model fails to explain the observed total nuclear angular momentum for several other nuclei used in NMR (e.g., ^{19}F and ^{23}Na). For example, for the ^{23}Na nucleus, which satisfies $Z = 11$ and $N = 12$, the net angular momentum quantum number from the model should be $i = 5/2$. This is because eight out of the eleven protons distribute in the 1st level (2) and 2nd level (6), for a magic number 8. The remaining three protons will be distributed in the lowest energy state ($i = 5/2$) of the 3rd level, with two of them paired and one non-paired for a final $i = 5/2$. The contribution from the twelve neutrons is zero. However, the observed $i = 3/2$ disagrees with the model prediction for ^{23}Na. Different factors might lie beneath these failures, e.g., the model does not take into account the proton-neutron coupling for ^{19}F and it assumes central potentials with spherical symmetry for ^{23}Na. In general, it can be stated that only nuclei containing odd numbers of nucleons (A) could have a net nuclear angular momentum.

The magnetic moment of the nucleus can be calculated using the approximation (1.3) for the unpaired proton/neutron. However, it is convenient to express it as a function of the gyromagnetic ratio $\gamma = g_J e/2m$ for each nucleus, with m the mass of the unpaired nucleon.

$$\vec{\mu} = \gamma \vec{I} \tag{1.4}$$

TABLE 1.1
List of useful stable isotope nuclei in NMR

Nucleus	Z	N	i	$2I/\hbar$	p	$I_z/\hbar$	$\gamma/2\pi$ (MHz/T)
Predicted by the Nuclear Shell Model							
^1H	1	0	1/2	$\sqrt{3}$	+1	$\pm 1/2$	42.577
^{13}C	6	7	1/2	$\sqrt{3}$	−1	$\pm 1/2$	10.708
^{17}O	8	9	5/2	$\sqrt{35}$	+1	$\pm 1/2, \pm 3/2, \pm 5/2$	≈ -5.772
^{31}P	15	16	1/2	$\sqrt{3}$	+1	$\pm 1/2$	17.235
^{15}N	7	8	1/2	$\sqrt{3}$	−1	$\pm 1/2$	−4.316
Not Predicted by the Nuclear Shell Model							
^{19}F	9	10	1/2	$\sqrt{3}$	+1	$\pm 1/2$	40.052
^{23}Na	11	12	3/2	$\sqrt{15}$	+1	$\pm 1/2, \pm 3/2,$	11.262

Notes: Acronyms: Z- Atomic number; N- Neutron number. The sign of the gyromagnetic ratio defines the direction of spin precession ($\gamma > 0$ – clockwise and $\gamma < 0$ – counterclockwise)

To illustrate the procedure, let us calculate the gyromagnetic ratio for ^1H and ^{17}O. For the ^1H nucleus, $Z = 1$ and $N = 0$, respectively. The only (unpaired) proton is in the 1st energy level (i.e., $l = 0$). Therefore, the Landé factor $g_J = 5.586$, resulting in a gyromagnetic ration for ^1H of $\gamma / 2\pi = 42.577$ MHz / T. For the ^{17}O nucleus, $Z = 8$ and $N = 9$. The unpaired neutron is the 3rd energy level ($l = 2$), for a $g_J \approx -0.765$. The gyromagnetic ratio for ^{17}O is $\gamma / 2\pi \approx -5.772$ MHz / T. Table 1.1 summarizes the gyromagnetic ratios for most useful NMR nuclei.

1.3 T_1/T_2 RELAXATIONS FOR BRAIN TISSUES – BLOCH EQUATIONS

The ^1H atom is the simplest and more commonly used in NMR. Its nucleus consists of a single proton orbited by an electron. Because of the higher mass of the proton compared to that of the electron (m_e) ($m_p / m_e \cong 1836$), the gyromagnetic ratio of the latter is ~ 658 times higher than that of the proton. Therefore, excitation of the electron will require RF pulses in the microwave range, which could be hazardous to living organisms. The nuclear orbital momentum is $\vec{L} = 0$; hence, there is no spin-orbit coupling for that proton. As the nucleus does not have neutrons, proton-neutron coupling is not a concern, and the central potential has a spherical symmetry. Therefore, the use of the ^1H nucleus is referred as proton NMR and $\vec{\mu}$ is called the spin magnetic momentum. In this subsection, general concepts are illustrated for the ^1H nucleus. Any generalization to other nuclei requires more complex mathematical constructs. For simplicity, any static field is applied in the z-direction $B\hat{z}$ and RF pulses to excite the nuclei have components only in the xy plane.

For any nucleus, its net nuclear magnetic moment interacts with an externally applied magnetic field $\vec{B}(\vec{r},t)$. After exposing the spins to a constant magnetic field in the z-direction $B_z\hat{z}$, exchange interactions of them with the surrounding lattice at temperature T reach an equilibrium, for which quantized energy states ε_{m_i} above the ground state must follow a Fermi-Dirac distribution as nucleons have spin $1/2$. However, if the nucleons interact weakly, the probability of the quantized energy states could be accurately represented by the normalized Boltzmann factor: $P\left(\epsilon_{m_i}\right) = e^{-\epsilon_{m_i}/kT}/Z$, with the partition function $Z = \sum_{m_i} e^{-\epsilon_{m_i}/kT}$. The Boltzmann constant is defined as $k = 1.380649 \times 10^{-23} J/K$, and T stands for the absolute temperature (K). Based on the definition of the potential energy $U = -\vec{\mu} \cdot \vec{B} = -\mu_z B_z$, the interaction potential for the nuclear magnetic moment in the m_i-th state can be expressed as $\epsilon_{m_i} = -\gamma m_i \hbar B_z$, which represents quantized energy states according to the quantum number m_i associated with each quantized direction of the nuclear angular momentum I_z. The energy difference between two consecutive states $\Delta\epsilon_{m_i, m_i'} = (m_i - m_i')\gamma\hbar B_z$ is also quantized. By definition, the state m_i has a higher interaction potential than m_i'. Equivalently, the z-components of the nuclear magnetic moments corresponding to each energy state distribute also from those having a low "ground" energy state ($m_i = i$) to those with highest energy state ($m_i = -i$), with the number of spin in m_i energy state defined as $N_{m_i} = \rho P(\epsilon_{m_i})$. For example, the two stable values for the ^{1}H nuclear magnetic moments in this situation are $\mu_z = \pm\gamma\hbar/2$. The energy difference between the two stable states (low-energy "*up*", $m_i = 1/2$; high-energy "*down*", $m_i' = -1/2$) is $\Delta\epsilon_{-\frac{1}{2},\frac{1}{2}} = \gamma\hbar B$. At the equilibrium distribution, the splitting of spins satisfies $N_+/N_- = e^{-\gamma\hbar B_z/kT}$, with $N_\pm$ standing for the number of spins in the $m_i = \pm 1/2$ states, respectively. At body temperature ($T = 310\ K$) in a typical MRI scanner field of $B_z = B_0 = 1.5\ T$, the splitting radio is $N_-/N_+ = 0.999990112756669$, for an excess of 9.8872×10^{-6} spins in the low-energy state. In contrast, there are six stable values for the [17]O nuclear magnetic moments: $\mu_z = \pm\gamma\hbar/2, \pm\gamma 3\hbar/2, \pm\gamma 5\hbar/2$, which distribute according to $P(\epsilon_{m_i})$.

The density $\rho = N/V$ of unpaired nucleons contributing to μ_z in a volume V, with N being the total number of these unpaired nucleons in that volume. Therefore, an excess of spins takes place in the z-direction at the thermodynamic equilibrium, which causes a net magnetization M_0 along the z-direction:

$$M_0 = \rho \sum_{m_i=-i}^{i} P(\epsilon_{m_i})\mu_z(m_i) \cong \rho \frac{i(i+1)(\gamma\hbar)^2}{3kT} B_z \tag{1.5}$$

For the ^{1}H nucleus ($i = 1/2$), $M_0 = \rho(\gamma\hbar)^2 B_z/4kT$. Expression (1.5) constitutes a quantum mechanics derivation of the phenomenological Curie law (Pierre Curie, 1859–1906) for the dependency of magnetization with the temperature and magnetic field strength $M_0 \sim B_z/T$.

As a result of torque $\vec{\tau} = \vec{\mu} \times B_z\hat{z}$, the nuclear magnetic moment precesses around the magnetic field direction at the Larmor frequency ($\omega_L = \gamma B_z$). If the nuclear magnetic moments are exposed to an RF stimulation with energy $\Delta E = \hbar\omega$ matching the quantized energy differences (i.e., $\omega = \omega_L$), nuclear magnetic moments in the states with lower interaction potential will absorb that quantum energy as they jump to the states with the higher interaction potential; therefore, absorption of an RF energy occurs on resonance. The $\vec{\mu}$ precession along axis z must satisfies $d\vec{\mu}/dt = \gamma\vec{\tau}$. From the quantum mechanics viewpoint, it is appropriate to use the expected value $\langle\vec{\mu}\rangle = \langle\psi|\vec{\mu}|\psi\rangle$ of the nuclear magnetic moments inside volume V, with $\psi(\vec{r},t)$ the linear solution of the Schrödinger equation $H\psi(\vec{r},t) = i\hbar\partial\psi(\vec{r},t)/\partial t$. If the Hamiltonian operator is defined as $H = -\hbar^2\nabla^2/2m_p - \vec{\mu}\cdot\vec{B}$, the solution for the ^{1}H nucleus takes the form $\psi(\vec{r},t) = C_{1/2}\psi_{1/2}e^{-i\epsilon_{1/2}t/\hbar} + C_{-1/2}\psi_{-1/2}(\vec{r})e^{-i\epsilon_{-1/2}t/\hbar}$. Note that for this particular case, symbol $i = \sqrt{-1}$ represents the imaginary number. The wave function $\psi(\vec{r},t)$ does not depend explicitly on $\vec{r}$ because of the absence of orbital angular momentum for the proton in the ^{1}H nucleus. Then, it can be demonstrated that $\langle\vec{\mu}\rangle$ satisfies the equations (1.6).

$$\langle\mu_x\rangle = \frac{\gamma\hbar}{2}sin(\theta)cos(\phi_0 - \omega_L t)$$
$$\langle\mu_y\rangle = \frac{\gamma\hbar}{2}sin(\theta)sin(\phi_0 - \omega_L t) \qquad (1.6)$$
$$\langle\mu_z\rangle = \frac{\gamma\hbar}{2}cos(\theta)$$

Parameters $\theta = a_+/a_-$ and $\phi_0 = \alpha_- - \alpha_+$ are defined as functions of the radial $a_\pm$ and angular $\alpha_\pm$ polar coordinates of the coefficients $C_{\pm1/2} = a_\pm e^{i\alpha_\pm}$, the latter to be determined by the initial conditions. Therefore, the ^{1}H nuclear magnetic moments precess at the Larmor frequency of $\omega_L = 63.86\ MHz$ (Figure 1.1). An RF stimulation with exact energy $\Delta E = \hbar\omega_L = 0.6735\times10^{-26}\,J$ will make nuclear magnetic moments in the lower energy "up" state to jump to the down state.

The rotating reference system – The introduction of an x-y rotating reference system is an appropriate mathematical framework to model RF stimulation. Assuming a new reference system $x'y'$ is rotating clockwise at an angular velocity $\vec{\Omega} = -\gamma B_r\hat{z} = -\omega_r\hat{z}$ respect to the fixed laboratory reference system xy. (Figure 1.2), the nuclear magnetic moments on that new reference system must satisfy: $(d\vec{\mu}/dt)' = \gamma\vec{\mu}\times\vec{B}_{eff}$, with $\vec{B}_{eff} = (B_z - B_r)\hat{z}$. Vector $\vec{\mu}$ is constant for the particular case that $\omega_L = \omega_r$.

The application of a *left-circulating polarized* RF magnetic field $\vec{B}_1(t) = B_1(\hat{x}\cos(\omega_1 t) + \hat{y}\sin(\omega_1 t))$ is represented in the $x'y'$ rotating reference system by a simple expression: $\vec{B}_1 = B_1\hat{x}'$. Therefore, the nuclear magnetic moment $\vec{\mu}$ performs a precession around a more general effective magnetic field (Figure 1.3); i.e., $(d\vec{\mu}/dt)' = \gamma\vec{\mu}\times\vec{B}_{eff}$.

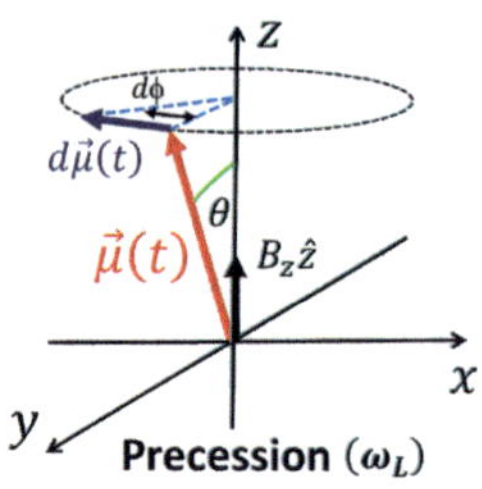

FIGURE 1.1 Illustration of the precession of the nuclear magnetic moment around the $\hat{z}$ axis.

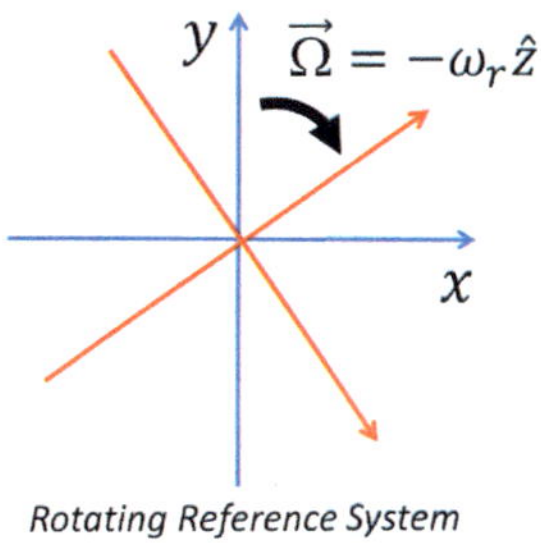

FIGURE 1.2 Definition of the rotating reference system.

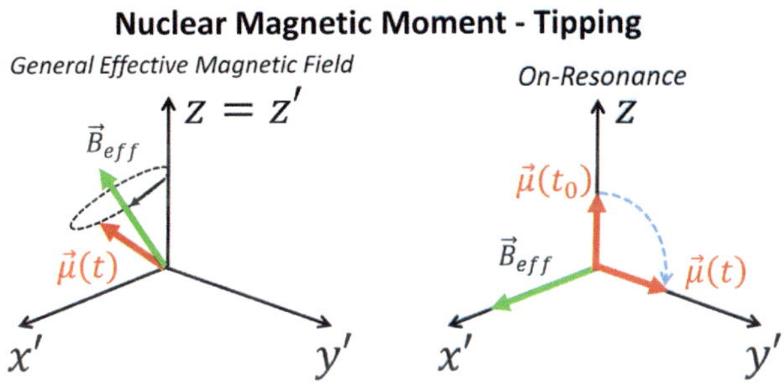

FIGURE 1.3 The nuclear magnetic moment $\vec{\mu}$ – tipping around a general effective magnetic field (left) and on-resonance (right).

$$\vec{B}_{eff} = \left((\omega_L - \omega_r)\hat{z}' + \omega_1 \hat{x}' \right) / \gamma \tag{1.7}$$

Note that $\hat{z}' = \hat{z}$. Selecting the angular frequency of the $x'y'$ rotating reference system as the Larmor frequency for perfectly static uniform B_z; i.e., $\omega_r = \omega_L$, a simplified equation $(d\vec{\mu} / dt)' = \omega_1 \vec{\mu} \times \hat{x}'$ is obtained for the moment tipping as a result of the polarized magnetic field. This particular condition is called on-resonance, a property underlying the NMR acronym. The solution of this equation in the $x'y'$ rotating reference system is:

$$\begin{aligned} \langle \mu_{x'} \rangle &= \langle \mu_{x'}(0) \rangle \\ \langle \mu_{y'} \rangle &= \langle \mu_{y'}(0) \rangle cos(\omega_1 t) + \langle \mu_z(0) \rangle sin(\omega_1 t) \\ \langle \mu_z \rangle &= -\langle \mu_{y'}(0) \rangle sin(\omega_1 t) + \langle \mu_z(0) \rangle cos(\omega_1 t) \end{aligned} \tag{1.8}$$

$\langle \vec{\mu}(0) \rangle$ represents the expected value of the initial condition of the nuclear magnetic moment.

Under the on-resonance condition, the polarized magnetic field $\vec{B}_1(t)$ can produce an optimal tipping of $\vec{\mu}$ around the $\hat{x}'$ axis (Figure 1.3). A RF pulse is the application of $\vec{B}_1(t)$ on-resonance for a finite time τ, for which $\vec{\mu}$ rotates by an angle of $\Delta\theta = \gamma B_1 \tau$. For the ^{1}H nucleus, 90° RF pulse is achieved by applying a polarized magnetic field of amplitude $B_1 = 18.45\mu T$ for $2ms$. If a 180° pulse is desirable using the same τ, the amplitude must be twice larger, $B_1 = 36.89\mu T$. From a quantum mechanics perspective, flipping $\vec{\mu}$ a 90° angle is equivalent to transfer exceeded spins in the up-state to the down-state. A 180° pulse will flip all proton spins to the down-state.

The MRI technique comprises a variety of imaging modalities all based on NMR principles. The "voxel" constitutes the elemental imaging unit in the 3D spaces. Therefore, it is useful to introduce the concept of bulk magnetization $\vec{M} = \sum_{i \in V} \vec{\mu}_i / V$ in a voxel as the sum of the nuclear magnetic moments of a similar chemical species in it, which is normalized by the volume of the voxel V. Consequently, this bulk magnetization must follow the equation:

$$d\vec{M} / dt = \gamma \vec{M} \times (B_z \hat{z} + \vec{B}_1(t)) \tag{1.9}$$

However, the bulk magnetization $\vec{M}$ decays because of interactions of the nuclear magnetic moments with: **a)** the surrounding lattice and **b)** with themselves. The first relaxation has a rate $R_1 = 1/T_1$ and occurs as any nuclear magnetic moment tends to align with the magnetic field to minimize the internal energy $U = -\vec{\mu} \cdot \vec{B}$. The T_1 value represents the time it takes the longitudinal (z-axis) component of $\vec{M}$ to decay ~ 63 % toward its equilibrium value M_0 (1.5) after a 90° RF pulse (Figure 1.4, top). The second relaxation has a rate $R_2 = 1/T_2$ and occurs as precessing nuclear magnetic moments interact with each other causing variations in the precession frequency. These variations produce dephasing; hence a decay of the magnetization in the xy plane. Similarly, the T_2 value represents the time it takes the transversal (xy plane) component of $\vec{M}$ to decay ~ 63 % toward zero after a 90° pulse (Figure 1.4, bottom). Therefore, dynamic equation for the evolution of $\vec{M}$ must also include these two relaxation mechanisms (1.10).

$$\frac{d\vec{M}}{dt} = \gamma \vec{M} \times (B_z \hat{z} + \vec{B}_1(t)) + \frac{(M_0 - M_z)}{T_1} - \frac{\vec{M}_\perp}{T_2} \tag{1.10}$$

The phenomenological relationship (1.10) is known as the Bloch equation. In the rotating reference system, equations (1.10) take the simplified form of equations (1.11).

$$\begin{aligned} \left(\frac{dM_{x'}}{dt}\right)' &= (\omega_L - \omega_r) M_{y'} - \frac{M_{x'}}{T_2} \\ \left(\frac{dM_{y'}}{dt}\right)' &= -(\omega_L - \omega_r) M_{x'} + \omega_1 M_z - \frac{M_{y'}}{T_2} \\ \left(\frac{dM_z}{dt}\right)' &= -\omega_1 M_{y'} + \frac{(M_0 - M_z)}{T_1} \end{aligned} \tag{1.11}$$

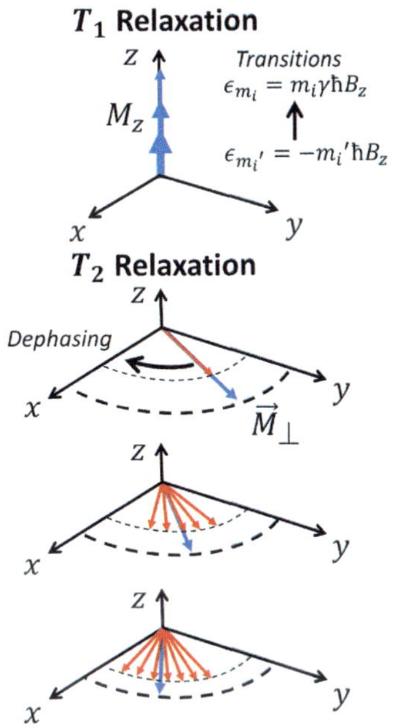

FIGURE 1.4 Representation of the T_1 and T_2 relaxations in the laboratory reference system.

Typically, the polarizing RF pulse $\vec{B}_1(t)$ is applied for a very short period (ms), which defines the flipping angle $\Delta\theta$. Hence, the explicit solutions in the laboratory reference system for the tangential x-y (1.12) and longitudinal z (1.13) components of $\vec{M}$, which is decaying in the presence of a static magnetic field are very useful. Note that at the thermodynamic equilibrium, the randomness of the $\langle\vec{\mu}\rangle$ phase (1.6) implies $\vec{M}_\perp = 0$.

$$\vec{M}_\perp(t) = \begin{bmatrix} M_x(t) \\ M_y(t) \end{bmatrix} = e^{-(t-t_0)/T_2} \begin{pmatrix} M_x(t_0)cos(\omega_L t) + M_y(t_0)sin(\omega_L t) \\ -M_x(t_0)sin(\omega_L t) + M_y(t_0)cos(\omega_L t) \end{pmatrix} \quad (1.12)$$

$$M_z(t) = M_z(t_0)e^{-(t-t_0)/T_1} + M_0\left(1 - e^{-(t-t_0)/T_1}\right) \quad (1.13)$$

Expression (1.12) can be rewritten in the complex form as: $\vec{M}_\perp(t) = e^{-(t-t_0)/T_2}e^{-i(\omega_L t-\phi_0)}\left|\vec{M}_\perp(t_0)\right|$, with ϕ_0 the phase at t_0. The initial magnetization vector $\vec{M}(t_0)$ is determined by the applied flipping angle. For 90° and 180° pulses, $\vec{M}(t_0) = [\vec{M}_\perp(t_0), 0]$ and $\vec{M}(t_0) = [0, 0, -M_0]$, respectively. For all applications in this book, the polarization fields are perfect RF pulses (i.e., turn on/off instantaneously, and are constant) and the interest is only on the evolution of $\vec{M}$ after the polarization field ceases.

The existence of time-varying inhomogeneities in the magnetic field $\Delta B(t)$ in the z-axis causes variations in the precession frequencies of the nuclear magnetic moments $\Delta\omega(t) = \gamma\Delta B(t)$, and therefore, a phase accumulation effect $\phi_A(t) = -\int_{t_0}^{t} \Delta\omega(t')dt'$. In

TABLE 1.2
The T_1 (spin-lattice) and T_2 (spin-spin) relaxations for the ^{1}H nucleus in five important human brain tissues/fluids

Brain Tissues/Fluids	T_1 Relaxation (*ms*)	T_2 Relaxation (*ms*)
Gray Matter	950	100
White Matter	600	80
Cerebral Spinal Fluid	4500	2200
Arterial Blood	1200	200
Venous Blood	1200	100

this case, the transversal component $\vec{M}_\perp(t)$ is described by a more general equation: $\vec{M}_\perp(t) = e^{-(t-t_0)/T_2} e^{-i(\omega_L t - \phi_A(t) - \phi_0)} \left|\vec{M}_\perp(t_0)\right|$. Inhomogeneities in the magnetic field are of two types: induced ΔB_G and intrinsic ΔB_I.

The T_1 and T_2 relaxations depend on the brain tissue properties. Table 1.2 shows these relaxations in five important human brain tissues/fluids for the ^{1}H nucleus.

1.4 NMR SIGNAL MEASUREMENTS

To perform MRI, scanners of different types and strengths are used. The strength of a scanner is defined by the constant magnetic field B_0 inside the imaging region, given in Tesla ($1T = 10^4$ *Gauss*). MRI scanners with $B_0 < 0.3\ T$, $0.3 \le B_0 < 1.0\ T$ and $1.0 \le B_0$ are considered of low, middle, and high field, respectively. There are three major types of MRI scanners: **a)** open bore, **b)** dipolar electromagnetic and **c)** close bore (cylindrical). In general, open bore scanners uses permanent magnet in a C-shape (or horseshoe) configuration and they operate at a field strength ranging from $0.2T$ to $1.0T$. Dipolar configuration scanners use opposite resistive/superconductive coils to generate a B_0 in the range of $0.5-1.2T$. This book focuses only on principles and methods for data collected with high-field close bore MRI scanners (Figure 1.5A), which generate a stable and homogeneous B_0 using a superconducting solenoidal design (i.e., the magnet) with coils bathed in liquid helium. These MRI scanners also use two other types of coils: **a)** RF coils (Figure 1.5B) and **b)** gradient coils (Figure 1.5C).

RF coils are used to excite the nuclei with a polarized RF pulses and to read the energy radiated by them as the transversal magnetization $\vec{M}_\perp$ decay to zero. The radiated energy is observed in the form of changes in the magnetic flux $\Phi_M = \oint_c d\vec{l} \cdot \vec{A}$ through the RF coil (c) due to a time-varying magnetic field $\vec{B} = \nabla \times \vec{A}$. A decaying $\vec{M}$ causes a magnetization current in the brain tissue $\vec{J}_M = \nabla \times \vec{M}$. The magnetic potential $\vec{A}$ is related to $\vec{J}_M$ via the Ampère's law (1.14, André-Marie Ampère, 1775–1836) under the Lorentz gauge $\nabla \cdot \vec{A} + c^{-2}\partial\phi / \partial t = 0$ (Ludvig Lorenz, 1829–1891). The symbol ϕ stands for the electric potential.

$$\vec{A}(\vec{r},t) = \frac{\mu_0}{4\pi} \iiint_V dV' \frac{\vec{J}_M(\vec{r}\,',t)}{\left|\vec{r}-\vec{r}\,'\right|} \tag{1.14}$$

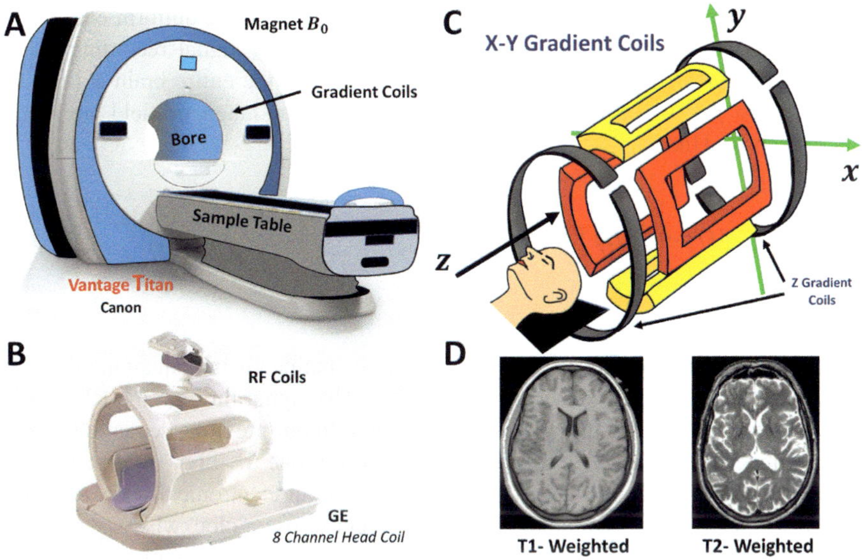

FIGURE 1.5 A. Drawing of a 1.5 T close bore MRI scanner (Vantage Titan, Canon). B. A picture of the 8-channels head coils (GE). C. Illustration of the three gradient coils used in MRI. D. T_1 and T_2 – weighted images from two different subjects.

$\mu_0 = 1.256637 \times 10^{-6} H/m$ is the magnetic permeability of free space and $c = 299792458 m/s$ the speed of light in the vacuum. By applying the reciprocity theorem for magnetic fields, it can be demonstrated that the signal $s = -d\Phi_M / dt$ detected by a RF coil from an inhomogeneous brain $T_2(\vec{r})$ inside an MRI scanner, after a RF stimulation ending at time t_0, and in the presence of a spatially dependent magnetic field, is given by equation (1.15).

$$s(t) \sim \iiint_V dV(\omega_L + \gamma\Delta B_I(\vec{r}))e^{-t/T_2(\vec{r})}B_c(\vec{r})\left|\vec{M}_\perp(\vec{r},t_0)\right|e^{i((\omega_r-\omega_L)t+\phi_A(\vec{r},t)+\phi_0(\vec{r})+\theta_c(\vec{r}))} \tag{1.15}$$

The phase accumulation is defined as $\phi_A(\vec{r},t) = -\int_{t_0}^{t} \Delta\omega(\vec{r},t')dt' = -\gamma\Delta B_I(\vec{r})(t-t_0) - \vec{k}_{xy}(t)\cdot\vec{r}_{xy}$. Here, it has been considered that induced variations in the magnetic field are typically caused by linear gradients in the xy plane $\Delta B_G(t) = \vec{G}_{xy}(t)\cdot\vec{r}_{xy} = G_x(t)x + G_y(t)y$ useful for the xy image encoding in the spatial frequency domain $\vec{k}_{xy}(t) = \gamma\int_{t_0}^{t}\vec{G}_{xy}(t')dt'$ discussed later. These gradients switch on/off depending on the strategy used to obtain an image. The field amplitude $B_c(\vec{r})$ and angle $\theta_c(\vec{r})$ are determined from the xy projection of the following integral in the complex form:

$$\frac{\mu_0}{4\pi}\oint_C \frac{(\vec{r}-\vec{r}')}{\left|\vec{r}-\vec{r}'\right|^3}\times d\vec{l}' = B_c^x(\vec{r})\hat{x} + iB_c^y(\vec{r})\hat{y} = B_c(\vec{r})e^{i\theta_c(\vec{r})}$$

For a better interpretation, several simplifications are made on the general equation (1.15). First, it is assumed that transmitter/receiver RF coils are uniform enough to ensure $B_c(\vec{r}) = B_c$ and $\theta_c(\vec{r}) = \theta_c$. Second, we assume that signal recording starts at $t_0 = 0$ and that at that time the initial phase is zero in the entire brain $\phi_0(\vec{r}) = 0$. Finally, $\omega_L \gg \gamma \Delta B_I(\vec{r})$ as intrinsic inhomogeneities in the magnetic field are much smaller than B_0. Under these assumptions:

$$s(t) \sim \omega_L A \iiint_V dV e^{-t/T_2(\vec{r})} \left|\vec{M}_\perp(\vec{r},0)\right| e^{i\left((\omega_r - \omega_L)t - \gamma \Delta B_I(\vec{r})t - \gamma \vec{k}_{xy}(t)\cdot\vec{r}_{xy}\right)} \tag{1.16}$$

The factor A is defined as $A = B_c e^{i\theta_c}$.

T_1 and T_2 weighted images (Figure 1.5D) can be obtained using equation (1.16) in combination with methods described henceforth. The signal through the RF coil is measured via an analog-to-digital converter (ADC). In our previous deduction, it has been assumed that $\omega_L \gg 1/T_1$ and $\omega_L \gg 1/T_2$. To remove the high frequency $\sim \omega_L$ components from signal s, a demodulation is carried out electronically. From a theoretical perspective, this demodulation is equivalent to use a clockwise x-y rotating reference system at frequency $\omega_r \approx \omega_L$. If further electronic amplification factors are ignored, the signal amplitude is defined as $|s| = \omega_L M_0 A V \sim B_0^2$. Note that as measurement noise increase linearly with B_0, the final signal-to-noise ratio (SNR) is proportional to B_0. Intrinsic magnetic field inhomogeneities in the z-axis (e.g., by internal susceptibility effects) cause an additional suppression of the signal s as nuclear magnetic moments will precess at different frequencies. Under certain circumstances, it can be demonstrated that the addition of these inhomogeneities within a small volume (e.g., a voxel) could be approximated by a transversal relaxation time T_2' : $\int e^{-i\gamma\Delta B_I(\vec{r}')t} dV' \approx e^{-t/T_2'(\vec{r})}$. The dephasing due to the combination of intrinsic field inhomogeneities (T_2') and nuclear magnetic moment interactions (T_2) is described by the use of a decay T_2^*, which satisfies $R_2^* = 1/T_2^* = 1/T_2 + 1/T_2' = R_2 + R_2'$. Therefore, the NMR signal in the laboratory reference system is finally written as:

$$s(t) \sim \omega_L A \iiint_V dV e^{-t/T_2^*(\vec{r})} \left|\vec{M}_\perp(\vec{r},0)\right| e^{i((\omega_r - \omega_L)t - \vec{k}_{xy}(t)\cdot\vec{r}_{xy})} \tag{1.17}$$

To have a better understanding of signal s, let us simplify equation (1.17) by assuming that: **1)** only nuclear magnetic moments inside a single voxel in the brain at location $\vec{r}_v$ contribute to signal s (i.e., $T_2^*(\vec{r}_v) = T_2^*$, **2)** xy imaging linear gradients of magnetic fields are not turned on $\vec{k}_{xy}(t) = 0$, **3)** a 90° RF pulse is applied along the x-axis to ensure $\left|\vec{M}_\perp(0)\right| = M_0$ at the end of the pulse, and **4)** the demodulation is perform at $\omega_r = \omega_0$. Under these assumptions, (1.17) transforms into (1.18).

$$s(t) \sim \underbrace{\omega_L M_0 V_v A}_{s_0} e^{-t/T_2^*} \tag{1.18}$$

The simplest experiment, called free induced decay (FID), involves the measurement of signal $s(t)$ for a period T_s after the 90° RF pulse has ceased and before totally

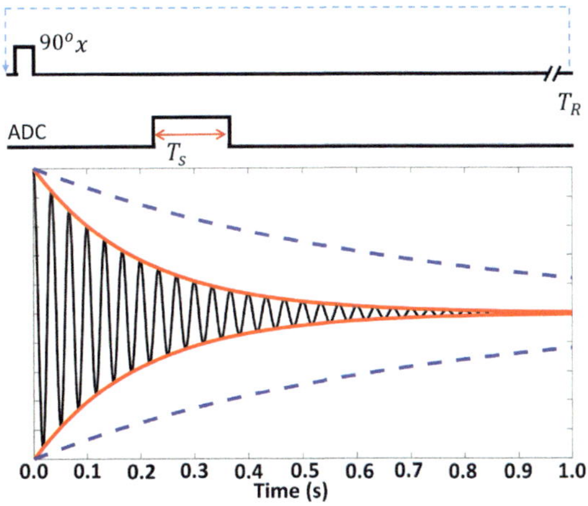

FIGURE 1.6 The free induction decay (FID) imaging sequence in MRI.

Note: The RF and ADC pulses are shown with the respective $s(t)$ signal (top). The T_2 (dashed blue) and T_2^* (solid red) envelops (a demodulated signal) are shown in the bottom panels.

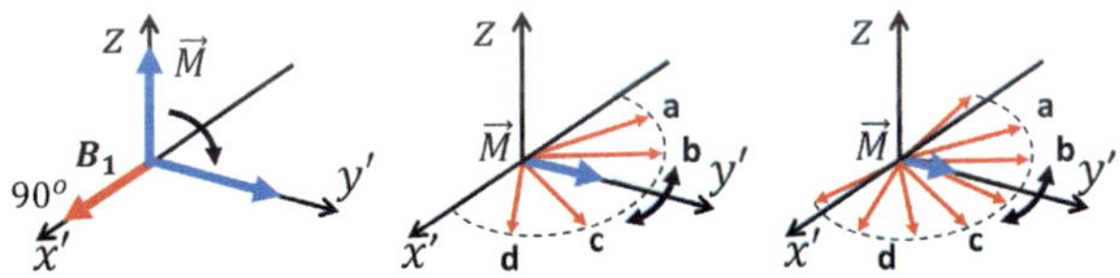

FIGURE 1.7 $\vec{M}$ tipping for the FID sequence in the rotating reference system.

losing the transversal magnetization. The FID sequence diagram and the nuclear magnetic moment evolution for this sequence is shown in Figure 1.6. Note that a measurement, represented by a finite set N_s of discrete observations $T_s = (N_s - 1)\Delta t$, are indicated by an ADC on-gradient. In many situations, T_2^* is small due to the existence of inhomogeneities in the magnetic field, which makes observations with an FID sequence challenging. Also, separation of the T_2 and T_2' effects using an FID sequence is difficult. Figure 1.7 shows the particular M tipping for the FID sequence.

To obtain T_2 measurements, a spin-echo (ES) sequence is used instead (Figure 1.8). An orthogonal (y-axis) 180° RF pulse of double amplitude is applied a period (τ) after the original 90° RF pulse along the x-axis. The latter RF pulse flips 90° the nuclear magnetic moments around the y-axis in the rotating reference system, which causes a rephasing (or refocus) after the same period τ. Therefore, at $T_E = 2\tau$ there is an echo in the signal s. Note that phase lost due to nuclear magnetic moment interactions (T_2) is irreversible. The phase of the nuclear magnetic moments can be predicted by the following equation:

$$\Delta\phi(t) = \begin{cases} -\gamma\Delta B_I t & 0 \le t \le \tau \\ -\gamma\Delta B_I \left(t - T_E\right) & t > \tau \end{cases} \tag{1.19}$$

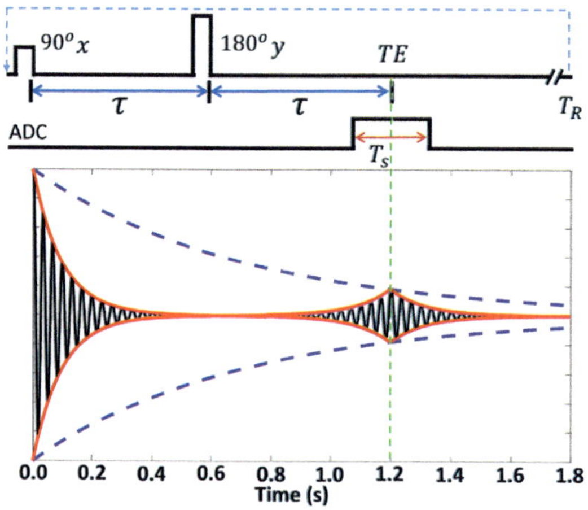

FIGURE 1.8 The spin-echo (SE) imaging sequence in MRI.

Note: The RF and ADC pulses are shown with the respective $s(t)$ signal (top). The T_2 (dashed blue) and T_2^* (solid red) envelops (a demodulated signal) are shown in the bottom panel.

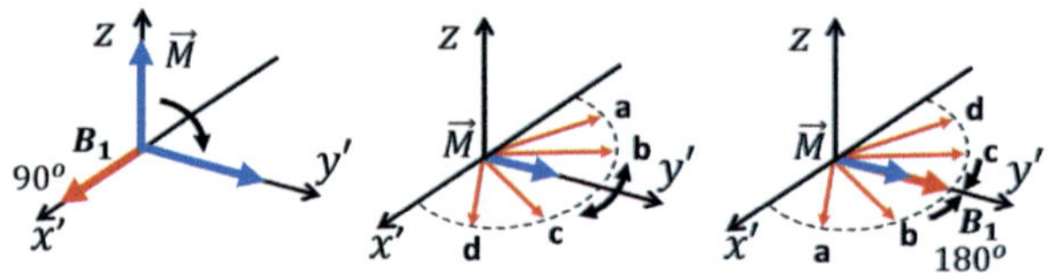

FIGURE 1.9 $\vec{M}$ tipping for the SE sequence in the rotating reference system.

As s_0 in the equation (1.18) is unknown, T_2 is estimated by repeating the SE sequence to measure the signal for two different echo times (a short T_{E_1} and a long T_{E_2}); i.e., $T_2 = (T_{E_1} - T_{E_2}) / ln(s(T_{E_2}) / s(T_{E_1}))$. Figure 1.9 shows the particular $\vec{M}$ tipping for the SE sequence.

The double SE sequence is very useful as it typically provides the *nucleus density* with the first echo (short T_{E_1}) and a T2-weighted contrast with the second echo (long T_{E_2}). In a more modern approach, T_2 is estimated using multiple 180° RF pulses ($m = 1, 2, \cdots$) at times $m\tau$ after a single 90° RF pulse as long as the T_2-relaxation has not been completely vanished. The multi-echo SE sequence is fast, reducing scan time in medical examinations. The transversal magnetization (rotation reference system) during the different phases of the SE sequence can be modeled by the equation (1.20).

$$\vec{M}_\perp(t) = \left|\vec{M}_\perp(t_0)\right| \begin{cases} e^{-t/T_2^*} & 0 < t < \tau \\ e^{-t/T_2} e^{-(T_E - t)/T_2{}'} & \tau < t < T_E \\ e^{-t/T_2^*} e^{T_E/T_2{}'} & t > T_E \end{cases} \tag{1.20}$$

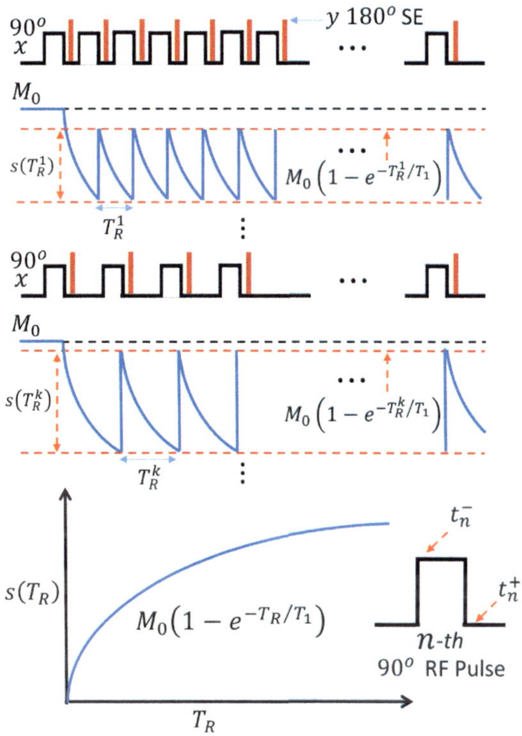

FIGURE 1.10 First method to estimate T_1: Multiple repetitions of the FID (or SE) sequence with different T_R values is used to fit T_1.

Note: The additional 180° RF pulse for the SE sequence is indicated by red bars.

To reduce noise-related estimation errors, FID and SE sequences are repeated several times (Figures 1.6 and 1.8, top panels, *blue dashed lines indicating sequence repetition*), with the repetition time T_R being a critical parameter. While choosing T_R, researchers/clinicians must take into consideration two aspects: **a)** the full recovery of the longitudinal magnetization M_0 and **b)** efficiency/cost in terms of the total scan time. Ignoring the first 90° RF pulse at t_0 in both sequences and using equation (1.13), it can be stated that the recovered longitudinal magnetization at the time t_n^- of applying subsequent ($n = 1,2\cdots$) 90° RF pulses ($t_n^- - t_0 = nT_R$) is: $M_z(t_n^-) = M_0(1 - e^{-T_R/T_1})$. This is the total magnetization to be projected to the xy-plane by any subsequent 90° RF pulse (Figure 1.10). Keep in mind that at the time the 90° RF pulse has ceased t_n^+, $M_z(t_n^+) = 0$. $\vec{M}$ dynamics within the time RF pulses are on are not considered in this book; hence, $t_n^+ - t_n^- \approx 0$ (i.e., instantaneous RF pulses). Note that T_1 from a sample can be measured by the following steps: **1)** perform the FID (or SE) sequence with at least two repetitions ($n > 1$) for a set of particular T_R^k values ($k = 1,2\cdots$), **2)** measure the signal after the k-th 90° RF pulse $s(T_R^k) = s(t_k^+)$, and **3)** estimate T_1 by fitting $s(T_R^k) = M_0(1 - e^{-T_R^k/T_1})$. The evolution of

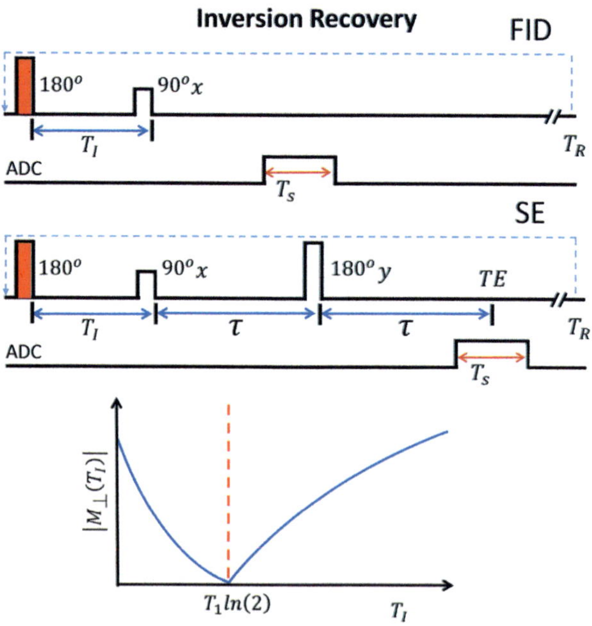

FIGURE 1.11 Second method to estimate T_1: The inversion recovery sequence for the FID (top) and SE (bottom) paradigms can be also used to estimate T_1 as the transversal magnetization is expected to be zero at $T_1 ln(2)$.

the transversal magnetization after the n-th 90° RF pulse ($t_n^+ < t_n < t_{n+1}^-$) is described by the following equations:

$$\vec{M}_\perp(t_n) = \begin{cases} M_o(1-e^{-T_R^k/T_1})e^{-t_n/T_2^*} & FID \\ M_o(1-e^{-T_R^k/T_1})e^{-t_n/T_2} & SE \end{cases} \tag{1.21}$$

The inversion recovery sequence constitutes another useful method to measure T_1 (Figure 1.11). This sequence consists of applying an early 180° RF pulse followed by a 90° RF pulse. The 180° RF pulse flips totally the longitudinal magnetization, i.e., $M_z(t_0=0) = -M_0$. Therefore, equation (1.13) describing the temporal evolution of $M_z(t)$ transforms into:

$$M_z(t) = M_0(1-2e^{-t/T_1}) \tag{1.22}$$

At the particular time $t = T_I = T_1 ln(2)$, named the inversion time, the longitudinal magnetization vanishes $M_z(T_I) = 0$ for any tissue with a T_1 relaxation value. If at this particular time a 90° RF pulse is applied, a transversal component $\vec{M}_\perp$ is not developed in brain tissues with that particular T_1 value; hence, this particular tissue does contribute to NMR signal s. In other words, the signal from this particular brain tissue has been suppressed. Considering equation (1.12), but in the rotation reference system (demodulation) and considering $\phi_0 = 0$, the transversal magnetization satisfies:

$$\vec{M}_\perp(t) = M_0 \left|(1-2e^{-T_I/T_1})\right| e^{-(t-T_I)/T_2^*} \quad t > T_I \tag{1.23}$$

The inversion recovery sequence can be performed under either an FID or a SE paradigm (also referred to as sequence). If a SE sequence is used, the transversal magnetization at the time of the echo $T_I + T_E$ satisfies $\vec{M}_\perp(T_I + T_E) = M_0 \left|(1-2e^{-T_I/T_1})\right| e^{-T_E/T_2}$. The T_1 value for a particular tissue (e.g., fat) is estimated by performing the inversion recovery sequence with FID (or SE) for many values of T_I to determine that for which the signal s is null (Figure 1.11, *red vertical dashed line*). If multiple repetitions of the inversion recovery are performed with a T_R separation, the signal s is zero at $T_I = T_1 ln(2/(1+e^{-T_R/T_1}))$.

1.5 SPATIAL ENCODING FOR MRI

The classical approach to create an image based on NMR principles consists of two steps: **1)** brain slices (Figure 1.5D) are selected using a linear gradient in one of the axes, and **2)** a high-resolution image for each slice is created by applying gradients in the two remaining axes followed by a Fourier transform. The selection of the x, y an z axis for step **(1)** generates sagittal, coronal, and axial images, respectively. In this book, only the case of axial (z-axis) imaging is considered. The Larmor frequency as a function of the z-axis is modeled as $\omega_L(z) = \omega_0 + \gamma G_z(t)z$, with $\omega_0 = \gamma B_0$ the Larmor frequency associated with the constant magnetic field B_0. The $G_z(t)$ gradient is turned on and off to ensure an axial slice selection with minimum signal lost, given by equation (1.24).

$$G_z(t) = \begin{cases} G_{SS} & 0 \le t \le \tau_{RF} \\ -G_{SS} & \tau_{RF} < t \le 3\tau_{RF}/2 \\ 0 & 3\tau_{RF}/2 < t \end{cases} \tag{1.24}$$

The application of the 90° RF pulse with band frequency $\Delta\omega_L = \omega_0 + \gamma G_{SS}(z_s \pm \Delta z/2)$ in the time window $0 \le t \le \tau_{RF}$ excites nuclei in an axial slice center at z_s with region of support $z_s - \Delta z/2 \le z \le z_s + \Delta z/2$. Therefore, the slice thickness Δz is a function of the bandwidth $\Delta\omega_L$ and the gradient γG_{SS}.

$$\Delta z = \Delta\omega_L / \gamma G_{SS} \tag{1.25}$$

To cancel $\vec{\mu}$ dephasing caused by the slice selection gradient, a negative gradient $-G_{SS}z$ is applied in the time window $\tau_{RF} < t \le 3\tau_{RF}/2$. Any signal s recorded using FID (or SE) paradigm is only originated from nuclei inside the selected slice. Figure 1.12 shows the SE paradigm used to reconstruct the xy image, which is based on the spatial 2D Fourier transform.

For convenience, the position in the 3D space $\vec{r}$ is decomposed into the 2D $\vec{r}_{xy}$ and 1D z_s components: $\vec{r}=[\vec{r}_{xy},z_s]'$. $\vec{r}_{xy}=[x,y]'$. As the signal will only be originated from the selected slice, let us define the time-varying effective nuclear density in the selected slice as $\rho(\vec{r}_{xy},t)=\omega_L(z_s)\Delta z A e^{-t/T_2^*(\vec{r}_{xy})}\left|\vec{M}_\perp(\vec{r}_{xy},0)\right|$. Then, equation (1.17) for a demodulated signal at $\omega_L(z_s)$ can be written as:

$$s(t)\sim\iint_{XY} dxdy\rho(\vec{r}_{xy},t)e^{-i\vec{k}_{xy}(t)\cdot\vec{r}_{xy}}=\iint_{XY} dxdy\rho(x,y,t)e^{-i(k_x(t)x+k_y(t)y)} \tag{1.26}$$

Taking into consideration the total magnetization to be projected to the xy-plane by consecutive 90° RF pulses separated by T_R intervals, the effective nuclear density at the time of maximal echo signal $s(T_E)$ takes the form:

$$\rho(x,y)=\underbrace{\omega_L(z_s)\Delta z A M_0(x,y)}_{\rho_0(x,y)}e^{-T_E/T_2^*(x,y)}(1-e^{-T_R/T_1(x,y)}) \tag{1.27}$$

By definition, the 2D continuous Fourier transform of $\rho_{z_s}(x,y)$ is:

$$\rho(k_x,k_y)=\iint dxdy\rho(x,y)e^{-i2\pi(k_x x+k_y y)} \tag{1.28}$$

Equations (1.26) and (1.28) suggest that $\rho(x,y)$ can be reconstructed by applying the inverse Fourier transform to a 2D arrangement $\bar{s}(k_x,k_y)$ of observed signal $s(t)$.

$$\rho(x,y)=\iint dk_x dk_y \bar{s}(k_x,k_y)e^{i2\pi(k_x x+k_y y)}$$

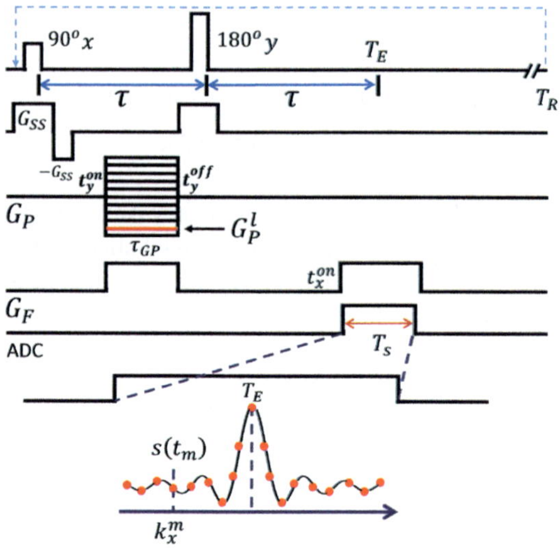

FIGURE 1.12 A very popular SE sequence.

Note:The FID sequence is similar, but without the need for a 180° RF pulse and its associate G_{SS} gradient.

To that end, the time dependency of the magnetic field gradients in the xy plane $\vec{G}_{xy}(t)$ must be defined in a way that an entire $\{k_x, k_y\}$ space (K-space) is covered while observing $s(t)$. There are multiple strategies to fill the K-space with collected $s(t)$ data. The most popular strategy (SE) is defined by equations (1.29) and (1.30), also illustrated in Figure 1.12.

$$k_x^m = \gamma \int_{t_x^{on}}^{t_m} G_x(t)dt = \gamma G_F(t_m - t_x^{on}) = \gamma G_F(2m+1)\Delta t \qquad m = -n_x, \cdots, n_x - 1 \quad (1.29)$$

Under this definition, $n_x = N_s / 2$; hence, $min(k_x) = -\gamma G_F T_s / 2$ and $max(k_x) = \gamma G_F T_s / 2$.

$$k_y^l = \gamma \int_{t_y^{on}}^{t_y^{off}} G_y(t)dt = \gamma G_P^l(t_y^{off} - t_y^{on}) = \gamma \Delta G_P l \tau_{GP} \qquad l = -n_y, \cdots, n_y - 1 \quad (1.30)$$

According to equation (1.30), a k_y^l line is selected by applying a phase-encoding gradient with a specific magnitude G_P^l during a fixed period τ_{GP}. Then, this k_y^l line is filled by observing $s(t)$ (ADC) in a discrete number of time instants t_m after the application of the frequency-encoding gradient (1.29). The 2D arrangement $\bar{s}$ is constructed as $\bar{s}(k_x^m, k_y^l) = s(t_x^{on} + (m + n_x)\Delta t)$ for each k_y^l line. The resolution of the K-space is defined as $\Delta k_x = \gamma G_F \Delta t$ and $\Delta k_y = \gamma \Delta G_P \tau_{GP}$. If the field of view (FOV) of the observed 2D area has a $L_x \times L_y$ dimension in mm, the resolution of the image will be $\Delta x = L_x / 2n_x$ and $\Delta y = L_y / 2n_y$. The total acquisition time to obtain N_z brain slices is: $T_{acq} = 2n_y N_z T_R$. The SNR per voxel (v) is $SNR / v \propto \Delta x \Delta y \Delta z \sqrt{2n_y T_s}$. A few

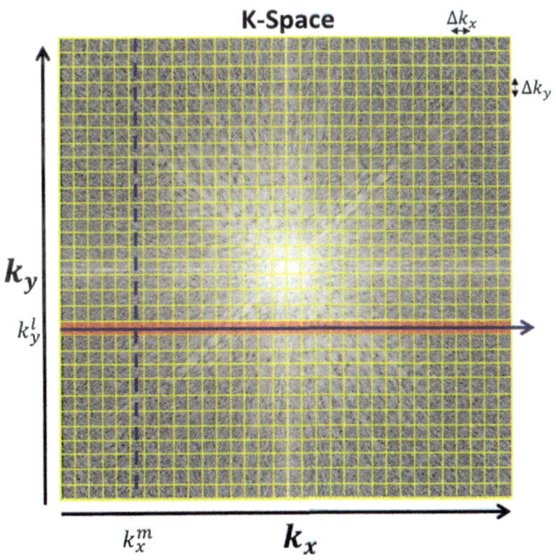

FIGURE 1.13 An illustration of the K-space filling for the SE sequence.

Note: The red line in the K-space is related to the G_P gradient and the ADC reading in Figure 1.12.

TABLE 1.3
Characteristics of the three most important MRI contrasts for anatomical evaluation

Contract	T_E	T_R	Formula
Proton Density	short	long	$C_{AB}=\rho_{0,A}(x,y)-\rho_{0,B}(x,y)$ $e^{-T_E/T_2^*}\rightarrow 1$ and $e^{-T_R/T_1}\rightarrow 0$
T_1-Weighting	short	short	$C_{AB}=\rho_{0,A}(1-e^{-T_R/T_{1,A}})-\rho_{0,B}(1-e^{-T_R/T_{1,B}})$ $e^{-T_E/T_2^*}\rightarrow 1$
T_2-Weighting	long	long	$C_{AB}=\rho_{0,A}e^{-T_E/T_{2,A}^*}-\rho_{0,B}e^{-T_E/T_{2,B}^*}$ $e^{-T_R/T_1}\rightarrow 0$ Using a SE sequence $T_2^*\rightarrow T_2$

whole-sequence repetitions N_{rep} is typically used to increase the SNR/v. The process to fill the K-space for the SE paradigm is shown in Figure 1.13.

We referred readers to a more specialized book (Brown et al., 2014) for a deeper understanding of: **a)** other strategies to cover the K-space, **b)** oversampling to avoid aliasing, **c)** types of image distortions and motion artifacts, and **d)** SNR improvement with multiple acquisitions.

According to (1.27), the contrast between brain tissues A and B is defined as:

$$C_{AB}=\hat{\rho}_A-\hat{\rho}_B=\rho_{0,A}e^{-T_E/T_{2,A}^*}(1-e^{-T_R/T_{1,A}})-\rho_{0,B}e^{-T_E/T_{2,B}^*}(1-e^{-T_R/T_{1,B}}) \tag{1.31}$$

Therefore, by properly selecting T_E and T_R, at least three main image contrasts can be obtained from estimated images $\hat{\rho}(x,y)$ (Table 1.3).

To enhance T_1–weighting contrast, it is recommended to use the following optimal T_R value:

$$T_R^{opt}=\frac{ln(\rho_{0,B}/T_{1,B})-ln(\rho_{0,A}/T_{1,A})}{1/T_{1,B}-1/T_{1,A}} \tag{1.32}$$

To enhance T_2–weighting contrast, the following optimal T_E value is recommended:

$$T_E^{opt}=\frac{ln(\rho_{0,B}/T_{2,B}^*)-ln(\rho_{0,A}/T_{2,A}^*)}{1/T_{2,B}^*-1/T_{2,A}^*} \tag{1.33}$$

1.6 ECHO-PLANAR IMAGING (EPI)

There are several strategies to reduce the total acquisition time ($T_T=N_{rep}T_{acq}$) for a whole-brain image reconstruction by: **a)** decreasing T_R without compromising the

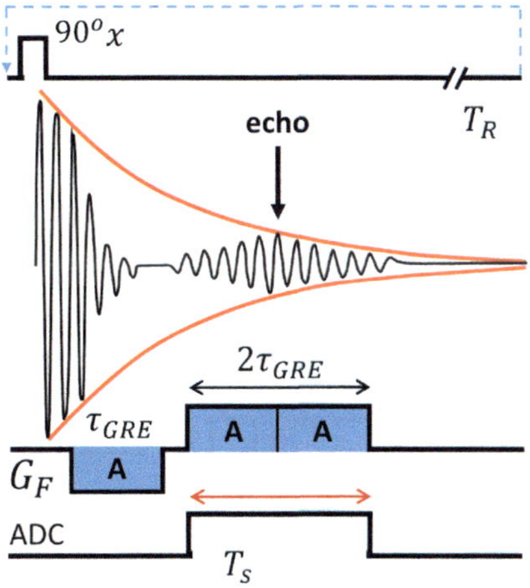

FIGURE 1.14 The gradient-recalled-echo sequence.

image type, **b)** reducing N_{rep} even to a value less than 1 by using a partial Fourier reconstruction, and **c)** lowering y-axis image resolution by a factor $\Delta y \rightarrow \lambda \Delta y$ ($0 < \lambda < 1$). Unfortunately, these strategies lead to some sort of deterioration in image quality. For measurements associated to phase-shifting (T_2), multiple 180° RF pulses after the initial 90° RF pulse in a SE sequence can be used to measure about 8–16 additional echoes in a single T_R as intrinsic T_2 dephasing continues, a sequence named rapid acquisition with relaxation enhancement (RARE – known as TURBO/FAST SE). However, segmenting the K-space by phase-encoding multiple k_y lines per RF excitation constitutes the most useful way to reduce T_T. This type of MRI sequence is named echo-planar imaging (EPI).

A simple gradient-recalled-echo (GRE) sequence consists of an accelerated FID signal decay/recover by using dephasing/rephasing frequency gradients (G_F) of same amplitude but opposite polarity (Figure 1.14). Nuclear magnetic moments scrambled by the initial dephasing gradient of duration τ_{GRE} are phase-refocused causing a final echo after applying the rephasing gradient for the same period, without affecting the undergoing T_2 and T_2^* relaxation mechanisms. The rephasing gradient is used concurrently with the ADC to measure the echo; hence, typically $T_s = 2\tau_{GRE}$. In a GRE sequence, the echo is recorded much more quickly than in a SE sequence. Low-flip-angle excitations are also used in GRE sequences, which allow further reductions in T_R.

The gradient-echo (GE) principle is the fundamental building-block for EPI. In an EPI, multiple negative and positive G_F gradients are alternatively turned on after a magnetization preparatory component, a strategy allowing the coverage of the K-space in a time efficient way. The EPI principle can be performed with the preparatory sequences discussed earlier: **a)** gradient-echo (90° RF pulse, GE-EPI), **b)** spin-echo ($90^o + 180^o$ RF pulses, SE-EPI), and **c)** inversion-recovery ($180^o + 90^o$ RF pulses, IR-EPI). In this book, an EPI with a gradient-echo preparation (yellow block) is the only one discussed: sequence (Figure 1.15) and K-space filling (Figure 1.16).

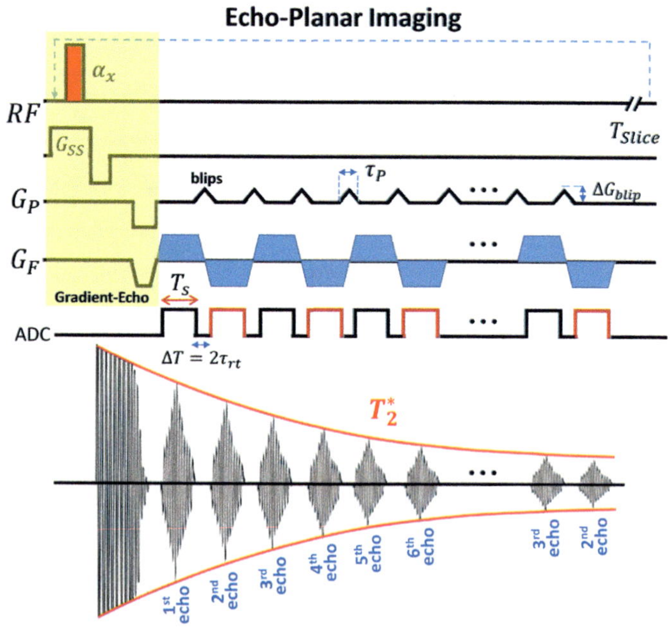

FIGURE 1.15 An EPI sequence with gradient-echo (yellow) preparation (GE-EPI).

FIGURE 1.16 K-space coverage for a classical zig-zagging under the GE-EPI sequence shown in Figure 1.15.

Under the most classical GE-EPI sequence "*blipped*", a first G_P gradient is applied to select the most negative k_y line $(-n_y\Delta k_y)$ in the K-space. This G_P gradient is performed simultaneously with an initial dephasing G_F gradient. Then, the k_y line-by-line "zig-zag" phase encoding is achieved by applying short (τ_P) gradient pulses (blips) between $\pm G_F$ gradients to switch the selected k_y line up in the K-space: $-n_y\Delta k_y \rightarrow -(n_y-1)\Delta k_y \rightarrow \cdots \rightarrow 0 \rightarrow \cdots \rightarrow (n_y-2)\Delta k_y \rightarrow \left(n_y-1\right)\Delta k_y$. The time needed to collect a single k_y line T_s adds to the time required to switch to the following line via the blip $\Delta T = 2\tau_{rt}$, with τ_{rt} representing the rise time of each readout. The EPI is initiated by a nuclear magnetic moment preparation with a RF pulse at a particular flip angle $\alpha_x < 90^o$. In summary, a single brain z-slice is obtained in an EPI sequence by applying the α_x RF pulse followed by strong switched frequency-encoding gradients simultaneously applied an intermittent "blipped" phase-encoding gradient. The total time needed to complete a 2D brain slice is $T_{Slice} = 2n_y(T_s + \Delta T)$. If a certain number of repetitions N_{rep} are used, the whole-brain image reconstruction under an GE-EPI sequence is $T_T = N_{rep}N_zT_{Slice}$. L_x and L_y are determined by the 2D FOV; hence $L_x = L_y = 250\,mm$ in typical brain imaging. For a triangle-shaped blips, their amplitude at the peak must be $\Delta G_{blip} = 2/(\gamma L_y \tau_P)$. The duration of the blips is usually selected as $\tau_P = 2\tau_{rt}$. Common ^{1}H MRI scanners can produce gradients of $30-45\,mT/m$ in peak amplitude, for a range in blip duration of $4.17\,\mu s < \tau_P < 6.26\,\mu s$. The time resolution during the read-out is $\Delta t = 1/(\gamma max(G_F)L_x)$. Therefore, for ^{1}H MRI scanners, Δt could take any value between $2.09\,\mu s$ to $3.13\,\mu s$ for $L_x = 250\,mm$. In EPI, an average slice xy (in plane) resolution is $2.5\,mm \times 2.5\,mm$, which implies $n_x = n_y = 50$. Assuming the shortest blips of $\tau_P = 4.17\,\mu s$ and the highest temporal resolution $\Delta t = 2.09\,\mu s$, the total time to obtain a single z-slice image of the brain is $T_{Slice} = 21.3\,ms$. Therefore, a whole-brain image consisting of 64 z-slices (single repetition) can be created in $T_T < 1.4\,s$. This temporal resolution is critical for functional brain imaging. Therefore, MRI scanners are nowadays created with the possibility to generate larger gradient $> 45\,mT/m$. Having a high temporal resolution is also critical to reduce the effect of movement artifacts. In fact, the EPI sequence was initially proposed due to the need for cardiac imaging. The first EPI sequence was used at Mansfield's laboratory to measure a beating rabbit heart in 1981 and an infant human heart in 1983. Non-blipped EPI sequences (e.g., spiral) are not discussed in this book.

Spoiling techniques are needed to avoid signal saturation in fast imaging sequences, e.g., EPI. Spoiling refers to a destruction of any remaining coherence ($\vec{M}_\perp = 0$) before the application of the RF pulse in GRE sequences with multiple cycles. There are three main spoiling methods: **a)** natural spoiling by using a long T_R, **b)** gradient spoiling by turning on G_{SS} (or G_F) with variable amplitudes at the end of the cycle, and **c)** RF-spoiling using variable phases in the RF pulse. The RF-spoiling method is the most used. Zur et al. (1991) proposed the following recurrent formula for the change in RF-pulse phase between GRE cycles: $\phi_i = \phi_0(i^2 + i + 2)/2$. At the echo time, the signal s from a perfectly spoiled-GRE sequence, assuming a longitudinal steady-state, is represented by the equation (1.34).

$$s(T_E) = K\rho_0 \frac{sin(\alpha_x)(1 - e^{-T_R/T_1})e^{-T_E/T_2^*}}{(1 - cos(\alpha_x)e^{-T_R/T_1})} \tag{1.34}$$

At a particular flip angle $\alpha_x = \alpha_E$, signal s is maximal ($\partial s/\partial\alpha_x = 0$). This optimal α_x is known as the Ernst's angle, $\alpha_E = arccos(e^{-T_R/T_1})$. The use of α_E in EPI helps reducing signal loss. T_1 and T_2 -weighting images can also be obtained by selecting appropriated α_E. The EPI sequence discussed above is called single-shot as the entire K-space is covered with a single preparation. Currently, the use of segmented coverage of the K-space using a small number of preparations are named multi-shot EPI. parallel imaging (PI) (multiband-excitation) is also used to reduce whole-brain image reconstruction time. There are two main techniques proposed to date: **1)** image domain PI (SENSE – sensitivity encoding) and **2)** K-space PI (GRAPPA – generalized auto-calibrating partial parallel acquisition). SENSE has been the most used. In general, PI can be combined with EPI under different preparations.

Phase shifting due to the following three mechanisms are not recovered in a GRE sequence: **1)** magnetic field inhomogeneities, **2)** tissue susceptibility gradients, and **3)** molecule diffusion. In modern MRI scanners, the homogeneity of the magnetic field is in the range of parts per million (ppm). Hence, other than those intentionally created by gradient coils, phase shifts by magnetic field inhomogeneities can be ignored. However, two important MRI techniques are based on the non-canceling effect at the center of the echo of the phase shifts due to: tissue susceptibility gradient (fMRI) and molecule diffusion (DWI).

1.7 FUNCTIONAL MRI (FMRI) – THE 90S REVOLUTION

Magnetic susceptibility originates from the interaction of electrons and nuclei with an externally applied magnetic field. As discussed earlier in this chapter, a nucleus could have associated a total magnetic moment, which is the foundation of NMR. Similarly, electrons orbiting around the nucleus could have associate orbital and angular (spin) magnetic moments. A theoretical understanding of the electron magnetic moment of atoms in biological tissues is not an objective of this book. For paramagnetic substances, these magnetic moments align with the external magnetic field augmenting its strength. A particular type of electron magnetic moment originates from unpaired electrons in an atom/molecule. Substances with unpaired electrons are called paramagnetic, as their magnetic moments align with any external magnetic field. In contrast, a substance is called diamagnetic when its intrinsic magnetic moments orient in opposite direction to the external magnetic field, canceling part of it. Contribution to the susceptibility by the nuclear magnetic moments is negligible. Contributions to the susceptibility from electron magnetic momentum are of two types: **a)** Langevin diamagnetism through the angular momenta of those electrons in filled orbitals (*very weak*), and **b)** Curie paramagnetism through the spins of unpaired electrons (*very strong*). The effect of Curie paramagnetism on magnetic susceptibility could be a thousand times larger than that from Langevin diamagnetism. In Table 1.1, every atom was associated to its particular atomic number (Z), which represents the number of electrons orbiting the nucleus. For example, ^{17}O has eight electrons ($Z = 8$). Atoms are filled with electrons according to an organization in energy levels (shells, with energy quantum number $n = 1,\cdots$), each formed by combinations of subshells ($s,p,d,f,\cdots$): **1)** the first energy level ($n = 1$) – one subshell s (s^1), **2)** the second energy level ($n = 2$) – two subshells (s^2, p^2), **3)** the third energy level ($n = 3$) – three subshells (s^3, p^3, d^3), and **4)** the fourth energy level ($n = 4$) – four subshells (s^4, p^4, d^4, f^4), and continue $\cdots$. According to the orbital quantum number pair $\{l,m\}$ with ($l = 0,\cdots$ and $-l \le m \le l$), each subshell could have a different number of orbitals, e.g., $s - 1$ orbital ($\{0,0\}$) and $p - 3$ orbitals ($\{1,-1\}$, $\{1,0\}$,$\{1,1\}$). Due to Pauli exclusion principles for Fermions, up to a total of two electrons ($s = \pm 1/2$) can co-exist in each orbital. Orbitals are first filled with electrons having the same spin quantum number s. Therefore, the eight electrons of ^{17}O are distributed from the lowest to the highest energy levels as follow: $2s^1 2s^2 4p^2$. The first three electron in the subshell p^2 are filled with $s = +1/2$. One of these orbitals is completed with the fourth electron ($s = -1/2$). Therefore, there are two unpaired electrons, making ^{17}O a paramagnetic substance. The Curie paramagnetism is very important in the development of contrast substance for MRI, with gadolinium (Gd – 7 unpaired electrons in the f^4 shell – Problem 1.7) being the most popular in clinical practice and animal research. For readers interested in a deeper understanding of this topic, a book about electron paramagnetic resonance (EPR) spectroscopy is recommended (Wertz and Bolton, 1986). In a groundbreaking study for bioinorganic chemistry, Pauling and Coryell (1936) demonstrated the paramagnetic properties of deoxygenated hemoglobin (deoxy-Hb), due to the four unpaired electrons per Fe atom. The hemoglobin protein comprises four one-iron subunits. In theory, the four unpaired spins per Fe should have an associated spin magnetic momentum of 4.90 Borh Magnetom (BM, $\mu_B = e\hbar / 2m_e = 9.27 \times 10^{-24} Am^2$). By using susceptibility measurements, it was verified that the spin magnetic momentum for the deoxy-Hb was 5.46 BM. In contrast, the oxygenated hemoglobin (oxy-Hb) was diamagnetic. Hence, deoxy-Hb must create magnetic field gradients nearby brain vessels, thus reducing the s signal from the ^1H nuclei (abundant in the brain) via an accelerated phase shifting. The differences in the magnetic susceptibility properties of the oxy-Hb and deoxy-Hb is the fundamental principle underlying functional MRI (fMRI). The feasibility of the fMRI technique was first demonstrated at the AT&T Bell Laboratories (Ogawa et al., 1990). It is known that increases in neuronal activity induce local changes in cerebral blood flow (CBF) and oxygen tissue pressure (pO_2), a phenomenon called functional hyperemic response. Local changes in CBF are a result of the activation of a variety of neurovascular coupling mechanisms (Chapter 10). Local changes in pO_2 are caused by the oxygen metabolism of glucose (Glc) to reestablish ion gradients associates with action potentials and postsynaptic graded potentials (Chapters 5–9). The cerebral metabolic rate of oxygen ($CMRO_2$) is commonly used to quantify Glc oxidative metabolism. Therefore, a blood-oxygen-level-dependent (BOLD) contrast is obtained by comparing changes in signal s at different time instants by using EPI sequences. A revolution in different areas of neuroscience was triggered by the development of this high-resolution brain imaging modality to study neuronal activity. The physiological mechanisms underlying the interplay between CBF and $CMRO_2$ in the temporal variations of signal s are discussed later in more detail (Chapters 9 and 10). There have been several theoretical models proposed to account for the BOLD effect. In 1997, a first model with extra and intra -vascular compartments and a CBV exchange effect at the post-capillary venous level was proposed by Buxton et al. (1998), which was extended later (Obata et al., 2004) to include an independent time-dependent oxygen extraction fraction (OEF); i.e., the Buxton-Obata's model. A simpler model was proposed a year later (Davis et al., 1998), which considered only an extravascular compartment and assumed uniform cerebral blood volume (CBV) changes in the whole vasculature. Despite its simplicity, Davis's model can explain a variety of BOLD signal data. Davis's model was optimized by Griffeth and Buxton (2011) using a more realistic general model for the BOLD response. In this section, the general Griffeth and Buxton's model is introduced. According to it, one extravascular and three intravascular compartments filled with ^{1}H proton spins contribute to signal s coming from a brain voxel at T_E of a GRE sequence (1.35).

$$s = (1 - v) s_E + \sum_{k=\{A,C,V\}} s_k v_k \tag{1.35}$$

v is the relative normalized CBV, which includes arteries (A), capillaries (C) and venous (V) inside the voxel occupying volume fractions v_A, v_C and v_V, respectively. The intra and extra-vascular signals are modeled as: $s_E = s_{0,E} e^{-T_E R^*_{2,E}}$ and $s_k = s_{0,k} e^{-T_E R^*_{2,k}}$ $(k = \{A, C, V\})$, respectively. The magnitudes $s_{0,E} = \rho_E (1 - e^{-T_R / T_{1,E}})$ and $s_{0,k} = \rho_I (H_T)(1 - e^{-T_R / T_{1,k}})$ correspond to their respective baseline values of the respective vascular signal. The intra and extra vascular spin densities are ρ_E and $\rho_I(H_T)$, respectively. $\rho_I(H_T)$ must depend on blood hematocrit H_T. These spin densities, which are related to water concentrations in tissue and blood, are assumed activation-independent. As T_1 relaxation is not affected by changes in local deoxy-Hb concentrations, it must be homogeneous inside the voxel ($T_1 = T_{1,E} = T_{1,\forall_k}$) and constant (i.e., activation-independent). Therefore, the following intrinsic signal rates are defined:

$$\varepsilon_k = \frac{s_{0,k}}{s_{0,E}} = \lambda(H_T) \frac{e^{-T_E R^*_{2,k}}}{e^{-T_E R^*_{2,E}}} \tag{1.36}$$

The effective extra-to-intra vascular spin density ratio $\lambda(H_T) = \rho_I(H_T) / \rho_E$ was estimated as 1.15 for $H = 0.44$ (Griffeth and Buxton, 2011). This ratio might absorb differences in the T_1 relaxation between intra and extra vascular tissue, and/or any dependence of the intravascular $T_{1,k}$ with H_T. It can be demonstrated that the activation-dependent variation in signal $\delta s = (s - s_0) / s_0$ satisfies (1.37).

$$\delta s = \Omega \left[(1 - v) e^{-T_E \Delta R^*_{2,E}} + \sum_{k=\{A,C,V\}} \varepsilon_k v_k e^{-T_E \Delta R^*_{2,k}} \right] - 1 \tag{1.37}$$

With $\Omega = 1 / (1 - v_0 + \sum_{k=\{A,C,V\}} \varepsilon_k v_{0,k})$

The baseline volume fractions are defined as v_0, and $v_{0,A}$, $v_{0,C}$ and $v_{0,V}$ for arteries, capillaries and venous, respectively. By definition, $\Delta R^*_{2,E} = R^*_{2,E} - R^*_{2,0,E}$ and $\Delta R^*_{2,k} = R^*_{2,k} - R^*_{2,0,k}$. Uludağ et al. (2009) provides linear regressions for the dependency of $R^*_{2,0,E}$ with B_0. The R_2 relaxation rates do not depend on deoxy-Hb concentration; hence, $\Delta R^*_{2,E} = \Delta R'_{2,E}$ and $\Delta R^*_{2,k} = \Delta R'_{2k}$ $(\forall_k)$. Changes in these transversal relaxations are modeled as:

$$\Delta R'_{2,E} = \sum_{k=\{A,C,V\}} \Delta R'_{2,E(k)}$$

$$\Delta R'_{2,E(k)} = \begin{cases} \frac{4\pi}{3}\Delta\chi H_T \gamma_H B_0 \left(v_k \left| S_{O_2}^{off} - S_{O_2,k} \right| - v_{0,k} \left| S_{O_2}^{off} - S_{O_2,0,k} \right| \right), k = \{A,V\} \\ 0.04\left(\Delta\chi H_T \gamma_H B_0\right)^2 \left(v_\text{k} \left| S_{O_2}^{off} - S_{O_2,\text{k}} \right|^2 - v_{0,k} \left| S_{O_2}^{off} - S_{O_2,0,\text{k}} \right|^2 \right), \text{k=\{C\}} \end{cases}$$

$$\Delta R'_{2,k} = C_{B_0}\left(H_T\right)\left[\left(S_{O_2}^{off} - S_{O_2,k}\right)^2 - \left(S_{O_2}^{off} - S_{O_2,0,k}\right)^2\right]$$

The dependence of coefficient $C_{B_0}(H_T)$ with H_T varies for different B_0. Griffeth and Buxton (2011) provided the dependence for $B_0 = 3T$.

$$C_{3T}(H_T) = 302.06H_T + 41.83 \tag{1.38}$$

Under certain theoretical considerations (Li and van Zijl, 2020), the dependence for other B_0 values can be obtained from bovine blood sample data using a multi-echo sequence at different oxygenation levels more available in a more recent study (Grgac et al., 2017). $S_{O_2,k}$ and $S_{O_2,0,k}$ represent oxygen saturation and its respective baseline level for each vascular compartment $k = \{A,C,V\}$. $\Delta\chi$ is the difference in magnetic susceptibility between the fully deoxygenated and oxygenated blood. γ_H is the gyromagnetic ratio of ^{1}H protons. Finally, $S_{O_2}^{off}$ represents the oxygen saturation that produces a zero difference in the magnetic susceptibility of blood and tissue. The dependency of these magnitudes with neuronal activity and other physiological variables are discussed in further chapters. Simplified versions of this model are presented in Part 2.

1.8 DIFFUSION WEIGHTED IMAGING (DWI)

Theoretical considerations about ion diffusion are discussed in Chapter 5. Thereafter, the concept of diffusion of both ions and large molecules are used to describe different physiological processes (e.g., neurotransmission, intracellular signaling, and transmembrane currents). As diffusion is a crucial aspect in brain physiology, having a technique to create images with the diffusional patterns of specific ions/molecules inside the brain is very valuable. Diffusion is referred to the microscopic movement of ions/molecules caused by their random interactions (collisions) inside the intra and extracellular compartments, a process known as Brownian motion. However, cellular microarchitecture in the brain imposes preferential directions and magnitudes to such random movements at each location. In principle, you can create images of these directions/magnitudes by quantifying voxel-wide shifts in the phase of the nuclear angular moment $\vec{\mu}$ caused by diffusion. A generalization of the Bloch's equation to include attenuation of $\vec{M}$ due to nucleus diffusion (1.39), initially proposed by Torrey (1956), constitutes the fundamental theoretical framework to develop diffusion-based imaging techniques.

$$\frac{d\vec{M}}{dt} = \gamma\vec{M}\times(B_z\hat{z}+\vec{B}_1(t))+\overbrace{\left[\frac{(M_0-M_z)}{T_1}-\frac{\vec{M}_\perp}{T_2}\right]}^{Relaxation}-\overbrace{\nabla\cdot(\vec{\vec{D}}\cdot\nabla\vec{M})}^{Diffusion} \tag{1.39}$$

The 2nd-rank tensor $\vec{\vec{D}}$ comprises different components related to diffusion. The diagonal has the diffusion coefficients along the principal direction x, y and z. Off-diagonal components are related to correlations in random motions along pair directions. Hence, $\vec{\vec{D}}$ is by definition symmetric. As the interest is to model ion/molecule diffusion inside a voxel to create a discrete image, $\vec{\vec{D}}$ is assumed to be homogeneous inside it, i.e., space-independent. Although equation (1.39) is general, this section is focussed only on imaging water molecule diffusion, which constitutes the foundation for the diffusion weighted imaging (DWI) technique. Because of a diversity of cellular organization/morphology in the brain, water molecule diffusion at each location is restricted according to the geometrical characteristics of plasma/organelles membranes at the microscopic level. DWI allows researchers to visualize the magnitude of bulk diffusional processes in brain voxels. Preferential directions due to anisotropic diffusion, for example restricted directions inside neuronal axons, are quantified by means of the diffusion tensor imaging. In DWI, the diffusion tensor is assumed diagonal $\vec{\vec{D}}=\{D_{ii}\}$; hence, the Bloch-Torrey general equation reduces to:

$$\frac{d\vec{M}}{dt} = \gamma\vec{M}\times(B_z\hat{z}+\vec{B}_1(t))+\overbrace{\left[\frac{(M_0-M_z)}{T_1}-\frac{\vec{M}_\perp}{T_2}\right]}^{Relaxation}-\left(D_{xx}\frac{\partial^2\vec{M}}{\partial x^2}+D_{yy}\frac{\partial^2\vec{M}}{\partial y^2}+D_{zz}\frac{\partial^2\vec{M}}{\partial z^2}\right) \tag{1.40}$$

Solving (1.40) given known the diffusion coefficient D_{ii} in each direction is straightforward. Defining an MRI sequence to estimate them is more challenging. There are several sequences developed over the years to that end. In this book, the classical SE sequence under a Stejskal-Tanner bipolar pair pulse is used to illustrate general principles underlying DWI (Stejskal and Tanner, 1965). Under this diffusion weighted SE sequence, the RF and G_{SS} gradients are turned on as in the SE classical sequence (Figure 1.12). However, a diffusion gradient in a particular principal direction (x, y or z) is turned on during two consecutive periods symmetrically allocated at both sizes of the 180° RF refocusing pulse. These diffusion gradients must have same duration (δ) and amplitude ($G_D^i, i=(x,y,z)$). Their onsets are separated by an interval Δ. Image encoding in DWI could be performed using the EPI protocol with the G_P and G_F gradients (DW-EPI). If the diffusion gradient is applied along a principal direction i (G_D^i), phase-shifting due to diffusion in that particular direction is not recovered at the echo; hence, $s(T_E)$ must be further attenuated. However, diffusion in the other two orthogonal principal directions has no effect on s at the echo. The solution of equation (1.40) for the transversal magnetization $\vec{M}_\perp(t)$ under this SE-DWI sequence is:

$$\vec{M}_\perp(t)=M_0e^{-i(\omega_Lt-\phi_0)}e^{-t/T_2}e^{-(b_xD_{xx}+b_yD_{yy}+b_zD_{zz})} \tag{1.41}$$

$$b_i = \gamma_H^2 \int_0^{T_E} \left(\int_0^t G_D^i(t')dt' \right)^2 dt = (\gamma_H G_D^i \delta)^2 \left(\Delta - \frac{\delta}{3} \right) \quad (1.42)$$

In this case, it has been considered that $t_0 = 0$. $|\vec{M}_\perp(t_0)| = M_0$ due to the preparation with a 90° RF pulse. Note that equation (1.42) can be used to obtain analytical solutions for b_i under other diffusion-sensitizing gradient configurations $G_D^i(t)$ (e.g., sinusoidal or trapezoidal pulses). For the particular case of a multi-echo SE sequence $b_i = (\gamma_H G_D^i)^2 T_E^3 / 12n^2$, with n the number of echos. For a given b_i value, the NMR signal at the echo is defined as: $s_{b_i}(T_E) = s_0 e^{-b_i D_{ii}}$. s_0 is the NMR signal at the echo obtained with the diffusion gradient off. Therefore, the diffusion coefficient in that particular principal direction is estimated by the equation: $D_{ii} = -ln(s_{b_i}(T_E)/s_0)/b_i$. The DWI is formed by applying equation (1.43).

$$s_{DWI} = \sqrt[3]{s_{b_x}(T_E)s_{b_y}(T_E)s_{b_z}(T_E)} = s_0 e^{-b(D_{xx}+D_{yy}+D_{zz})/3} \quad (1.43)$$

With $s_0 = \rho_0(1 - e^{-T_R/T_1})e^{-T_E/T_2}$. DWI uses a grey scale to quantify water diffusion restriction, with a bright white for those areas with total 3D restricted diffusion and a dark black for those in which water molecules move freely in all directions. To avoid T_2 shine-through cause by the dependency of s_0 with T_2, the apparent diffusion coefficient (ADC, Eq. 1.44) is used instead in many applications.

$$ADC = (D_{xx} + D_{yy} + D_{zz})/3 \quad (1.44)$$

1.9 PROBLEMS AND EXERCISES

Problem 1.1 – If the molarity of ^{1}H atoms in the gray matter is $88M$, calculate the total number of ^{1}H spins in the up-state (low energy) at the equilibrium inside a $2 \times 2 \times 2\,mm$ neocortical voxel after the person's head is placed inside a $3T$ MRI scanner at $309K$.

Problem 1.2 – Considering that the ionization potential for an organic compound is approximately $\sim 10^{-19} J$, demonstrate why the energy absorbed by a ^{1}H nucleus in a 1.5 T magnetic field is not ionizing. Compare it to the energy to an $5 \times 10^{19}\,Hz$ x-ray photon.

Problem 1.3 – If a material has a longitudinal relaxation time $T_1 = 200\,ms$, what will the longitudinal magnetization be $400\,ms$ after an applied 180° RF pulse has ceased, and the system is returning to its equilibrium magnetization?

Problem 1.4 – The net magnetization of a material inside a $3T$ MRI scanner decreased after increasing the temperature. Determine the relative change in temperature (ΔT in K) if the net magnetization decreased 10% from its initial value at room temperature ($T = 297\,K$).

Problem 1.5 – What the inversion time (T_I) of an adipose sample is if it has a T_1 relaxation of 0.75 seconds?

Problem 1.6 – In average, brain gray matter has longitudinal and transversal relaxation $T_1 = 950\,ms$ and $T_2 = 100\,ms$, respectively. Calculate the absolute value of the transversal magnetization $|M_{xy}(t)|$ at the particular time t for which the longitudinal magnetization has decayed half of the total gray matter magnetization $|M_z(t)| = M_0 / 2$ for a 90° RF pulse.

Problem 1.7 – Demonstrate that Gd is paramagnetic.

Problem 1.8 – Draw the EPI-DWI sequence with a SE preparation.

REFERENCES

Abragam A. *The Principles of Nuclear Magnetism*. International Series of Monographs on Physics 32, Oxford Science Publication, Oxford Press, 1961.

Brown RW, Cheng Y-CN, Haacke EM, Thompson MR, Venkatesan R. *Magnetic Resonance Imaging: Physical Principles and Sequence Design* (2nd Edition). Wiley Blackwell, Hoboken, New Jersey, 2014.

Buxton RB, Wong EC, Frank LR. Dynamics of blood flow and oxygenation changes during brain activation: The balloon model. *Magn Reson Med* 39, 855–864, 1998.

Cavanagh J, Skelton N, Fairbrother W, Rance M, Palmer A-III. *Protein NMR Spectroscopy* (2nd Edition). Elsevier Academic Press, Burlington, Massachusetts, 2007.

Davis TL, Kwong KK, Weisskoff RM, Rosen BR. Calibrated functional MRI: Mapping the dynamics of oxidative metabolism. *Proc Natl Acad Sci USA* 95, 1834–1839, 1998.

Ernst RR, Bodenhausen G, Wokaun A. *Principles of Nuclear Magnetic Resonance in One and Two Dimensions*. International Series of Monographs on Chemistry 14. Oxford Science Publication, Oxford Press, 1987.

Grgac K, Li W, Huang A, Qin Q, van Zijl PCM. Transverse water relaxation in whole blood and erythrocytes at 3T, 7T, 9.4T, 11.7T and 16.4T; Determination of intracellular hemoglobin and extracellular albumin relaxivities. *Magn Reson Imaging* 38, 234–249, 2017.

Griffeth VEM, Buxton RB. A theoretical framework for estimating cerebral oxygen metabolism changes using the calibrated-BOLD method: Modeling the effects of blood volume distribution, hematocrit, oxygen extraction fraction, and tissue signal properties on the BOLD signal. *NeuroImage* 58, 198–212, 2011.

Li W, van Zijl PCM. Quantitative theory for the transverse relaxation time of blood water. *NMR Biomed* 33, e4207, 2020.

Obata T, Liu TT, Miller KL, Luh WM, Wong EC, Frank LR, Buxton RB. Discrepancies between BOLD and flow dynamics in primary and supplementary motor areas: Application of the balloon model to the interpretation of BOLD transients. *NeuroImage* 21, 144–153, 2004.

Ogawa S, Lee TM, Kay AR, Tank DW. Brain magnetic resonance imaging with contrast dependent on blood oxygenation. *Proc Natl Acad Sci USA* 87, 9868–9872, 1990.

Pauling L, Coryell CD. The magnetic properties and structure of hemoglobin, oxyhemoglobin and carbonmonoxyhemoglobin. *Proc Natl Acad Sci USA* 22(4), 210–216, 1936.

Stejskal EO, Tanner JE. Spin-diffusion measurements: Spin echoes in the presence of a time-dependent field gradient. *J Chem Phys* 42, 288–292, 1965.

Torrey HC. Bloch equations with diffusion. *Physi Rev* 104(3), 563–565, 1956.

Uludag K, Muller-Bierl B, Ugurbil K. An integrative model for neuronal activity induced signal changes for gradient and spin echo functional imaging. *NeuroImage* 48, 150–165, 2009.

Wertz JE, Bolton JR. *Electron Spin Resonance: Elementary Theory and Practical Applications*. Chapman and Hall, New York – London, 1986.

Zur Y, Wood ML, Neuringer LJ. Spoiling of transverse magnetization in steady state sequences. *Magn Reson Med* 21, 251–263, 1991.

2 Large-Scale Electrophysiological Techniques

2.1 OVERVIEW

The activation of neurons causes electric currents to flow across their plasma membrane (Chapters 5 and 6). As neurons are suspended in a highly conductive extracellular medium, these primary transmembrane currents induce secondary ohmic currents that propagate inside the brain. The understanding of these primary and secondary electric currents inside the brain is the main objective of electrophysiologists. As a result of these currents, an electric potential difference can be observed between a pair of electrodes placed at arbitrary locations inside the brain. To increase spatial resolution and specificity in the type of neuronal activity recorded, these intracranial electric potentials are normally recorded with high-resolution microelectrodes arrays (MEA). Depending on the filtering frequency bands, intracranial electric potentials can be classified as either local field potentials (LFP, $0 \leq \omega \leq 500\,Hz$) or multiunit activity (MUA, $500\,Hz < \omega < 20\,kHz$). Electrical potential differences can be also recorded using a grid of electrodes placed on the cortical sheet, a technique referred as electrocorticogram (ECoG). A finer classification might also include subcutaneous, epidural, and subdural recordings, which are not discussed in this book. Secondary ohmic currents can also flow through the less conductive skull reaching the outer surface of the scalp. Because of the parallel-like organization of principal neurons in the brain facilitating electric current summation, potential differences on the scalp can been observed in particular experimental situations. These scalp electric potentials are referred as electroencephalogram (EEG). Activation of neurons cause also intracellular currents, whose directions are defined by the neuronal morphology. These microscopic intracellular currents, in conjunction with the secondary ohmic currents generate observable magnetic fields outside the head, a technique referred to as magnetoencephalogram (MEG). The amplitude of the magnetic field is determined by the extension of the brain activated area and the morphological characteristics of the activated neurons. In this chapter, the physical foundations of electrophysiological recording are presented. The particulars for large-scale electrophysiological techniques (EEG and MEG) are developed in detail. Applications of the theoretical framework developed in this chapter to understand intracranial electric recordings (LFP/MUA) and ECoG are discussed in Part 2 of this book series. In addition, this theoretical framework is also useful to understand other types of electrophysiological

DOI: 10.1201/9780367901509-2

recordings not discussed in the book (e.g., electro/magneto-cardiogram – heart, electromyogram – muscles, electrooculography – eyes, and electroretinography – retina). A primer review of the EEG and MEG techniques was published by Lopes da Silva (2013).

2.2 THE ELECTROENCEPHALOGRAM – ELECTRODE SYSTEMS AND REFERENCES

A brief history of the EEG – Luigi A Galvani (1737–1798) was the first scientist to recognize the connection between electricity and life. For that reason, he is considered the father of the bioelectricity. In 1786, he found that the leg of a dead frog twitched if touched with a metal knife, which prompted him to suggest the existence of a "nervo-electrical substance" activating muscles in living organisms. Galvani's suggestion was in line with the one-fluid theory proposed by Benjamin Franklin (1706–1790) thirty-nine years earlier. As Galvani also studied the relationship between chemical reactions and electricity, he is also referred to as the father of electrochemistry. By 1792, Alessandro GAA Volta (1745–1827), who disagree with Galvani's suggestion, associated the finding to electricity created when a moisture is placed between two metals, in this case, the knife and the metal plate upon which the frog was placed. This general principle led Volta to develop the first electrochemical cell (battery). Applications to study the electrical activity in muscles and the heart advanced quickly in the field of bioelectricity. First, stimulating electrodes were used in the 1780s to study muscle contractions by the measurement of spring extension. The Lippmann capillary electrometer was invented around 1872 by Jonas FG Lippmann (1845–1921), which allowed measuring voltage changes in a nerve. The first electrocardiography (ECG/EKG) machine was created in 1887 by Augustus D Waller (1856–1922), which consisted of a projector and a train toy. The creation of string galvanometers by Willem Einhoven (1860–1927) triggered the introduction of the ECG into practice from 1901, a work for which he received the Nobel Prize in Medicine in 1924. Emil du Bois-Reymond (1818–1896) discovered the nerve action potential (AP) in electric fish around 1848. For this work, he is considered the father of experimental electrophysiology. Edouard Hitzig (1838–1927) and Gustav Fritsch (1838–1907) induced focal seizures in dogs by applying strong electrical stimulation on their cerebral cortex (1870). They were trying to study the Jacksonian march, first clinically described by John Hughling Jackson (1835–1911). Five year later, Richard Caton (1842–1926) registered spontaneous and evoked electrical activity from the brains of rabbits and monkeys using a sensitive galvanometer. He reported the existence of fluctuations in brain activity during sleep and absence after death. Advances in instrumentation allowed Napoleon Cybulski (1854–1919) to report in 1914 the first ECoG during an epileptic seizure induced in a dog by electrical stimulation. The first spontaneous and evoked electrical activity from the brain of dogs was reported in 1913 by Vladimir Pravdich-Neminsky (1879–1952). However, the most transcendental finding for clinical electrophysiology, a brain electrical rhythmic activity from the human brain, was reported by Hans Berger (1873–1941) from 1929 to 1930. The Berger's rhythm in humans is known today as the α-band rhythm (7.812–13.28 Hz), which exists also in other mammals. The finding by Berger triggered the massive application of EEG to study the normal and pathological human brain. Although there were many contributions in the post Berger period, only a mention is made here to four of the most relevant lines of clinical and research applications: **1)** epileptiform type of activity were reported in several studies in the 1930s, **2)** EEG studies during sleep gained interest in the 1940s, **3)** the kappa rhythm linked to thinking was reported in 1948, and **4)** psychiatric patients were treated with chronically implanted electrodes via a frontal leucotomy in the 1950s, which were also used to characterize seizures in epileptic patients. A more detailed historical overview of this EEG golden period can be found in these two reviews (Collura, 1993; Stone and Hughes, 2013).

EEG equipment – Together with these applications, technological developments were quickly incorporated to improve the quality of EEG data and to allow hardware-based EEG post-processing. Apparatus to observe EEG evolved from the smoked drum kymograph to the polygraphs (i.e., ink-writing amplifiers) with roll and folding-chart papers, which was followed by the Grass EEG recorders (models I-III). Further technological advances in the fabrication of electron tubes, transistors with memory and microprocessors led to the development first of cathode-ray oscilloscopes and later of the digital EEG amplifiers (1980s). The definition of spontaneous (spectral analysis) and evoked potentials (time-locked averaging) from periphery stimulation or mentalizing states, stimulated early the respective construction of EEG recording systems with Fourier analyzers (e.g., the Grass-Gibbs analyzer) leading to compressed spectral arrays (CSA) in the 1940s and the George Dawson technology (i.e., the storage capacitor) for noise cancellation during repetitive synchronized responses in the 1950s. The development of multi-channel EEG systems was facilitated by the advances in sub-miniature vacuum tubes and silicon solid-state devices.

Currently, commercial EEG systems, comprising dozens-to-hundreds of electrodes, are made of differential amplifiers. Typical differential amplifiers are AC-coupled, i.e., the offset EEG signal is eliminated. However, DC-coupled amplifiers have been used in some studies. Figure 2.1 shows a particular DC amplifier (BrainAmp DC) with a 64 channel EEG caps (EasyCap). Each differential amplifier records a pair of electrodes, which can be set using either a sequential (bipolar) or a referential (common reference, e.g., the linked ears) montage. A 3D position/orientation tracking system (Isotrack II, Polhemus) is typically used to localize electrodes with respect to MRI visible landmarks (e.g., vitamin E or fish oil capsules). Differential amplifiers could achieve $60-100\,dB$ of voltage gain. EEG signal is first anti-aliasing band-pass/notch filtered and then digitalized using analog-to-digital converter (ADC) at sampling rate of up to $512\,Hz$. Because of the high impedance of microelectrodes, differential amplifiers used to record LFPs and MUA required very high input impedance ($T\Omega$). Also, sampling rates of up to $20\,kHz$ is required to detect/reconstruct neuronal spikes from MUA. Referential EEG data post-processing includes special operations to remove the ambiguity in the selection of the common reference, e.g., **a)** average reference, **b)** REST, and **c)** Laplacian. Wet electrodes are typically small disks made of a high-conductive material (e.g., gold, stainless steel, platinum, titanium, or silver. Electroplating is also a very popular technique to reduce electrode impedance. Wet electrodes can be also coated using chemical procedures (e.g., silver-chloride and iridium-oxide). These wet electrodes connect to the scalp through a conductive gel. However, recent

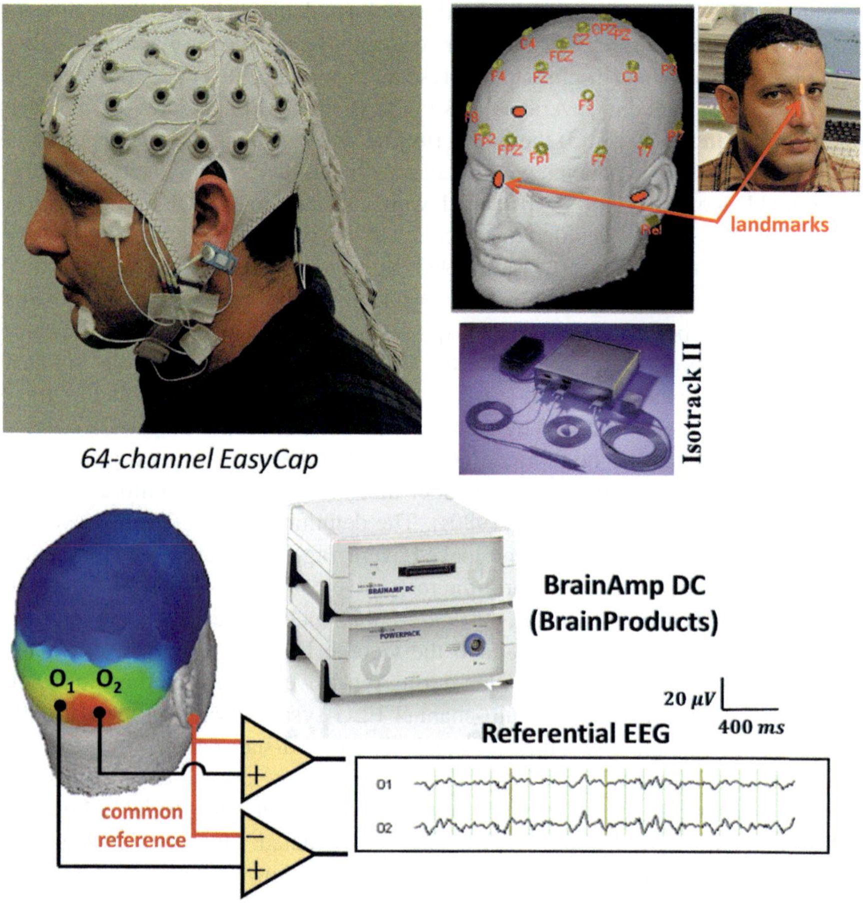

FIGURE 2.1 Technical aspects for EEG data recording: 1) a picture of myself with the 64-channel EEG caps (BrainCap MR, EasyCap) as a postdoctoral fellow at the Ryuta Kawashima's lab (2003), 2) the 10/20 electrode positions are co-registered to the individual MRI using the Isotrack II system, 3) the BrainAmp DC amplifiers is shown on the middle panel (from BrainProducts website), and 4) time series from two occipital electrodes (O_1 and O_2) under a closed-eyes experimental condition and the α-peak topographic map are on the bottom.

advances in conductive materials and integrated electronic technologies have led to the generation of gel-free (dry and/or active) EEG electrodes (Lopez-Gordo et al., 2014), which has boosted portable EEG applications for long-lasting EEG studies. There are many EEG electrode configurations. However, the 10/20 and 10/10 reference systems are the most used in clinical studies and research (Figure 2.2). Highly dense EEG electrode configurations have been also introduced in the past (e.g., the Geodesic Sensor Net and the 10/5 system; Oostenveld and Praamstra, 2001). These reviews discussed important aspects of electrodes (Im and Seo, 2016; Shad et al., 2020).

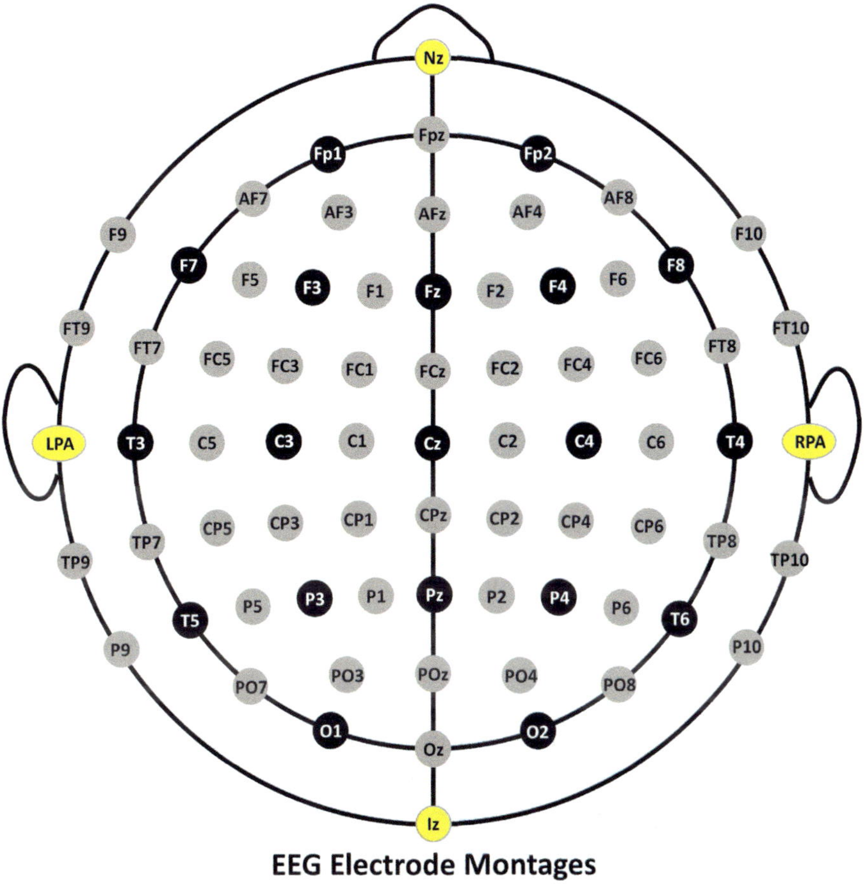

FIGURE 2.2 The 10/20 (black) electrode system, with the respective extension to the 10/10 system (grey).

Note: The 10/10 system renames four electrodes of the 10/20 system (*T3 → T7, T4 → T8, T5 → P7 and T6 → P8*). Anatomical landmarks (yellow) are used for positioning the EEG electrodes (Nz – Nasion, Iz – Inion, RPA (LPA) – Right (Left) Preauricular.

2.3 THE MAGNETOENCEPHALOGRAM – MAGNETOMETER/ GRADIOMETERS

A brief history of the MEG – The history of MEG begins in 1963 with the discovery of magnetic signals from the heart using a gradient detection technique (two adjacent/identical coils) (Baule and McFee, 1963), i.e., named the magnetocardiography (MCG). This gradient technique was essential to remove ambient background magnetic noise, including the earth's magnetic field. Environmental magnetic noise can be of the order of $10^8\ fT - 0.1\mu T$; hence it is important to remove it. Using multiple recording locations, chest maps of magnetic potentials were created. These observations were confirmed later even with a single coil but with the subject inside a shielding room to remove background noise (Cohen 1967). Measuring a magnetic signal from the brain associated with the human α-rhythm was more challenging as this signal is in the order of $10^3\ fT$. The first observation of the MEG signatures of the α-rhythm in humans, concurrently recorded with a pair of EEG electrodes, was obtained by David Cohen in a multilayer (moly-permalloy and aluminum) magnetically shielded chamber constructed at MIT (Cohen, 1968). However, the most important advancement in MEG data acquisition was linked to the use of the RF superconducting quantum interference device (SQUID) magnetometer developed by James E Zimmerman at Ford Motor Company. By using such a magnetometer installed inside the shielded chamber, Cohen (1972) was able to measure MEG signals with a high sensitivity ($< 4\times10^{-10}\ gauss$), which allowed him to record spontaneous cortical activity of about $10\,fT$ in amplitude. MEG systems based in SQUID technology have been commercially available since the 1980s. The CTF systems were among the first (Figure 2.3). Current magnetically shielded rooms, known as a passive method, consist of nested layers made of an aluminum sublayer and multi-sheets of high-permeability ferromagnetic materials. Active methods consist of fluxgate magnetometers for noise cancellation from specific directions (Platzek et al., 1999). A fluxgate magnetometer is composed of an excitation coil and a pick-up coil each on a ferromagnetic core material. With this method, attenuation for DC field is reached in more than $60\,dB$. Optically pumped magnetometers (OPM) have been also proposed in the early 2000s as a next generation of non-cryogenic MEG sensing technology. OPM are based on the spin-exchange relaxation-free (SERF) principle proposed by Happer (1972). Combined with flux guides (FG), OPM can achieve sensitivity of up to $8.1\,pT\,/\,Hz$ (Kim and Savukov, 2016).

Principles of the SQUID – A general description about the RF SQUID working principle is provided here, with no specification to most of the compensation/amplification electronics underlying MEG signal measurements. RF SQUID are based on

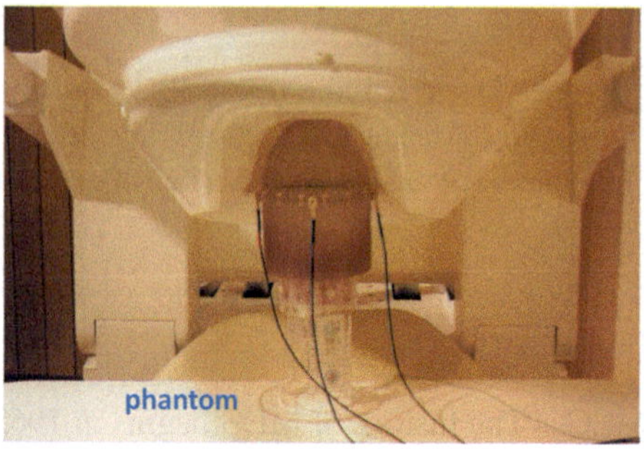

CTF Systems Inc

FIGURE 2.3 Whole-head Omega biomagnetometer (CTF Systems) and calibration phantom at the Brain Science Institute, RIKEN, Japan (Andreas Ioannides's lab).

Note: The head position is dynamically monitored using external coils as those shown on the phantom.

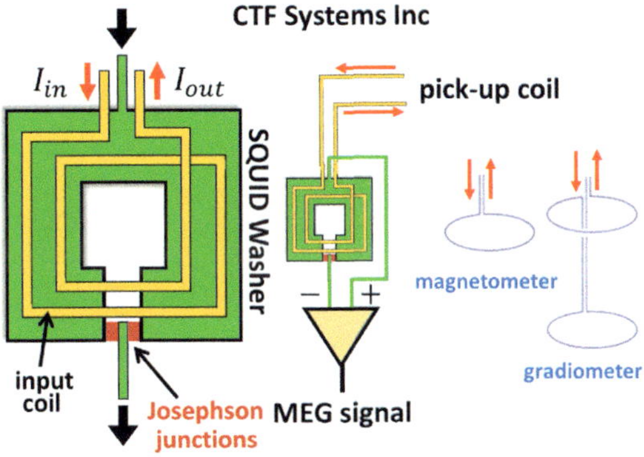

FIGURE 2.4 Illustration of the physical principles of MEG sensing with SQUID.

the AC Josephson's effect (Figure 2.4). Changes in magnetic fluxes through pick-up coils are induced by time-varying primary/secondary electric currents in the brain. Depending on design, there are two classical pick-up coils: magnetometers and gradiometers. These coils are placed in cryostat cylinders (liquid helium), which are arranged in a way that cover most part of the head pointing perpendicular to the scalp. Pick-up coils are connected to input coils, which are placed on top of a SQUID Washer made of a superconducting material. Electric currents circulating along the input coils induce supercurrents in the SQUID Washer. The washer comprises weak links "Josephson junctions" through which a voltage drops. A relationship between the supercurrent $I(t)$ and the voltage across the weak link $V(t)$ was proposed by Brian D Josephson in 1962, which consists of two equations: $I(t) = I_c sin(\varphi(t))$ and $\partial\varphi(t) / \partial t = K_J V(t)$. The junction acts as a voltage-to-frequency converter. The critical current I_c depends on the superconductor. $\varphi(t)$ is known as the Josephson's phase. Electrons flow through the junctions by quantum tunneling, an effect that led him to receive the Nobel Prize in 1973. Measurements of the losses in the voltage allows the estimation of the magnetic flux across the pick-up coils. Theoretical details for the AC SQUID effect can be found in Erné et al. (1976).

As an example, Figure 2.5 shows these magnetic flux (MEG signal) measurements from an epileptic patient. In this case, magnetometers were co-registered with the patient's MRI. The topography color map corresponding to an inter-ictal spike (dashed vertical blue line) is also plotted.

2.4 THE THREE FUNDAMENTAL LAWS OF ELECTROMAGNETISM – THE INTEGRAL FORMS

The history of electricity has been tightly linked to important progresses in electrophysiology. Therefore, this section presents a description of major discoveries leading to the current theoretical construct for electromagnetism. It all starts with the first discovery by the Greek philosopher Thales of Miletus (c.624 BC–c.546 BC) about the property of amber to attract lightweight objects after being rubbed against cloth,

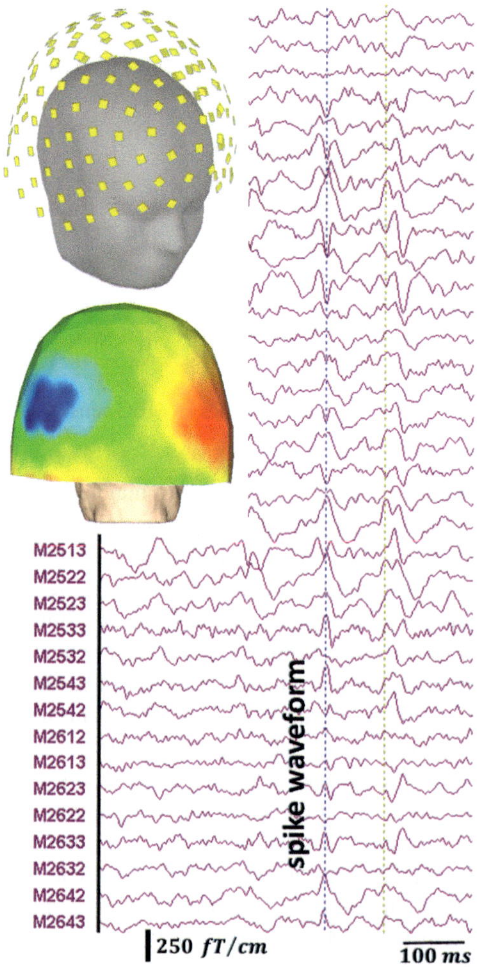

FIGURE 2.5 MEG time series from an epileptic patient (Tohoku University).

which is known today as the triboelectric effect. The triboelectric effect is the basis of static shock, which might have been experienced by many humans for centuries. The interest for the understanding of the triboelectric effect grew after Otto von Guericke (1602–1686) created the first electric generator in 1663, which led to the two-fluid theory. In 1747, Franklin proposed a one-fluid theory for the charging and discharging of a body. Although not yet known at that time, the concept of the elementary charge $q_e = 1.602176634 \times 10^{-19}\, C$ (single proton/electron) underlay the referred fluid. The quantization of the charge into multiples of 1/3 is still debated as quarks are not considered isolated particles. Using a torsion balance, the force between two charged spheres separated by a distance r was measured by Charles A de Coulomb (1736–1806) by 1785, which is known as the Coulomb law. Assuming q and q' the respective charges of these spheres, the Coulomb law can be mathematically written as $\vec{F}(\vec{r}) = \frac{k_e qq'}{r^2}\hat{r}$, with $k_e = 8.99899 \times 10^9 Nm^2 / C_{-2}$ representing the Coulomb's constant. By definition, $k_e = 1/4\pi\varepsilon_0$, with $\varepsilon_o = 8.854 \times 10^{-12} F/m$ representing the electric permittivity for the vacuum. Note that the force is a vector pointing along the line connecting the charges (Figure 2.6A) and its direction depends on the signs of these charges (i.e., repulsion – equal sign and attraction – opposite sign). The Coulomb force could be represented in terms of a radially distributed electric field $\vec{E}(\vec{r}) = q\hat{r}/4\pi\varepsilon_0 r^2$, leading to $\vec{F}(\vec{r}) = q'\vec{E}(\vec{r})$, associated later to the concept of electric line of force.

The debate between Galvani and Volta in the following years must have injected fuel into the research community, leading to the three most important physical laws describing electromagnetic phenomena proposed in a seventeen-year period (1813–1831). First, further development and use of the electrochemical cells led to the discovery of the electrolysis and electroplating. These two concepts are the fundamental building block of any electrophysiological observation. In 1832, Faraday's experiments (Michael Faraday, 1791–1867) suggested the discrete character of the charge according to his two laws of electrochemistry relating the mass *m*, the molar mass *M*, the valence *v* and the charge *Q* of a substance, which could be mathematically formulated as: $m = MQ / Fv$. $F = 96485.33212\ C / mol$ is the Faraday's constant. However, it was not until the observation by Robert A Millikan (1868–1953) in 1910 that the elementary charge was finally verified, a work leading him to receive the Nobel Prize in 1923. Magnetic phenomena have been observed also for many years, especially due to the natural existence of permanent magnets in the form of rock minerals (e.g., lodestone). However, it was the discovery by Hans C Ørsted (1777–1851) in 1820 on the magnetic effect of electric currents what iconized the term magnetism in the scientific community. He noticed that a needle of a compass oriented perpendicular to a nearby wire if an electric current was flowing along it. Based on this result, André-Marie Ampère (1775–1836) proposed the same year a law to describe the attraction/repulsion force between two parallel current-carrying wires. The first quantitative description of the Øersted's effect was carried out by Jean-Baptiste Biot (1774–1862) and Félix Savart (1791–1841) by observing the timing of damped oscillations of a magnetic needle placed perpendicular to and at different distances *r* of a long (~ ∞) wire carrying a constant current *I* (C / s). Based on the observed period of these oscillations, they proposed a model for the differential force exerted by the wire on the needle. In modern mathematical terms, the Biot-Savart law can be represented as $\vec{B}(\vec{r}) = \mu_0 I(\hat{l} \times \hat{r}) / 2\pi r$ (Figure 2.6B). Using the differential action/wire concept proposed by Pierre-SM de Laplace (1749–1827), a more general formulation of this law for a wire with an arbitrary shape *C* and an observation location $\vec{r}$ in the 3D space is $\vec{B}(\vec{r}) = \frac{\mu_0}{4\pi}\int_C \frac{I(\vec{r}')d\vec{l}' \times (\vec{r} - \vec{r}')}{(\vec{r} - \vec{r}')^3}$.

In 1831, Michael Faraday (1791–1867) used a galvanometer to measure the electric current generated in a wire wrapped to a side of an iron ring (torus) when another independent wire wrapped into the opposite side of the torus was connected/disconnected to/from a battery (Figure 2.6C). He called this current the *wave of electricity* and associated it to a new concept, the "*line of force*". Transient currents were also reported by Faraday whenever a magnet was moved quickly in and out of a coil. In 1834, Emil K Lenz (1804–1865) proposed the concept of "magnetic flux" through an area S limited by a close wire ($\phi(t) = \int_S \vec{B}(\vec{r},t) \cdot d\vec{S}$), and associated it to an induce electromotive force (EMF): $\varepsilon_f(t) = -d\phi(t)/dt$.

Although the quantum character of the charge is important to describe phenomena at the subatomic levels, it is convenient to work with a continuous distribution of charge density $\rho(\vec{r})$ when modeling electromagnetic effects in large bodies. The total charge Q_T inside a volume V is defined as $Q_T = \int_V \rho(\vec{r})dV$. Under this representation, the particular case of a point charge q (e.g., of an electron) at location $\vec{r}_q$ is represented as $\rho(\vec{r}) = q\delta(\vec{r} - \vec{r}_q)$, with δ the Dirac delta function. Based on the principle of superposition of electric fields $\vec{E}(\vec{r})$ and the Coulomb law, Johann CF Gauss (1777–1855) proposed in 1813 that the flux of the electric field across a closed surface S is proportional to the total electric charge Q_T in the volume V limited by S. The Gauss's law (2.1) constitutes the first of the four fundamental laws of electromagnetism.

$$\oint_S \vec{E}(\vec{r}) \cdot d\vec{S} = \frac{Q_T}{\varepsilon_0} \tag{2.1}$$

Electric currents inside a volumetric region are better represented by the electric current density $\vec{J}(\vec{r})$ (A/cm^2). Trivially, an electric current through a thin wire of radius R and oriented in the $\hat{l}$ direction can be represented as a function of a volumetric current $\vec{j}$ as $I = \pi R^2 \vec{j} \cdot \hat{l}$. In general, the total flux I_S of the electric current density through a surface S is defined $I_S = \int_S \vec{j}(\vec{r}) \cdot d\vec{S}$. The Ampère's circuital law (2.2) proposed in 1823 constitutes the second fundamental law of electromagnetism. It stablishes that the total flux I_S is proportional to the line integral of the magnetic field on the closed-loop wire L limiting surface S.

$$\oint_L \vec{B}(\vec{r}) \cdot d\vec{l} = \mu_0 I_S \tag{2.2}$$

Note that equations (2.1) and (2.2) imply these cause-effect phenomena occur instantaneously, i.e., a volumetric charge Q_T and a surface current flux I_S cause static electric and magnetic fields, respectively. The Faraday's law of induction constitutes the third fundamental law of electromagnetism and the one involving temporal dynamics for the first time. Several years after discovered, a mathematical expression for the induction law was provided:

$$\oint_L \vec{E}(\vec{r},t) \cdot d\vec{l} = \varepsilon_f(t) \tag{2.3}$$

The close-loop wire L encloses the area S through which the magnetic field flux $\phi(t)$ is varying. Because of the Lenz's law $\varepsilon_f(t) = -\int_S \partial\vec{B}(\vec{r},t)/\partial t \cdot d\vec{S}$, the induction law implies a non-instantaneous relationship between $\vec{E}$ and $\vec{B}$. Therefore, whenever convenient a time variable t will be incorporated. As electric and magnetic field depends on space $\vec{r}$ and time t, the total derivative d/dt will be substituted by a partial derivative $\partial/\partial t$.

Although many theoreticians are still searching for a free magnetic pole, its existence has not been demonstrated experimentally. Based on a preliminary work by William Gilbert (~1544–1603), Faraday proposed the final fundamental law, the Gauss's law of magnetism for the absence of magnetic monopoles:

$$\oint_S \vec{B}(\vec{r}) \cdot d\vec{S} = 0 \tag{2.4}$$

Electric change continuity law: Assuming $Q_T(t)$ is the total change inside a volume V at a giving time instant t, which is distributed as a charge density $\rho(\vec{r},t)$ (Figure 2.6D). Because of mass conservation, changes in $Q_T(t)$ over time $(I(t) = dQ_T(t)/dt)$ must be reflected in a flux of charge $I(t) = \oint_S \vec{j} \cdot d\vec{S}$ across the surface S limiting V. Hence, the charge $\rho(\vec{r},t)$ and current $\vec{j}(\vec{r},t)$ densities hold a continuity relationship, mathematically represented as:

$$\int_V \frac{\partial \rho(\vec{r},t)}{\partial t} dV = \oint_S \vec{j}(\vec{r},t) \cdot d\vec{S} \tag{2.5}$$

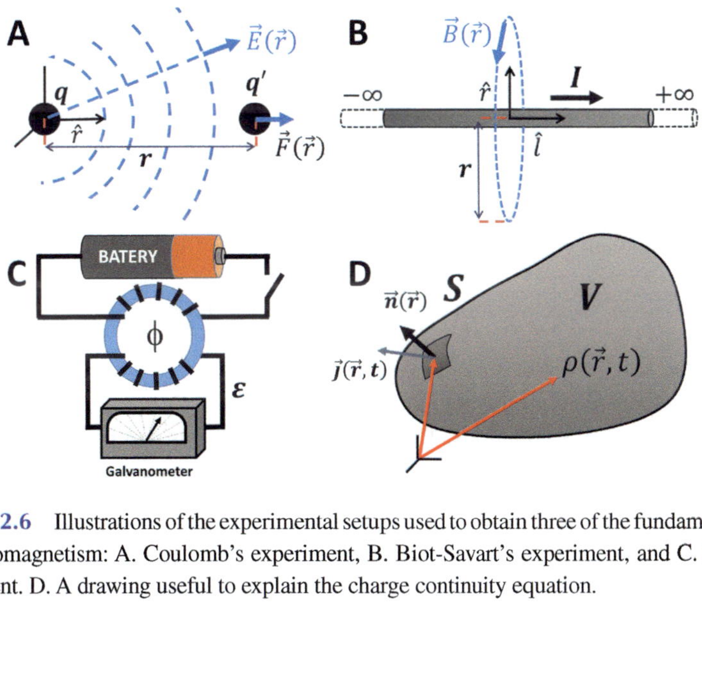

FIGURE 2.6 Illustrations of the experimental setups used to obtain three of the fundamental laws of electromagnetism: A. Coulomb's experiment, B. Biot-Savart's experiment, and C. Faraday's experiment. D. A drawing useful to explain the charge continuity equation.

2.5 THE MICROSCOPIC MAXWELL EQUATIONS – THE DIFFERENTIAL FORMS

To transform all these laws from their integral to differential forms, it is necessary to introduce the gradient (∇), the divergence ($\nabla\cdot$) and the rotational ($\nabla\times$) operators. Readers are referred to appendix A for a summary of most important operators and identities in vector calculus and needed in this book. Using the Gauss's (A.1) and Stoke's (A.2) theorems, the fundamental laws can be written in their differential forms as:

$$\nabla\cdot\vec{E}(\vec{r},t)=\frac{\rho(\vec{r},t)}{\varepsilon_0} \quad \textit{Gauss's law}-\textit{electric field} \tag{2.6}$$

$$\nabla\cdot\vec{B}(\vec{r},t)=0 \quad \textit{Gauss's law}-\textit{magnetic field} \tag{2.7}$$

$$\nabla\times\vec{E}(\vec{r},t)=-\frac{\partial\vec{B}(\vec{r},t)}{\partial t} \quad \textit{Faraday's law} \tag{2.8}$$

$$\nabla\times\vec{B}(\vec{r},t)=\mu_o\vec{j}(\vec{r},t) \quad \textit{Ampere's law} \tag{2.9}$$

For the deduction of these three equations, it was considered that volume and surfaces are always positives, i.e., $dV\geq 0$ and $d\vec{S}\geq 0$. Also using the Gauss's theorem, the differential form of the charge continuity law is defined as:

$$\nabla\cdot\vec{j}(\vec{r},t)+\frac{\partial\rho(\vec{r},t)}{\partial t}=0 \tag{2.10}$$

In its current form, the Ampere's law does not comply with the charge continuity equation as applying the fundamental identity (A.5) of vector calculus to equation (2.9) results in $\nabla\cdot\vec{j}(\vec{r},t)=0$, which disagrees with the charge continuity equation (2.10). In his work on the dynamical theory for the electromagnetic field (1864), James C Maxwell (1831–1879) resolved the inconsistency in Ampere's law by adding to it the displacement current $\varepsilon_0\partial\vec{E}(\vec{r},t)/\partial t$.

$$\nabla\times\vec{B}(\vec{r},t)=\mu_0\vec{j}(\vec{r},t)+\mu_0\varepsilon_0\frac{\partial\vec{E}(\vec{r},t)}{\partial t} \tag{2.11}$$

The set of partial differential equations (2.6–2.8), and (2.11) are known as the microscopic Maxwell's equations. They describe the genesis and propagation of the electromagnetic field in the vacuum. The velocity of this propagation is $c=1/\sqrt{\mu_0\varepsilon_0}$; i.e., the speed of light. For simplicity, the dependency with $\vec{r}$ and t (or ω for the frequency domain) of physical magnitudes will be sometimes omitted in this book. Maxwell's system of equation is overdetermined as it comprises eight differential equations and six unknown variables, i.e., $\vec{E}$ and $\vec{B}$. Therefore, these equations are not fully independent. For example, equation (2.7) can be obtained from taking the divergence at both sides of equation (2.8), using fundamental identity (A.5), and imposing the condition of being zero at the infinity (radiation). Similarly, equation (2.6) is a consequence of equations (2.11) and (2.10), also using fundamental identity (A.5) and imposing a radiation condition.

The microscopic Maxwell's equations are not fully valid while describing the behavior of the electromagnetic field in biological tissues, which are composed of atoms, molecules, and cells. These elements can react to a time-varying electromagnetic field, affecting its mechanism of genesis and propagation. In brain tissues, there are three major mechanisms that can potentially contribute to the electric currents $\vec{j}$: **1)** convective, **2)** diffusive, and **3)** primary. Convective currents originate from the physical movement of the head with respect to the observation sensors; hence, they are negligible as EEG and MEG recordings are performed using electrodes and magnetometers/gradiometers that are fixed respect to the head. Contributions by diffusive currents have been represented by two different theoretical frameworks. The Poisson-Nernst-Planck equation for electrodiffusion is a theoretical framework that can be used to model the contribution from a specific ion to electro/magneto-physiological data. There is substantial evidence to ignore this type of contribution in the particular case of EEG and MEG data. Nevertheless, major implications of this assumption are discussed in Part 2 of this book series. A second theoretical framework models the effect of bulk ion diffusivity in the electromagnetic field. In this case, the brain is assumed to react to the application of an external electric field by a collective movement of ions, which will result in a secondary electricity current $\vec{j}_{ohm} = \ddot{\sigma}\vec{E}$. This electric current is also referred as ohmic because of the linear constitutive relationship with $\vec{E}$ (Georg S Ohm, 1789 –1854). The 2nd order tensor $\ddot{\sigma}(\vec{r})$ is used to represent anisotropies and inhomogeneities in brain tissues as they respond to an applied $\vec{E}$. Complex numerical methods are needed to model EEG and MEG data in the case of an anisotropic brain, e.g., finite element methods (FEM). For an isotropic brain, $\ddot{\sigma}(\vec{r}) = \sigma(\vec{r})\ddot{I}$, with $\ddot{I}$ representing the identity 2nd order tensor. Boundary element methods (BEM) are used instead to model EEG and MEG data for an isotropic piecewise homogenous brain, discussed later in this chapter. The principles lying behind the electric conductivity are discussed in Chapter 5. The contribution to $\vec{j}$ due to the activation of neurons in the brain is denominated the primary current density $\vec{j}^{(p)}$. To understand $\vec{j}^{(p)}$, it is convenient to introduce the concept of primary volumetric current source $s^{(p)}$, such that $s^{(p)} = \nabla \cdot \vec{j}^{(p)}$. The magnitude $s^{(p)}$ represents the total trans-membrane current from neurons inside an elementary volume at a particular position in the brain. Under these considerations, the total electric current takes the form:

$$\vec{j}(\vec{r},t) = \sigma(\vec{r})\vec{E}(\vec{r},t) + \vec{j}^{(p)}(\vec{r},t) \tag{2.12}$$

There are two other properties in brain tissues that can affect the genesis/propagation of electromagnetic field: **a)** polarization $\vec{P}$, and **b)** magnetization $\vec{M}$. For biological tissues, the constitutive relationship between $\vec{B}$ and $\vec{M}$ is typically linear $\vec{M} = \chi_m \vec{B} / \mu_0$. However, as the volume magnetic susceptibility of biological tissue is very small ($|\chi_m| \ll 1$), magnetization in the brain has been traditionally neglected. The effect of polarization of brain tissues in electro/magneto-physiological data is still under debate. Differences in MEG and EEG power-scaling have been associated with non-resistive properties of brain tissue (Dehghani et al., 2010). However, a more recent result suggested that difference in the spatiotemporal coupling as results of geometrical considerations might also explain the discrepancy in the power scaling even for a purely resistive brain (Bénar et al., 2019).

2.6 TISSUE POLARIZATION – MULTIPLE RELAXATION MECHANISMS

To understand tissue polarization in the brain, it is necessary to introduce the three fundamental dispersion mechanisms based on their frequency range: a) α-dispersion ($10^0 - 10^4\ Hz$), b) β-dispersion ($10^4 - 10^7\ Hz$), and c) γ-dispersion ($> 10^7\ Hz$). An important type of tissue polarization in the brain occurs from a microscopic separation of positive and negative ionic charges as tissue reacts to a time-varying electromagnetic field at different physical levels. Such a charge separation creates dipoles that can absorb electromagnetic energy. One form of charge separation occurs due to ion displacements in a solution, an effect amplified in the brain due to the existence of channel selectivity to particular ions. Note that the mechanisms underlying ionic polarization in crystals are different. Shilov and Borkovskaya (2010) were the first to propose a biophysical model for this selective ionic polarization mechanism and associated it to a *diffusion-controlled* α-dispersion. The charging of interfaces separating different cellular compartments could also create tissue polarization in the brain. A type of interfacial polarization has been associated with the presence of thin double-layer lipid membranes, which have a very low electric conductivity. Pauly and Schwan (1959) proposed a biophysical model to describe this type of interfacial polarization. Due to the high capacitance and small (*nm*) thickness of the cell membrane, the Pauly and Schwan's dispersion has a critical frequency in the β range. Reorganization of the charge on the surface of particles suspended in a solution (colloids) in response to an electromagnetic incident wave cause a type of bulk interfacial polarization, described by the *Maxwell-Wagner* γ-dispersion model.

There are several other dispersion mechanisms in the brain, which are not discussed here because they are irrelevant in the frequency range for electric/magnetic recording. For example, molecular dipoles exist also naturally in biological tissue (e.g., the water molecule), which can contribute to polarization by dipole rotation (*orientation polarization*). Furthermore, as the brain is also formed by molecules/atoms, charge separation could also happen because of deformations in atoms (*electronic polarization*) and molecules (*atomic polarization*). Electronic/atomic polarization are referred to as resonance phenomena, while all other types are referred to as relaxation phenomena. Illustrations of the three most important mechanisms for tissue polarization in the brain (i.e., diffusion-controlled, Pauly-Schwan, and Maxwell-Wagner) are shown in Figure 2.7. A general formula for the induced dipolar moments in these three types of brain tissue polarization was proposed by Shilov and Borkovskaya (2010).

Polarization $\vec{P}(\vec{r},t)$ counteracts the effect of an applied $\vec{E}$, which results in a reduce electric field $\vec{D}$, named electric displacement field (2.13).

$$\vec{D}(\vec{r},t) = \varepsilon_0 \vec{E}(\vec{r},t) + \vec{P}(\vec{r},t) \tag{2.13}$$

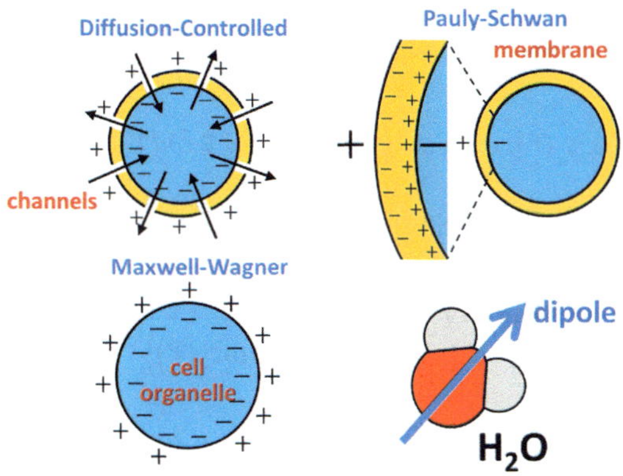

FIGURE 2.7 Illustrations of the four mean relaxation mechanisms in the brain.

Under the assumption of linearity, the polarization effect in the brain can be described by the following constitutive relationship in the frequency domain: $\vec{P}(\vec{r},\omega)=\chi_e(\vec{r},\omega)\vec{E}(\vec{r},\omega)$. The magnitude χ_e represents the electric susceptibility for an inhomogeneous brain, which in general could be a 2nd order tensor for anisotropic dielectrics. However, anisotropies in the electric polarization within the observation frequency range ($0 \le \omega \le 10\,kHz$) has been rarely referred to for brain tissues in the past. Equation (2.13) can be rewritten as a function of the electric permittivity $\varepsilon(\vec{r},\omega)=\varepsilon_0(1+\chi_e(\vec{r},\omega))$ as: $\vec{D}(\vec{r},\omega)=\varepsilon(\vec{r},\omega)\vec{E}(\vec{r},\omega)$. For a brain with polarization mechanisms, Maxwell's equations (2.6) and (2.11) transform into:

$$\nabla\cdot\left(\vec{D}(\vec{r},t)\right)=\rho_f(\vec{r},t) \tag{2.14}$$

$$\nabla\times\vec{B}(\vec{r},t)=\mu_0\left(\sigma(\vec{r})\vec{E}(\vec{r},t)+\vec{j}^{(p)}(\vec{r},t)+\frac{\partial\vec{D}(\vec{r},t)}{\partial t}\right) \tag{2.15}$$

ρ_f represents the free charge inside the brain. Please note that equation (2.12) was used to obtain (2.15).

Hence, equation (2.15) in the frequency domain is simplified to:

$$\nabla\times\vec{B}(\vec{r},\omega)=\mu_0\left(\left(\sigma(\vec{r})+i\omega\varepsilon(\vec{r},\omega)\right)\vec{E}(\vec{r},\omega)+\vec{j}^{(p)}(\vec{r},\omega)\right) \tag{2.16}$$

It is convenient to represent the electric permittivity as a function of the frequency-dependent complex relative electric permittivity $\varepsilon_r(\vec{r},\omega)$ as $\varepsilon(\vec{r},\omega)=\varepsilon_r(\vec{r},\omega)\varepsilon_0$. The architecture of the brain (i.e., the cellular organization) is highly inhomogeneous; hence, the relaxation mechanisms must be better represented by a parametric model for broadband dispersions (2.17).

$$\varepsilon_r(\vec{r},\omega)=\varepsilon_\infty(\vec{r})+\sum_{n=1}^{N}\int_0^\infty\frac{F_n(\tau,\vec{r})d\tau}{(1+i\omega\tau)} \tag{2.17}$$

By using the distribution $F_n(\tau) = \dfrac{sin(\alpha_n(\vec{r})\pi)}{2\pi(cosh((1-\alpha_n(\vec{r}))ln(\tau/\tau_n(\vec{r}))) - cos(\alpha_n(\vec{r})\pi))}$ with geometric means $\tau_n(\vec{r})$, each satisfying $\int_0^\infty F_n(\tau,\vec{r})d\tau = 1$ for each position $\vec{r}$, the Cole-Cole's model under multiple relaxation mechanisms is obtained:

$$\varepsilon_r(\vec{r},\omega) = \varepsilon_\infty(\vec{r}) + \sum_{n=1}^{N} \frac{\Delta\varepsilon_n(\vec{r})}{1+(i\omega\tau_n(\vec{r}))^{1-\alpha_n(\vec{r})}} \tag{2.18}$$

with $\Delta\varepsilon_n = \varepsilon_s^n - \varepsilon_\infty^n$.

The parameter α_n is called the Cole-Cole's exponent, which indicates how broad is the spectrum of relaxation times. For any *n-th* mechanism, the Cole-Cole's model reduces to the Debye's model (Peter JW Debye, 1884–1966) when $\alpha_n = 1$. ε_s^n and ε_∞^n define the static permittivity and its value at the high frequency limit for the *n*-th relaxation mechanism, respectively. The relaxation time τ_n determines how fast each mechanism responds to the electromagnetic wave. Defining the complex conductivity $\sigma^*(\vec{r},\omega) = \sigma(\vec{r}) + i\omega\varepsilon_0\varepsilon_r(\vec{r},\omega)$, equation (2.16) is rewritten as: $\nabla \times \vec{B}(\vec{r},\omega) = \mu_0(\sigma^*(\vec{r},\omega)\vec{E}(\vec{r},\omega) + \vec{j}^{(p)}(\vec{r},\omega))$. Characterization of the relaxation mechanisms in a particular brain tissue type is performed by measuring the induced total electric current in response to a time-varying electric field applied to a sample of the tissue, i.e., $\vec{J}(\omega) = \sigma^*(\omega)\vec{E}(\omega)$. In this case, tissue homogeneity and absence of primary electric currents are assumed.

In the general case of an inhomogeneous brain, it is useful to decompose the complex electric permittivity into its real and imaginary parts, i.e., $\varepsilon(\vec{r},\omega) = \varepsilon'(\vec{r},\omega) - i\varepsilon''(\vec{r},\omega)$. In this case, the complex electric conductivity expands into two main components.

$$\sigma^*(\vec{r},\omega) = \underbrace{\sigma(\vec{r}) + \omega\varepsilon''(\vec{r},\omega)}_{\sigma_e(\vec{r},\omega)} + i\omega\varepsilon'(\vec{r},\omega) \tag{2.19}$$

The effective conductivity $\sigma_e(\vec{r},\omega)$ includes contributions from the static electric conductivity $\sigma(\vec{r})$ and the imaginary part of the electric permittivity $\omega\varepsilon''(\vec{r},\omega)$. Equation (2.19) can be written as a function of the loss tangent, $tan(\delta_e(\vec{r},\omega)) = \sigma_e(\vec{r},\omega)/\omega\varepsilon'(\vec{r},\omega)$.

$$\sigma^*(\vec{r},\omega) = i\omega\varepsilon'(\vec{r},\omega)(1 - i tan(\delta_e(\vec{r},\omega))) \tag{2.20}$$

The term $tan(\delta_e(\vec{r},\omega))$ comprises losses from two mechanisms: **a)** ions interactions as $\sigma(\vec{r})/\omega\varepsilon'(\vec{r},\omega)$ and **b)** electromagnetic wave dissipation in the form of heat as $\varepsilon''(\vec{r},\omega)/\varepsilon'(\vec{r},\omega)$.

For each given position $\vec{r}$, $\varepsilon'(\vec{r},\omega)$ and $\varepsilon''(\vec{r},\omega)$ must satisfy the Kramers-Kronig causal relationship (linear assumption):

$$\varepsilon'(\vec{r},\omega) - \varepsilon_\infty(\vec{r}) = \frac{2}{\pi}\int_0^\infty \frac{x\varepsilon''(\vec{r},x)}{x^2 - \omega^2}dx$$

$$\varepsilon''(\vec{r},\omega)-\sigma(\vec{r})/\omega\varepsilon_0=\frac{-2\omega}{\pi}\int_0^\infty\frac{\varepsilon'(\vec{r},x)-\varepsilon_\infty(\vec{r})}{x^2-\omega^2}dx$$

Using these experimental methods, the electrical properties of brain tissue have been studied in the past by many. Based on previous experimental observations (Gabriel Gabriel, and Corthout, 1996) and on his own data (Gabriel et al., 1996a), Gabriel et al. (1996b) estimated the parameters (Table 2.1) of the Cole-Cole's model (2.18) for five important tissue type in the head: **a)** two elements of the bone (cancellous and cortical), **b)** the two most important components in the brain (gray matter and white matter), and **c)** the dry skin.

The authors needed to include four relaxation mechanisms in order to properly explain data from the three dispersion frequency ranges. Figure 2.8 shows separated curves of the four relaxation mechanisms for the particular case of gray matter. The critical frequency for the slowest mechanism was in the α-dispersion range (188.5 Hz). For the second mechanism, the critical frequency was $0.94\times10^4\,Hz$, which is right on the lower limit of the β-dispersion. The third mechanism has a critical frequency of $6.2814\times10^7\,Hz$, which could be either a high β- or low γ-dispersion. Although accounted for a very small portion of the relative electric permittivity ($\varepsilon_r<100$), a four relaxation was fitted with critical frequency ($2563\times10^{11}\,Hz$) in the high γ-dispersion range.

The latter mechanisms might be associated with orientation polarization associated with water molecules (Figure 2.7). Therefore, it is proposed a link between three of the reported relaxation mechanisms and the three previously described polarization types in the brain. Inhomogeneity in the electric permittivity of different brain tissues were more pronounced than that for their respective conductivities (Figure 2.9).

2.7 GAUGE – ELECTRIC AND MAGNETIC POTENTIALS

The general (macroscopic) Maxwell equations (2.7 and 2.8) and (2.14 and 2.15) for the electric and magnetic fields can be expressed as a function of the electric ϕ (scalar) and magnetic $\vec{A}$ (vector) potentials. Considering the fundamental identity (A.5) of vector calculus, the magnetic vector potential is defined from the Gauss's law for $\vec{B}$ (2.7) as:

$$\vec{B}(\vec{r},t)=\nabla\times\vec{A}(\vec{r},t) \tag{2.21}$$

Using (2.21) in the Faraday's law and considering another fundamental identity (A.6), the electric potential is defined as:

$$\vec{E}(\vec{r},t)=-\nabla\phi(\vec{r},t)-\frac{\partial\vec{A}(\vec{r},t)}{\partial t} \tag{2.22}$$

The minus sign $-\nabla\phi$ was added for convenience in future notation. The gradient operator ∇ is defined in appendix A (A.3). The concept of potentials in physics was introduced by Joseph L Lagrange (1736–1813) late in the XVIII century. However,

TABLE 2.1
Estimated parameters of the Cole-Cole's model

Tissue	ε_∞	Diffusion-Controlled			Pauly-Schwan			Maxwell-Wagner			
		$\Delta\varepsilon_1(\times 10^7)$	$\tau_1(ms)$	α_1	$\Delta\varepsilon_2(\times 10^5)$	$\tau_2(\mu s)$	α_2	$\Delta\varepsilon_3$	$\tau_3(ns)$	α_3	σ
Cancellous	2.5	2	15.915	0	0.2	159.15	0.20	300	79.58	0.25	0.07
Cortical	2.5	0.01	15.915	0	0.05	159.15	0.20	180	79.58	0.20	0.02
Gray	4	4.5	5.305	0	2	106.10	0.22	400	15.92	0.15	0.02
White	4	3.5	7.958	0.02	0.4	53.05	0.30	100	7.96	0.10	0.02
Skin	4	*NA*	*NA*	*NA*	0	*NA*	*NA*	1100	32.48	0.20	0.0002

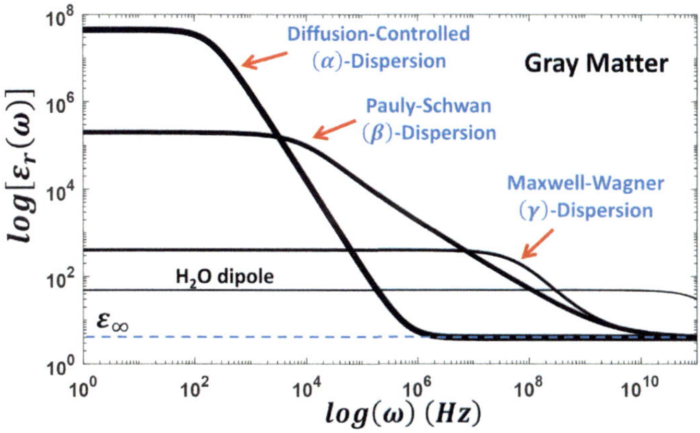

FIGURE 2.8 The frequency dependency of these four relaxation mechanisms for gray matter, with emphasis in the three most important for electromagnetic wave propagation in the α frequency range.

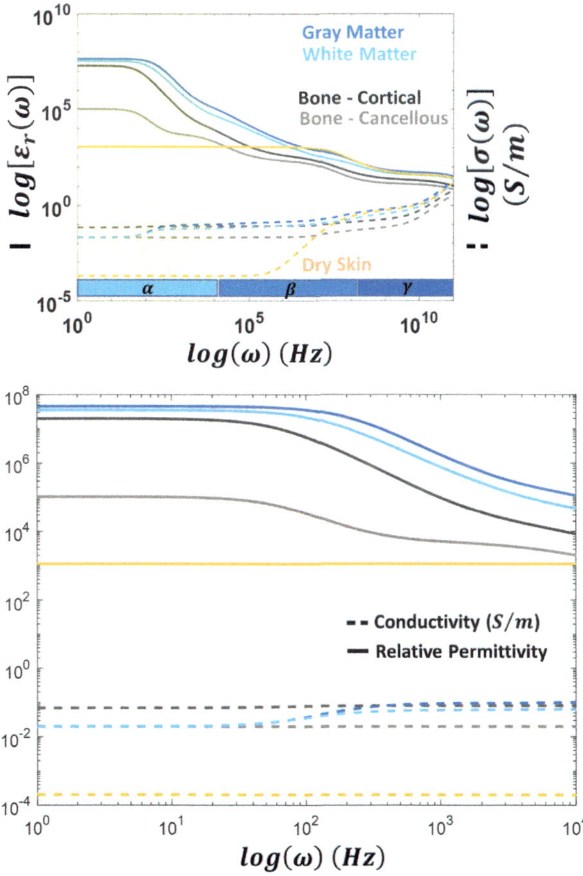

FIGURE 2.9 Comparison of the frequency dependency for the electric permittivity and conductivity of different tissue types in the head. Bottom: Inset for the α frequency range (yellow).

it was not until a few decades later (1813–28) that the theoretical frame was fully elaborated by George Green (1793–1841) and Johann CF Gauss. Siméon D Poisson (1781–1840) proposed an equation to represent the electric potential using a new element of vector calculus (i.e., the Laplacian differential operator ∇^2) developed by Pierre S Laplace (1749–1827) to describe celestial mechanics. The magnetic vector potential was proposed by William Thomson (Lord Kelvin, 1824–1907) in 1851. Potentials ϕ and $\vec{A}$ are undetermined by a scalar field ψ as defined by equations (2.23 and 2.24).

$$\vec{A}'(\vec{r},t) = \vec{A}(\vec{r},t) + \nabla\psi(\vec{r},t) \tag{2.23}$$

$$\phi'^{(\vec{r},t)} = \phi(\vec{r},t) - \frac{\partial\psi(\vec{r},t)}{\partial t} \tag{2.24}$$

Note that $\nabla\times\vec{A}' = \nabla\times\vec{A}$ and $\nabla\phi' - \partial\vec{A}'/\partial t = \nabla\phi - \partial\vec{A}/\partial t$. Therefore, certain gauges are required to solve this uncertainty. As the brain is both a conductor and a dielectric, it has been very useful to use the generalized Lorentz's gauge (Hendrik A Lorentz, 1853–1928): $\nabla\cdot\vec{A} + \mu_0\sigma^*\phi = 0$.

To understand the importance of the potentials, the particular solution of Maxwell equations for an infinite homogeneous conductor (σ) and dielectric ($\varepsilon(\omega)$) medium is obtained in this section. By using the generalized Lorentz's gauge, it is demonstrated that the scalar and vector potentials satisfy respective inhomogeneous Helmholtz equations in the frequency domain.

$$\nabla^2\vec{A}(\vec{r},\omega) + k^2(\omega)\vec{A}(\vec{r},\omega) = -\mu_0\vec{j}^{(p)} \tag{2.25}$$

$$\nabla^2\phi(\vec{r},\omega) + k^2(\omega)\phi(\vec{r},\omega) = \nabla\cdot\vec{j}^{(p)}(\vec{r},\omega)/\sigma^*(\omega) \tag{2.26}$$

The fundamental identity (A.7) of vector calculus was used to obtain equation (2.25). Note that the Laplacian of the magnetic potential $\vec{A}(\vec{r},\omega)$ is defined as the divergence of its gradient $\nabla\vec{A}(\vec{r},\omega)$. The gradient of a vector is a 2nd order tensor (A.4), and its divergence is a vector. To obtain equation (2.26), the fundamental identity (A.5) was used in (2.16) to find first a general equation for the electric field $\vec{E}$ in the frequency domain.

$$\nabla\cdot\left(\sigma^*(\vec{r},\omega)\vec{E}(\vec{r},\omega)\right) = -\nabla\cdot\vec{j}^{(p)}(\vec{r},\omega) \tag{2.27}$$

Note that in the context of electric and magnetic recording from the brain, equation (2.27) is used instead of the original Gauss's law for $\vec{D}$ (2.14). For the case of an infinite homogeneous conductor-dielectric medium, this equation trivially transforms to:

$$\nabla\cdot\vec{E}(\vec{r},\omega) = -\nabla\cdot\vec{j}^{(p)}(\vec{r},\omega)/\sigma^*(\omega) \tag{2.28}$$

The wave number k satisfies:

$$k^2(\omega) = -i\omega\mu_0\sigma^*(\omega) = \mu_0\omega^2\varepsilon(\omega)\left(1+\frac{\sigma}{i\omega\varepsilon(\omega)}\right) = -i\mu_0\omega\sigma\left(1+\frac{i\omega\varepsilon(\omega)}{\sigma}\right) \quad (2.29)$$

The general solutions of equations (2.25) and (2.26) are:

$$\vec{A}(\vec{r},\omega) = \frac{\mu_0}{4\pi}\int\frac{\vec{j}^{(p)}(\vec{r}',\omega)e^{-ik(\omega)|\vec{r}-\vec{r}'|}}{|\vec{r}-\vec{r}'|}\mathrm{dV}' \quad (2.30)$$

$$\phi(\vec{r},\omega) = \frac{1}{4\pi(\sigma+i\omega\varepsilon(\omega))}\int\frac{\nabla\cdot\vec{j}^{(p)}(\vec{r}',\omega)e^{-ik(\omega)|\vec{r}-\vec{r}'|}}{|\vec{r}-\vec{r}'|}\mathrm{dV}' \quad (2.31)$$

The exponential $e^{-ik(\omega)|\vec{r}-\vec{r}'|}$ represents the phase delay, which describe electromagnetic field propagation effects. The phase delay can be expanded by means of a Taylor's series as:

$$e^{-ik(\omega)|\vec{r}-\vec{r}'|} \approx 1 - ik(\omega)|\vec{r}-\vec{r}'| - \frac{(k(\omega)|\vec{r}-\vec{r}'|)^2}{2!} - \frac{i(k(\omega)|\vec{r}-\vec{r}'|)^3}{3!} - \ldots$$

2.8 QUASI-STATIC APPROACH OF THE MAXWELL EQUATIONS

Action potentials (AP) constitutes one of the fastest observable electrophysiological phenomena in the brain. As these potentials happen with a temporal scale of $1-3\,ms$, electric and magnetic recordings of brain activity are typically performed using low past-filters ($< 2\,kHz$). For researchers interested in a dynamic characterization of these APs, brain activity might be recorded at up to $20\,kHz$ using implanted microelectrodes for MUA. Therefore, equations (2.7 and 2.8) and (2.14 and 2.15) must be simplified for such a low frequency range using the quasi-static approximation, which is based on the negligibility of three effects: **1)** electromagnetic field propagation, **2)** inductive, and **3)** capacitive.

According to equations (2.30) and (2.31), the quasi-static approximation would be valid if $k(\omega)|\vec{r}-\vec{r}'| \ll 1$ and $i\omega\varepsilon(\omega) \ll 1$ for the referred low frequency range.

Based on experience, the longest distance between a voxel inside the brain ($\vec{r}'$) and an EEG/MEG recording sensor ($\vec{r}$) would be ½m; i.e., $R_{max} = max|\vec{r}-\vec{r}'| = 0.5\,m$. Assuming the average conductivity of gray matter $\sigma = 0.3\,S/m$ (Chapter 5) and $0 < \omega \leq 2\,kHz$, it can be demonstrated that the complex number $k(\omega)|\vec{r}-\vec{r}'|$ defined by equation (2.32) has both real/imaginary parts and modulus always smaller than 0.016. Errors in the magnitude and phase of $e^{-ik(\omega)|\vec{r}-\vec{r}'|}$ by assuming it equal the unity are less than 1.6 % (Figure 2.10A).

$$k(\omega)|\vec{r}-\vec{r}'| \leq k(\omega)R_{max} = \omega\sqrt{\mu_0\varepsilon(\omega)\left(1-i\frac{\sigma}{\omega\varepsilon(\omega)}\right)}R_{max} \quad (2.32)$$

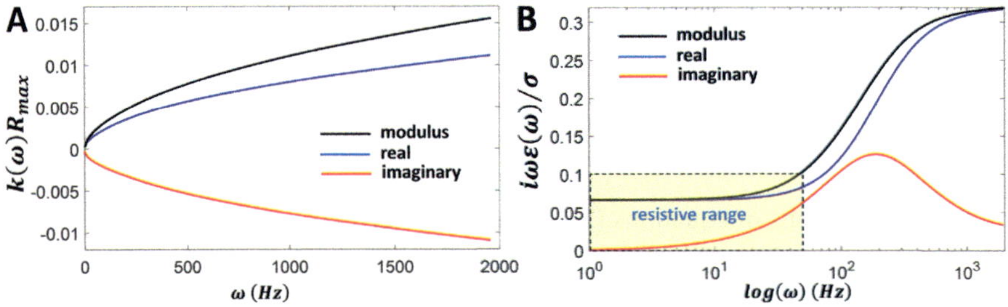

FIGURE 2.10 A. The dependency of the real/imaginary parts and modulus of $k(\omega)R_{max}$ with the frequency ω (linear scale). B. The dependency of the complex portion of the phasor $\sigma(1+i\omega\varepsilon(\omega)/\sigma)$ with the frequency ω (logarithm scale).

Note: Note that, these calculations were performed using the experimentally observed $\varepsilon(\omega)$ as described by (2.18) with the parameters for gray matter (Table 2.1).

Under certain assumptions, Plonsey and Heppner (1967) demonstrated that not only propagation effect are negligible but also inductive effects as they demonstrated that $\left|\omega\vec{A}\right|/\left|\nabla\phi\right| \cong \left(\left|k(\omega)\right|\left|\vec{r}-\vec{r}'\right|\right)^2 < 0.00026$. However, these authors pointed out that neglecting capacitive effects might be problematic and called it the weakest assumption of the quasi-static approximation. The capacitive effect is described by the complex electric conductivity $\sigma^*(\omega)=\sigma(1+i\omega\varepsilon(\omega)/\sigma)$ in equation (2.31). To consider that brain is purely resistive, the complex phasor must satisfy $i\omega\varepsilon(\omega)/\sigma \ll 1$ in the frequency range of interest. Figure 2.10B reveals that although this condition is valid for very low frequencies values ($\omega \le 50\,Hz$), it must be carefully used for frequencies higher than $50\,Hz$. The real part of this phasor is related to the electromagnetic wave dissipation by the brain in the form of heat $\omega\varepsilon''(\omega)/\sigma$ at a particular frequency (2.19 and 2.20). Such energy dissipation is large for frequencies higher than $200\,Hz$. This might explain why, compared to the LFPs, extracellular spikes propagate shorter distances. The imaginary part of the phasor, related to the degree of dephasing between the relaxation mechanism and the incident electromagnetic wave, can be ignored (up to a 10% error) in the entire frequency range ($< 2\,kHz$). Capacitive effects may also explain the differences in MEG and EEG power-scaling mentioned earlier. As the largest power of the EEG and MEG spectra is contained in frequencies $\le 50\,Hz$, simplified Maxwell's equations are obtained under the quasi-stationary condition (2.33–2.36).

$$\nabla\cdot\left(\sigma(\vec{r})\vec{E}(\vec{r},t)\right) = -\nabla\cdot\vec{j}^{(p)}(\vec{r},t) \tag{2.33}$$

$$\nabla\times\vec{E}(\vec{r},t) = 0 \tag{2.34}$$

$$\nabla\cdot\vec{B}(\vec{r},t) = 0 \tag{2.35}$$

$$\nabla\times\vec{B}(\vec{r},t) = \mu_0\left(\sigma(\vec{r})\vec{E}(\vec{r},t) + \vec{j}^{(p)}(\vec{r},t)\right) \tag{2.36}$$

The electric and magnetic potentials are defined as $\vec{E} = -\nabla\phi$ and $\vec{B} = \nabla \times \vec{A}$, with the following gauge relationship $\nabla \cdot \vec{A} + \mu_0 \sigma\phi = 0$. After some mathematical manipulations, the following scalar and vector Poisson's differential equations are finally obtained:

$$\nabla^2 \phi(\vec{r},t) = \left(\nabla \cdot \vec{j}^{(p)}(\vec{r},t) - \nabla\sigma(\vec{r}) \cdot \nabla\phi(\vec{r},t)\right) / \sigma(\vec{r}) \tag{2.37}$$

$$\nabla^2 \vec{A}(\vec{r},t) = -\mu_0 \left(\nabla\sigma(\vec{r})\phi(\vec{r},t) + \vec{j}^{(p)}(\vec{r},t)\right) \tag{2.38}$$

The fundamental identities (A.8) and (A.9) were used to obtain equations (2.37) and (2.38), respectively. Finally, the concept of a magnetic force is introduced in this section. Coulomb's electric force was formally extended by Hendrik A Lorentz (1853–1928) to the magnetic field contribution in 1895. For a single charged particle, the force is represented as: $\vec{F} = q\vec{E} + q\vec{v} \times \vec{B}$. A generalization of the force to a charge body is written as: $\vec{F} = \rho\vec{E} + \vec{j} \times \vec{B}$.

2.9 EEG AND MEG FORWARD PROBLEM

The EEG and MEG forward problem consists of solving equations (2.37 and 2.38) given known the conductivity profile $\sigma(\vec{r})$ and the primary current source density $\vec{j}^{(p)}(\vec{r},t)$ in the entire brain space. Based on knowledge of the physical system, there are several simplifications that can be made. First, the conductivity (air) outside the head is practically zero and, by definition, primary currents are limited only to the brain region. Therefore, as the total electric current cannot flow out from the head, the following condition on the scalp must be satisfied: $\sigma(\vec{r})\vec{E}(\vec{r},t) \cdot \vec{n}(\vec{r})\big|_{Scal} = 0$. Considering equation (2.34), the electric potential satisfies boundary condition (2.39). Also, the magnetic potential $\vec{A}$ must satisfy a radiation condition (2.40).

$$\sigma(\vec{r})\nabla\phi(\vec{r},t) \cdot \vec{n}(\vec{r})\Big|_{S_{scalp}} = 0 \quad \text{Neumann boundary condition} \tag{2.39}$$

$$\lim_{\vec{r}\to\infty} \vec{A}(\vec{r},t) = \vec{0} \tag{2.40}$$

A FEM is used to solve numerically equations (2.37) and (2.38) with boundary conditions (2.39) and (2.40). However, an alternative isotropic piecewise homogenous volume conductor model has been traditionally used by researchers and clinicians to represent EEG and MEG data. Under this model, the head is considered as a set of embedded compartments of isotropic conductivity σ_k $(k = 0,\cdots,N)$. The innermost region V_0 is the brain. Let us assume that surface S_k separates compartments V_k and V_{k+1}. The projection of the electric current to the surface normal must be conserved at all positions on the surface: $\sigma_k \nabla\phi_k(\vec{r},t) \cdot \vec{n}(\vec{r})\big|_{S_k} = \sigma_{k+1}\nabla\phi_{k+1}(\vec{r},t) \cdot \vec{n}(\vec{r})\big|_{S_k}$. In this notation, ϕ_k and ϕ_{k+1} represent the electric potential on surface S_k as evaluated from inside and outside region, respectively. As pointed out by Plonsey and Heppner (1967), this condition might not be satisfied is the quasi-static approach is invalid.

Similarly, the potential must be continuous across the surface $\phi_k(\vec{r},t)\big|_{S_k} = \phi_{k+1}(\vec{r},t)\big|_{S_k}$. Under all these considerations, equation (2.37) for each individual compartment V_k takes the simplified form:

$$\nabla^2\phi_k(\vec{r},t) = \nabla\cdot\vec{j}^{(p)}(\vec{r},t)/\sigma_k \quad \vec{r}\in V_k \tag{2.41}$$

The brain, skull and scalp are the most popularly recognized compartments under this model. The inclusion of other compartments has been also evaluated in several previous studies, e.g., the cerebral spinal fluid, dura matter and the ventricles. In this chapter, the brain (V_0), skull (V_1), and scalp (V_2) are the only compartment assumed, which have the respective electric conductivity values σ_0, σ_1, and σ_2. Surfaces S_0, S_1 and S_3 limit brain-skull, skull-scalp, and scalp-air, respectively. The conductivity of the air is $\sigma_3 = 0$. The definition of a Green's function is very important to solve the EEG and MEG forward problem (Appendix B). Applying the Green's second identity (B.4) to each compartment V_k $(k = 1,\cdots,3)$, and using the Laplacian Green function for the Euclidian space (B.6), the following equation defines the electric potential ϕ_l at a specific position $\vec{r}$ inside any compartment V_l:

$$4\pi\phi_l(\vec{r},t) = 4\pi\frac{\sigma_0}{\sigma_l}\phi_\infty(\vec{r},t) + \sum_{k=0}^{2}\frac{(\sigma_{k+1}-\sigma_k)}{\sigma_l}\oint_{S_k}\frac{\phi_k(\vec{r}',t)(\vec{r}-\vec{r}')\cdot d\vec{S}'}{|\vec{r}-\vec{r}'|^3} \tag{2.42}$$

With $\phi_\infty(\vec{r},t) = -\dfrac{1}{4\pi\sigma_0}\displaystyle\int_{V_0}\frac{\nabla\cdot\vec{j}^{(p)}(\vec{r}',t)}{|\vec{r}-\vec{r}'|}dV'$

Note that the term $\vec{j}_k(\vec{r},t) = (\sigma_{k+1}-\sigma_k)\phi_k(\vec{r},t)\vec{n}_k(\vec{r})/\Delta l$ represents a secondary current density at any position $\vec{r}$ on surface S_k as the results of a change in the conductivity value. This secondary current density is oriented in the direction of the conductivity gradient, i.e., normal $\vec{n}_k(\vec{r})$ to the surface S_k at position $\vec{r}$. Δl represents the thickness of the surface, which is assumed to be the same for all surfaces. Therefore, defining an infinitesimal volume shell δV_k (the surface S_k of thickness Δl), equation (2.42) can be rewritten as:

$$4\pi\phi_l(\vec{r},t) = 4\pi\frac{\sigma_0}{\sigma_l}\phi_\infty(\vec{r},t) + \frac{1}{\sigma_k}\sum_{k=0}^{2}\int_{\delta V_k}\frac{\vec{j}_k(\vec{r},t)\cdot(\vec{r}-\vec{r}')dV'}{|\vec{r}-\vec{r}'|^3} \tag{2.43}$$

In equation (2.42), the potential ϕ_l is defined inside compartment V_l, while ϕ_k takes values only on the surface S_k. Therefore, to calculate ϕ_k in all surfaces given $\vec{j}^{(p)}$, the limit of equation (2.42) for $\vec{r}\to S_l$ must be obtained, which results in (2.44):

$$2\pi(\sigma_{l+1}+\sigma_l)\phi_l(\vec{r},t) = 4\pi\sigma_0\phi_\infty(\vec{r},t) + \sum_{k=1}^{2}(\sigma_{k+1}-\sigma_k)\oint_{S_k}\frac{\phi_k(\vec{r}',t)(\vec{r}-\vec{r}')\cdot d\vec{S}'}{|\vec{r}-\vec{r}'|^3} \quad \vec{r}\in S_l \tag{2.44}$$

BEM approximations are made to numerically calculate the surface integrals in equation (2.44). A general approach has been to consider for each surface S_k a set of N_k functions $h_j^{(k)}(\vec{r})$ and discretization points $\vec{r}_{j'}^{(k)}$, such that $h_j^{(k)}(\vec{r}_{j'}^{(k)})=\delta_{j,j'}$. The symbol $\delta_{j,j'}$ stands for the Kronecker delta function. The electric potential on S_k is represented as $\phi_k(\vec{r},t)=\sum_{j=1}^{N_k}\phi_j^{(k)}(t)h_j^{(k)}(\vec{r})$. A common method to construct $h_j^{(k)}$ is based on the use of the Delaunay triangulation to discretize surface S_k by a set of N_k triangles $\Delta_j^{(k)}$, defining $h_j^{(k)}=1$ on the j-th triangle and zero elsewhere. In this case, the discretization points $\vec{r}_j^{(k)}$ could be considered as the center of mass of triangle $\Delta_j^{(k)}$. Then, the integrals in (2.44) can be expressed as a sum of the weighted solid angles $\Omega_j^{(k)}(\vec{r})$ of triangles $\Delta_j^{(k)}$ from position $\vec{r}$. Inside a triangle $\Delta_j^{(k)}$, the time-varying electric potential is assumed to be constant $\phi_j^{(k)}(t)$, representing the weights in expansion (2.45).

$$\oint_{S_k}\frac{\phi_k(\vec{r}',t)(\vec{r}-\vec{r}')\cdot d\vec{S}'}{|\vec{r}-\vec{r}'|^3}\approx\sum_{j=1}^{N_k}\phi_j^{(k)}(t)\int_{\Delta_j^{(k)}}\frac{(\vec{r}-\vec{r}')\cdot d\vec{S}'}{|\vec{r}-\vec{r}'|^3}=\sum_{j=1}^{N_k}\phi_j^{(k)}(t)\Omega_j^{(k)}(\vec{r}) \tag{2.45}$$

In practice, these surfaces are obtained from individual anatomical images (Figure 2.11, *top-panel*), created from a T_1-weighted MRI or a head statistical atlas. In some studies, a T_2-weighted MRI has been used instead. Let $\vec{r}_{i,j}^{(k)}$ be the counter-clock vertices $(i=\{1,2,3\})$, looking from outside, of the j-th triangle $\Delta_j^{(k)}$ on surface S_k with normal $\vec{n}_j^{(k)}$ (Figure 2.11, *bottom-panel*). Defining the centered vertices $\vec{s}_{i,j}^{(k)}=\vec{r}_{i,j}^{(k)}-\vec{r}$, the solid angle is defined as:

$$\Omega_j^{(k)}(\vec{r})=2\,arctan\left[\frac{\vec{s}_{1,j}^{(k)}\cdot\left(\vec{s}_{2,j}^{(k)}\times\vec{s}_{3,j}^{(k)}\right)}{\left|\vec{s}_{1,j}^{(k)}\right|\left|\vec{s}_{2,j}^{(k)}\right|\left|\vec{s}_{3,j}^{(k)}\right|+\left(\vec{s}_{1,j}^{(k)}\cdot\vec{s}_{2,j}^{(k)}\right)\left|\vec{s}_{3,j}^{(k)}\right|+\left(\vec{s}_{1,j}^{(k)}\cdot\vec{s}_{3,j}^{(k)}\right)\left|\vec{s}_{2,j}^{(k)}\right|+\left(\vec{s}_{2,j}^{(k)}\cdot\vec{s}_{3,j}^{(k)}\right)\left|\vec{s}_{1,j}^{(k)}\right|}\right]$$

Using this definition, equation (2.44) transforms into an algebraic equation:

$$2\pi\left(\sigma_{l+1}+\sigma_l\right)\phi_m^{(l)}(t)=4\pi\sigma_0\phi_\infty^{(l),m}(t)+\sum_{k=0}^{2}\left(\sigma_{k+1}-\sigma_k\right)\sum_{j=1}^{N_k}\phi_j^{(k)}(t)\Omega_j^{(k)}\left(\vec{r}_m^{(l)}\right)$$

Therefore, considering the vectors $\boldsymbol{\phi}(t)=\{\phi^{(l)}(t),l=0,1,2\}$ and $\boldsymbol{\phi}_\infty(t)=\{\phi_\infty^{(l)}(t),$ $l=0,1,2\}$, with $\boldsymbol{\phi}^{(l)}(t)=\{\phi_m^{(l)}(t),m=1,\cdots,N_l\}$ and $\boldsymbol{\phi}_\infty^{(l)}(t)=\{\phi_\infty^{(l),m}(t),m=1,\cdots,N_l\}$, this algebraic equation system can be written in matrix form: $\boldsymbol{R}\boldsymbol{\phi}(t)=\boldsymbol{\phi}_\infty(t)+\boldsymbol{\Omega}\boldsymbol{\phi}(t)$.

$$\boldsymbol{R}=\frac{1}{2}\begin{bmatrix}\left(\sigma_1+\sigma_0\right)I_{N_0} & 0 & 0\\ 0 & \left(\sigma_2+\sigma_1\right)I_{N_1} & 0\\ 0 & 0 & \sigma_2 I_{N_2}\end{bmatrix}$$

$$\boldsymbol{\Omega}=\frac{1}{4\pi}\begin{bmatrix}\left(\sigma_1-\sigma_0\right)\boldsymbol{\Omega}^{(0,0)} & \left(\sigma_2-\sigma_1\right)\boldsymbol{\Omega}^{(0,1)} & \sigma_2\boldsymbol{\Omega}^{(0,2)}\\ \left(\sigma_1-\sigma_0\right)\boldsymbol{\Omega}^{(1,0)} & \left(\sigma_2-\sigma_1\right)\boldsymbol{\Omega}^{(1,1)} & \sigma_2\boldsymbol{\Omega}^{(1,2)}\\ \left(\sigma_1-\sigma_0\right)\boldsymbol{\Omega}^{(2,0)} & \left(\sigma_2-\sigma_1\right)\boldsymbol{\Omega}^{(2,1)} & \sigma_2\boldsymbol{\Omega}^{(2,2)}\end{bmatrix}$$

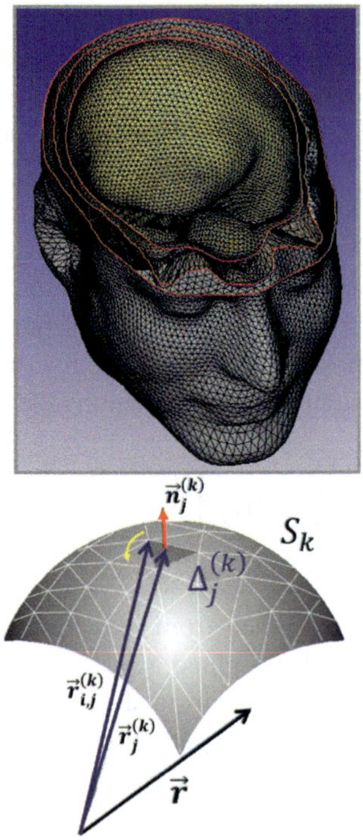

FIGURE 2.11 Triangulation of surfaces limiting different head tissues obtained from MRI tissue segmentation (top). An illustration of different triangulation parameters (bottom, modified from Riera and Fuentes, 1998).

With $$\mathbf{\Omega}(l,k) = \begin{bmatrix} \Omega_1^{(k)}\left(\vec{r}_1^{(l)}\right) & \cdots & \Omega_{N_k}^{(k)}\left(\vec{r}_1^{(l)}\right) \\ \vdots & \ddots & \vdots \\ \Omega_1^{(k)}\left(\vec{r}_{N_l}^{(l)}\right) & \cdots & \Omega_{N_k}^{(k)}\left(\vec{r}_{N_l}^{(l)}\right) \end{bmatrix}$$

Note that a correction by the auto solid angle is required. To improve accuracy and reduce computational burden, de Munck (1992) proposed a linear interpolation for the electric potential within each triangle. Electric potentials can only be defined up to an additive constant; hence, matrix $\boldsymbol{M} = \boldsymbol{R} - \Omega$ in the linear algebraic equation system $\boldsymbol{M}\boldsymbol{\phi}(t) = \boldsymbol{\phi}_\infty(t)$ is not inversible. Instead, the Moore-Penrose pseudoinverse $\boldsymbol{M}^+$ is used in iterative solvers, a method called multiple deflactions (Lynn and Timlake, 1968). As the conductivity of the skull is very small compared to that of the surrounding tissue (brain and scalp), $\boldsymbol{M}$ is also ill-conditioned. To deal with this issue, Hamalainen and Sarvas (1989) proposed the isolated problem approach, which consists on defining $\boldsymbol{\phi}(t) = \boldsymbol{\phi}'(t) + \boldsymbol{\phi}''(t)$, with $\boldsymbol{\phi}'(t)$ the electric potential produced by $\vec{j}^{(p)}$ on the surface of the brain with homogeneous σ_0 ignoring the existence of the skull and scalp; i.e.,

$\phi'(t)=(\phi^0_\infty,0,0)$. Then, $\phi''(t)=\boldsymbol{M}^{+}\phi'_\infty(t)$, with $\phi'_\infty(t)=\beta\phi_\infty(t)-\begin{pmatrix}2\phi^0_\infty/(\beta+1)\\0\\0\end{pmatrix}$ and $\beta=\sigma_0/\sigma_1$. In this method, it is also assumed that $\sigma_0\cong\sigma_2$.

Applying the Green's second identity (B.4) on the entire space ($\Re$ – 3D Euclidian linear space, i.e., $S\to\infty$) to each component of the magnetic potential $\vec{A}$ in (2.38), and using the Laplacian Green function for the Euclidian space and the radiation condition (2.40), it is demonstrated that:

$$4\pi\vec{A}(\vec{r},t)=\mu_0\int_{\Re}\frac{\left(\nabla\sigma(\vec{r}\,')\phi(\vec{r}\,',t)+\vec{j}^{(p)}(\vec{r}\,',t)\right)}{|\vec{r}-\vec{r}\,'|}dV' \tag{2.46}$$

Taking into consideration that $\nabla\sigma$ is different from zero only on the surfaces limiting the three head tissues, and that it can be approximated for any point $\vec{r}$ on S_k as $\nabla\sigma(\vec{r})=(\sigma_{k+1}-\sigma_k)\vec{n}_k(\vec{r})/\Delta l$, equation (2.46) transform into:

$$4\pi\vec{A}(\vec{r},t)=\mu_0\int_{V_0}\frac{\vec{j}^{(p)}(\vec{r}\,',t)}{|\vec{r}-\vec{r}\,'|}dV'+\mu_0\sum_{k=0}^{2}\left(\sigma_{k+1}-\sigma_k\right)\oint_{S_k}\frac{\phi_k(\vec{r}\,',t)}{|\vec{r}-\vec{r}\,'|}d\vec{S}' \tag{2.47}$$

Equivalent to (2.43), we can write (2.47) as a function of secondary current density $\vec{j}_k(\vec{r}\,',t)$, i.e., $4\pi\vec{A}(\vec{r},t)=\mu_0\int_{V_0}\frac{\vec{j}^{(p)}(\vec{r}\,',t)}{|\vec{r}-\vec{r}\,'|}dV'+\mu_0\sum_{k=0}^{2}\int_{\delta V_k}\frac{\vec{j}_k(\vec{r}\,',t)}{|\vec{r}-\vec{r}\,'|}dV'$. In (2.47), it was considered that primary current sources exist only inside the brain V_b. Using fundamental identity (A.10) and taking into consideration that $\vec{B}=\nabla\times\vec{A}$, a final equation for $\vec{B}$ is obtained (2.48).

$$4\pi\vec{B}(\vec{r},t)=4\pi\vec{B}_\infty(\vec{r},t)-\mu_0\sum_{k=0}^{2}\left(\sigma_{k+1}-\sigma_k\right)\oint_{S_k}\frac{\phi_k(\vec{r}\,',t)(\vec{r}-\vec{r}\,')}{|\vec{r}-\vec{r}\,'|^3}\times d\vec{S}' \tag{2.48}$$

With $\vec{B}_\infty(\vec{r},t)=-\frac{\mu_0}{4\pi}\int_{V_0}\frac{(\vec{r}-\vec{r}\,')\times\vec{j}^{(p)}(\vec{r}\,',t)}{|\vec{r}-\vec{r}\,'|^3}dV'$

Using the Stoke's Theorem with $\oint_{S_0}\frac{\vec{j}^{(p)}(\vec{r}\,',t)\cdot d\vec{S}'}{|\vec{r}-\vec{r}\,'|}=0$ (i.e., brain sources are only in the brain volume V_0) and fundamental identity (A.10), it is demonstrated that $\vec{B}_\infty(\vec{r},t)=\frac{\mu_0}{4\pi}\int_{V_b}\frac{\nabla\times\vec{j}^{(p)}(\vec{r}\,',t)}{|\vec{r}-\vec{r}\,'|}dV'$.

Once $\phi_k(\vec{r},t)$ is estimated in all surfaces, the EEG ($V_{er}(t)$, voltage difference) and MEG ($\phi_m(t)$, magnetic flux) observations are calculated by using following equations:

$$V_{er}(t) = \phi_2(\vec{r}_e, t) - \phi_2(\vec{r}_r, t)$$
$$\phi_m(t) = \vec{B}(\vec{r}_m, t) \cdot \hat{o}_m \Delta s_m$$

For EEG, $\vec{r}_e$ and $\vec{r}_e$ are the position of the *e-th* electrode and the common references, respectively. For MEG, $\vec{r}_m$, $\hat{o}_m$ and Δs_m are the position, normalized orientation, and coil area of the *m-th* magnetometer. An equivalent formula can be obtained for a gradiometer.

Alternatively, the use of the lead field (LF) formalism permits us to reformulate the EEG and MEG inverse problems by using vector Fredholm integral equations of the first kind.

$$V_{er}(t) = \int_{V_0} \vec{K}_e^E(\vec{r}) \cdot \vec{j}^{(p)}(\vec{r}, t) dV$$

$$\phi_m(t) = \int_{V_0} \vec{K}_m^M(\vec{r}) \cdot \vec{j}^{(p)}(\vec{r}, t) dV$$

The electric $\vec{K}_e^E(\vec{r})$ and magnetic $\vec{K}_m^M(\vec{r})$ LFs are proportional to the respective induced ohmic current ($\vec{j}_E(\vec{r})$ – EEG and $\vec{j}_M(\vec{r}, \omega)$ – MEG) inside the conductor after energizing a particular electrode/reference pair with a direct current (I_{er}) and a magnetometer coil with an alternative current $I_c(\omega)$ at any particular frequency ω. The LFs are consequence of the symmetry property of the Green's function, i.e., the reciprocity theorem (Figure 2.12).

$$\vec{K}_e^E(\vec{r}) = -\frac{\vec{j}_E(\vec{r})}{\sigma I_{er}}$$
$$\vec{K}_m^M(\vec{r}) = -\frac{\vec{j}_M(\vec{r}, \omega)}{i\omega \Delta S_m \sigma I_c(\omega)}$$

In this case, σ is the electric conductivity of the brain. V represents the head volume limited by the scalp. The flux and vortex sources of the LFs, as well as the specific

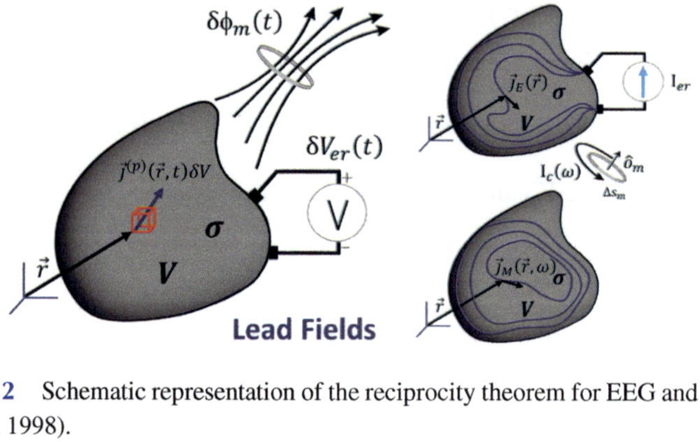

FIGURE 2.12 Schematic representation of the reciprocity theorem for EEG and MEG (Riera and Fuentes, 1998).

TABLE 2.2
Properties of the electric and magnetic LFs

Lead Field	$\nabla^2_{\vec{r}}$	$\nabla_{\vec{r}}\cdot$	$\nabla_{\vec{r}}\cdot$	$\vec{K}(\vec{r})\cdot\vec{n}(\vec{r})$, $\forall\vec{r}\in S$
Electric $\vec{K}_e^E(\vec{r})$	**Zero**	**Zero**	**Zero**	**Not zero**
Magnetic $\vec{K}_m^M(\vec{r})$	**Zero**	**Zero**	**Not zero**	**Zero**

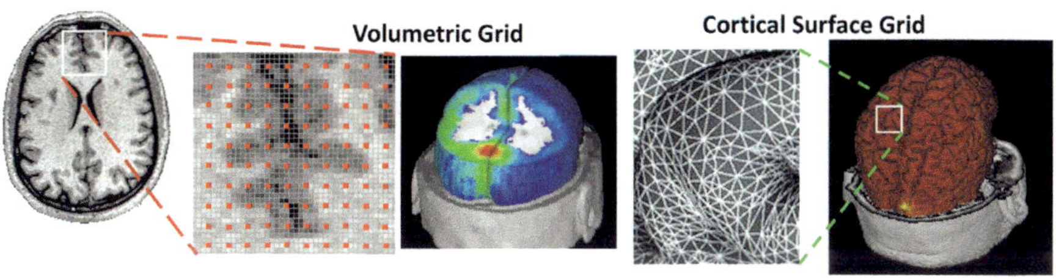

FIGURE 2.13 Grids (brain volume and cortical surface) used to construct functional images based on the EEG (and/or MEG) inverse solutions.

Note: The EEG inverse solution corresponding to my own a-rhythm is used to illustrate the technique.

boundary conditions associated with each of them are presented in Table 2.2. Note that both LFs have zero Laplacian inside the brain; hence, they are harmonic vector functions.

The electric LF for the isotropic piecewise homogeneous volume conductor model of the head was estimated in the past using a dyadic formalism (Riera and Fuentes, 1998). The case of three-concentric spheres was also obtained from the general dyadic formalism. An equivalent demonstration for the magnetic LF can be found in appendix C.

Brain images of neuronal electrical activity are created from EEG and MEG data by solving the abovementioned Fredholm integral equations of the first kind. This is known as the EEG and MEG inverse problem. To that end, grids covering the whole brain volume and/or cortical surface are used to discretize these integral equations (Figure 2.13). As an example, it is shown in this figure the brain volumetric and cortical sources of my own α-rhythm.

2.10 PROBLEMS AND EXERCISES

Problem 2.1 – Demonstrate the empirical equation $\vec{B}(\vec{r}) = \mu_0 I(\hat{l}\times\hat{r})/2\pi r$ from the general Biot-Savart law $\vec{B}(\vec{r}) = \frac{\mu_0}{4\pi}\int_C \frac{I(\vec{r}')d\vec{l}'\times(\vec{r}-\vec{r}')}{|\vec{r}-\vec{r}'|^3}$.

Problem 2.2 – Obtain the Kirchhoff equations (Gustav Kirchhoff, 1845) from Maxwell equations.

Problem 2.3 – Demonstrate equation (2.42).

Problem 2.4 – The node of Ranvier, at position $(10\,\mu m, 2\,\mu m, 10\,\mu m)$, of a single axon is behaving as a current source while action potentials travel along it. The radius and length of the node is $2\,\mu m$ and $1\,\mu m$, respectively. The volumetric current source generated by the action potential can be considered constant inside the node and equal $I = 5\,pA/\mu m^3$ at the signal peaks. A researcher is interested in recording the timing of these action potentials because this axon is codifying an important motor command from the brain. Because of the relative position of the axon inside the fascicle containing it, there is only one possible location for a parallel LIFE electrode used to record the extracellular potentials generated by this axon. The closest terminal of the electrode to the node is at position $(2\,\mu m, 15\,\mu m, z)$, with the z-component to be determined. The conductivity of the surrounding brain tissue is $3\,mS/cm$. The potential differences are recorded with respect to a bath electrode located far away from the nerve. With respect to the nerve and the electrode, the boundary enclosing this particular tissue area is far enough as to assume the medium infinite. The resolution of your amplifiers is $0.1\,\mu V$ and the signal-to-noise ratio is 3 when the electrode is at a particular position (z closest proximity and symmetric). The electrode can be considered as a perfect thin wire. The length of the electrode is $1\,mm$. Please, determine the z value (approaching from the right) from which the LIFE electrode will start reflecting the action potentials conducted by this axon. Please see Figure 2.14 for reference.

Problem 2.5 – The magnetic flow ϕ_B across a coil of surface Δ produced by the propagation of an action potential along an axon is measured by a researcher in an ideal experimental condition of homogeneous magnetic/electric medium. The magnetic permeability of the medium is not that different from the one for the vacuum. The transversal area of the coil is small compared to the distance to the axon; hence, the magnetic field can be assumed space-independent inside

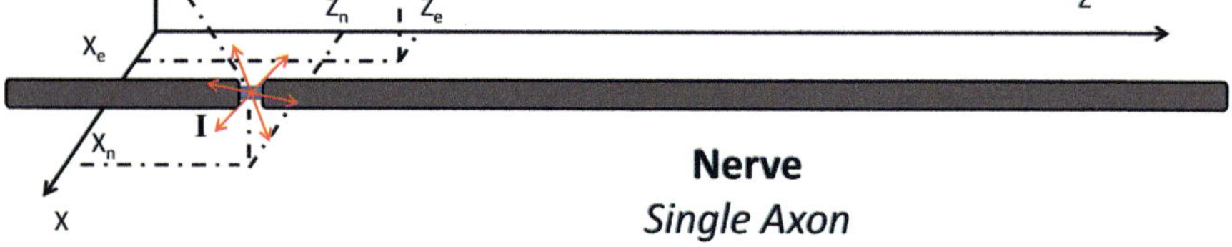

FIGURE 2.14 Illustration of the relative position of the LIFE electrode with respect to the node of Ranvier.

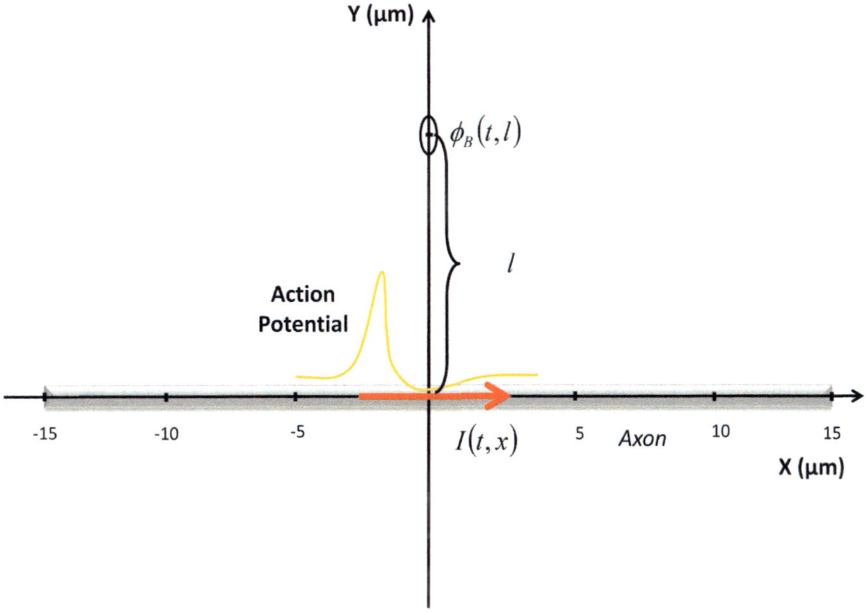

FIGURE 2.15 Illustration of action potential propagation through an axon and observed as magnetic flux on an external coil.

the coil. If the electric current flowing inside the axon during the action potential is $I(t,x)$ and the length of the axon is $30\,\mu m$, find a theoretical expression for $\phi_B(t,l)$ as a function of the distance to the axon l. The transversal area of the axon S is very small, so the electric current $I(t,x)$ can be assumed space-independent along the transversal position inside the axon. Please see Figure 2.15 for reference.

REFERENCES

Baule G, McFee R. Detection of the magnetic field of the heart. *Am Heart J* 66(1), 95–96, 1963.

Bénar CG, Grova C, Jirsa VK, Lina JM. Differences in MEG and EEG power-law scaling explained by a coupling between spatial coherence and frequency: A simulation study. *J Comput Neurosci* 47, 31–34, 2019.

Cohen D. Magnetic fields around the torso: Production by electrical activity of the human heart. *Science* 156(3775), 652–654, 1967.

Cohen D. Magnetoencephalography: Evidence of magnetic fields produced by alpha-rhythm currents. *Science* 161(3843), 784–786, 1968.

Cohen D. Magnetoencephalography: Detection of the brain's electrical activity with a superconducting magnetometer. *Science* 175(4022), 664–666, 1972.

Collura TF. History and evolution of electroencephalographic instruments and techniques. *J Clin Neurophysiol* 10(4), 476–504, 1993.

Dehghani N, Bedard C, Cash SS, Halgren E, Destexhe A. Comparative power spectral analysis of simultaneous elecroencephalographic and magnetoencephalographic recordings in humans suggests non-resistive extracellular media. *J Comput Neurosci* 29(3), 405–421, 2010.

de Munck JC. A linear discretization of the volume conductor boundary integral equation using analytically integrated elements. *IEEE Trans Biome Eng* 39(9), 986–990, 1992.

Erné SN, Hahlbohm H-D, Lübbig H. Theory of the RF biased Superconducting Quantum Interference Device for the non-hysteretic regime. *J Appl Phys* 47(12), 5440–5442, 1976.

Gabriel C, Gabriel S, Corthout E. The dielectric properties of biological tissues: I. Literature survey. *Phys Med Biol* 41, 2231–2249, 1996.

Gabriel S, Lau RW, Gabriel C. The dielectric properties of biological tissues: II. Measurements in the frequency range of 10 Hz to 20 GHz. *Phys Med Biol* 41, 2251–2269, 1996a.

Gabriel S, Lau RW, Gabriel C. The dielectric properties of biological tissues: III. Parametric models for the dielectric spectrum of tissues. *Phys Med Biol* 41, 2271–2293, 1996b.

Happer W. Optical pumping. *Rev Mod Phys* 44, 169–249, 1972.

Hamalainen M, Sarvas J. Realistic conductivity geometry model of the human head for interpretation of neuromagnetic data. *IEEE Transs on Biomed Eng* 36(2), 165–171, 1989.

Im C, Seo JM. A review of electrodes for the electrical brain signal recording. *Biomed Eng Lett* 6, 104–112, 2016.

Kim YJ, Savukov I. Ultra-sensitive magnetic microscopy with an optically pumped magnetometer. *Sci Rep* 6(24773), 1–7, 2016.

Lopes da Silva F. EEG and MEG: Relevance to neuroscience. *Neuro*n 80, 1112–1128, 2013.

Lopez-Gordo MA, Sanchez-Morillo D, Pelayo Valle F. Dry EEG electrodes. *Sensors* 14(7), 12847–12870, 2014.

Lynn M, Timlake W. The use of multiple deflations in the numerical solution of singular systems of equations with application to potential theory. *SIAM J Numer Anal* 5(2), 303–322, 1968.

Oostenveld R, Praamstra P. The five percent electrode system for high-resolution EEG and ERP measurements. *Clin Neurophysiol* 112, 713–719, 2001.

Pauly H, Schwan HPZ. Über die impedanz einer suspension von kugelförmigen teilchen mit einer schale. *Z Naturforsch A: Phys Sci* 14b, 125–131, 1959.

Platzek D, Nowak H, Giessler F, Röther J, Eiselt M. Active shielding to reduce low frequency disturbances in direct current near biomagnetic measurements. *Rev Sci Instrum* 70, 2465–2470, 1999.

Plonsey R, Heppner D. Considerations of quasi-stationary in electrophysiological systems. *Bull Math Biophys* 29, 657–664, 1967.

Riera J, Fuentes ME. Electric lead field for a piecewise homogeneous volume conductor model of the head. *IEEE Trans Biomed Eng* 45(6), 746–753, 1998.

Shad EHT, Molinas M, Ytterdal T. Impedance and noise of passive and active dry EEG electrodes: A review. *IEEE Sensors J* 20(24), 14565–14577, 2020.

Shilov VN, Borkovskaya YB. Theory of broadband dispersion of permittivity of biological cell suspensions. *Colloid J* 72(5), 686–695, 2010.

Stone JL, Hughes JR. Early history of electroencephalography and establishment of the American Clinical Neurophysiology Society. *J Clin Neurophysiol* 30(1), 28–44, 2013.

3 Transmission-Emission Imaging Modalities

3.1 OVERVIEW

Three important brain-imaging modalities rely on the propagation principles of high-energy photons inside biological tissues. Based on their generation mechanisms and energy ranges, these photons are classified as: **a)** hard x-rays (electron transitions, energy range $10-120\,keV$) and **b)** γ-rays (nuclear transitions, energy range $0.1-10\,MeV$). The hard x-rays can be absorbed (photo-absorption) and scattered (Compton and Rayleigh scattering) by biological tissues in an energy dependent manner, a property characterized by the linear attenuation coefficient μ. The computed tomography (CT) modality consists of a 3D reconstruction of μ inside any organ exposed to transmitting x-rays and has proven to be useful to image certain brain tissues. The γ-rays are produced by nuclear decays of unstable atoms (*radioisotopes*) that can be delivered to biological tissues through the blood circulation. Two brain-imaging modalities are based on the 3D reconstruction of the radioisotope density ρ from the emitted γ-photons, which are poorly absorbed by biological tissues. These two emission modalities differentiate by the nuclear-decaying mechanisms underlying the genesis of γ-photons: **a)** β^+-decay in positron emission tomography (PET) and **b)** γ-decay in single photon emission tomography. Because of the use of ionizing radiation, transmission (CT) and emission (PET/SPECT) imaging modalities are used mostly in both clinical practice (medical radiology) and pre-clinical animal studies. Under certain conditions, they have been applied in the past to advance neuroscience knowledge by using a low-level radiation in healthy volunteers. A common aspect of these three modalities is the use of computerized methods to obtain 3D images from 2D data collected with external sensors, named computer-assisted tomography (CAT).

3.2 TRANSMISSION X-RAY – THE VACUUM TUBES

The x-rays were discovered in 1895 by Wilhelm C Röntgen (1845–1923) using the early Lenard and Crookes tubes; hence, sometimes it is referred to as Röntgen radiation. For this work, considered the most impacting invention in medical imaging, he received the Nobel Prize in Physics in 1901. From its discovery, technological advances were introduced to improve x-ray measurements (e.g., the Coolidge x-ray tube) and coated screens. However, the field suffered quickly from retraction

DOI: 10.1201/9780367901509-3

because of the continuous reports of injuries associated with the use of the technique. Enthusiasm recovered after Marie S Curie (1867–1934) used a mobile x-ray machine in 1914 to evaluate injured soldiers in World War I. Today, x-rays are produced using a vacuum tube inside which electrons are accelerated from a hot cathode and projected at a high velocity on a metal target (i.e., the anode). The cathode consists of a wire filament along which an electric current is flowing to induce thermionic emission. The acceleration is caused by a high voltage between the cathode and the metallic anode. In medical imaging with hard x-rays, an alloy of tungsten (95 %) and rhenium (5 %) is typically used as anode target. The atomic number of tungsten ($Z = 74$) and rhenium ($Z = 75$) ensures a large electron cloud around the nucleus of atoms in the anode (Chapter 1). Therefore, the large majority of bombarding electrons interact elastically, i.e., via deflections in their original paths, with the outer orbital electrons in these clouds. This interaction causes a low-energy electromagnetic radiation (near infrared); and hence, the production of heat by the anode. A very small percentage of bombarding electrons collides with the nucleus and inner-orbital electrons. This latter form of interaction could be either elastic or inelastic. Under elastic collisions, the energy transferred from the bombarding electrons to the atoms in the anode is not significant. These elastic collisions result just in tortuous paths for the electrons inside the anode. However, an electron crossing in close proximity to the nucleus accelerates toward it because of a strong attractive electric force, potentially leading to an inelastic collision with the nucleus. Because of this interaction, the electron deaccelerates/deflects causing the emission of quantum of electromagnetic radiation. The energy of the quantum is equal to the reduction in the kinetic energy of the electrons (keV). The amount of transferred kinetic energy depends on the distance to the nucleus and inner-orbital electrons, creating a broad electromagnetic spectrum for the radiation, which is known as the *Bremsstrahlung radiation*. Therefore, approximately 5 % of the energy transferred from the bombarding electrons appears in the form γ-rays, while the remaining ~ 95 % is dissipated as heat.

Coated screens used to measure the transmitted radiation evolved over time, from photographic plates and films to digital detectors (e.g., image plates and flat panel detectors). Techniques for image reconstruction have also progressed. The focal-plane method was first introduced into x-ray transmission imaging by Bocage (1921). The method produced very blurred images in unwanted planes, an issue that Shoji Takahashi corrected in 1957 by aligning the photographic film and the x-ray source in the same plane (Doi et al., 2012). However, most important advances in the methods were linked to progresses in radioastronomy, electron microscopy and optics. The concept of back-projection underlies early reconstruction methods developed in the period of 1963–1971, yielding also very blurred images. In 1971, the CT technique was invented by Godfrey Hounsfield (1919–2004) at EMI Laboratories. For this work, he shared the Nobel Prize for Physiology or Medicine of 1979. He also created the Hounsfield number: $H = \frac{\mu_{tissue} - \mu_{H^2O}}{\mu_{H^2O}} \times 1000$, which depends primary on the atomic number and electron density in heavy atoms, but also on the photon energy. The number H describes the relative linear attenuation coefficient of a specific tissue in relation to that of water.

3.3 RADIOACTIVITY DECAY AND SPECIFIC ACTIVITY: RELEVANCE FOR SPECT AND PET

The conformational properties of the nucleus of an atom are determined by two forces: **a)** the attractive-repulsive strong nuclear force between the nucleons and **b)** the repulsive electromagnetic force between the positively charged protons. The attractive property of the strong nuclear force is very effective for nucleons within the distances of $1-2.5\,fm$. The force become repulsive for distances less than $\sim 0.7\,fm$ and the attraction properties quickly decay for distances $> 2.5\,fm$. The repulsive component determines the size of nuclei. Nucleons attraction due to the strong nuclear force compete with the electromagnetic repulsive force between the protons, leading to an equilibrium for which all nucleons are stably bound. Except for technetium ($_{43}Tc$) and promethium ($_{61}Pm$), all elements of the periodic table with $Z \leq 83$ have at least one stable isotope. Most known elements with $Z > 83$ are artificially produced by either neutron bombardment or fusion, with $^{235/238}$U and its decay products the notable exception. All known transuranic elements are artificial. Similarly, it is also possible to create radioactive isotopes for any natural element by forcibly adding neutrons or protons in reactors or particle accelerators, respectively. Adding more neutrons to an atom's nucleus increases its physical radius, and consequently the electromagnetic repulsive force becomes predominant as the strong nuclear force decays quickly with distance (Yukawa potential). Therefore, most isotopes of an atom are unstable. An unstable decaying isotope is called a radionuclide (or radioisotope). Hence, the mass number of an atom $A = Z + N$ (Chapter 1) determines not only the size of the nucleus but also how stable it is. The largest known completely stable isotope is ^{208}Pb, with 126 neutrons and 82 protons. Any unstable atomic nucleus decays to more stable ones in a process losing energy by radiation. There are three main mechanisms by mean of which unstable nuclei decay: **a)** α decay, **b)** β decay, and **c)** γ decay. While decaying, these nuclei emit certain energetic particles, a process that is stochastic in nature. The decaying nucleus is referred to as the parent. In α decay, an unstable parent breaks into two daughters representing different atoms. The parent and daughter atoms must satisfy the following relationship: $^{p+n}_{p}E_1 \rightarrow {}^{p+n-4}_{p-2}E_2 + {}^{4}_{2}H_e^{+2}$. Hence, during α decay an unstable parent atom $^{p+n}_{p}E_1$ transforms into a new element $^{p+n-4}_{p-2}E_2$, and a helium cation is created. Depending on the energetic particle emitted, there are two types of β decays: **1)** β^- decay – an electron is emitted and **2)** β^+ decay – a positron is emitted. In this type of decay, either a neutrino v_e (β^+ decay) or an antineutrino $\overline{v}_e$ (β^- decay) is also emitted, and in both cases, a nuclear transmutation takes place. The β^+ decay is also referred to as positron emission, and it underlies the PET imaging modality. The following relationships between the parent and daughters are preserved in β decays:

$$^{p+n}_{p}E_1 \rightarrow {}^{p+n}_{p+1}E_2 + e^- + \overline{\text{v}}_e \qquad \beta^- \text{ decay}$$

$$\textbf{Energy released}: Q(\beta^-) = \left(m\left({}^{p+n}_{p}E_1\right) - m\left({}^{p+n}_{p+1}E_2\right)\right)c^2$$

$$^{p+n}_{p}E_1 \rightarrow {}^{p+n}_{p-1}E_2 + e^+ + \text{v}_e \ \mathbf{2)} \qquad \beta^+ \text{ decay}$$

$$\textbf{Energy released}: Q(\beta^+) = \left(m\left({}^{p+n}_{p}E_1\right) - m\left({}^{p+n}_{p-1}E_2\right) - 2m_e\right)c^2$$

For any β decay to occur, $Q > 0$ must be satisfied. Therefore, the sum of the mass of the daughter ${}^{p+n}_{p+1}E_2$ and the particles emitted must be smaller than the mass of the parent ${}^{p+n}_{p}E_1$ in β^- decay. Interestingly, β^+ decay can only occur when the mass of the decaying parent ${}^{p+n}_{p}E_1$ exceeds that of the daughter ${}^{p+n}_{p-1}E_2$ by at least twice the mass of the electron. In γ decay, the nucleus of an atom makes a transitions from an excited (e) energy level to a ground (g) energy level, releasing a photon with large energy ($10^{18} - 10^{19}\ Hz$); i.e., a γ-ray: ${}^{p+n}_{p}E_1^{(e)} \rightarrow {}^{p+n}_{p}E_1^{(g)} + \gamma$. A nucleus can reach an excited energy level by: **a)** collisions with other nuclei/particles and **b)** as the result of either an α or a β^- decay. In SPECT, an unstable nucleus first decays either in the α or β^- form, leaving an excited daughter nucleus that then decays to the ground energy level by emitting a γ-ray photon. There are γ emitters that come from an electron capture (EC) instead of a regular decay; however, these are not discussed in this book. γ decay was discovered in 1900 by Paul U Villard (1860–1934) while studying the radiation emitted by natural radium (e.g., ${}^{226}Ra$). β^+ decay was discovered in 1934 by Frédéric (1900–1958) and Irène (1897–1956) Joliot-Curie. For this work, they received the Nobel Prize in Chemistry in 1935. They observed the emission of a positron when bombarding aluminum (${}^{27}Al$) with α particles (${}^{4}_{2}H_e^{+2}$). The daughter of this nuclear reaction was a non-natural radioactive element ${}^{30}P$; hence, their experiment was recognized as the first example of artificial radioactivity. During a β^+ decay, the nucleus emits a neutrino and a positron by changing one of its protons into a neutron. To understand this process, it is necessary to review how nucleons are formed from quarks. There are six types of quarks, also referred to "flavors": *up, down, top, bottom, strange*, and *charm*. Quarks forming a nucleon are kept together thanks to weak nuclear forces (electroweak theory), which are limited mostly to the nucleon diameters. According to Chapter 1, the neutron is formed by 1 *up* ($+2e/3$) and 2 *down* ($-1e/3$) quarks. On the other hand, protons are composed of 2 *up* ($+2e/3$) and 1 *down* ($-1e/3$) quarks. In β^+ decay, a proton is transformed into a neutron by means of weak force interactions, in which quarks in the proton can swap their flavors by the emission of a *W* boson. After its emission from a β^+ decay, the positron travels inside the biological tissue dispersing a bit of its kinetic energy until encountering an electron. Upon the electron-positron collisions, these forms of matter and antimatter are annihilated, resulting in the emission of two photons with energy $\sim 511\,keV$ in opposite ways ($\approx 180°$), i.e., γ-rays. The distance the positron travels before annihilation depends on its energy; however, it is in the $0-2\,mm$ range. The temporal evolution of the number of radioactive atoms is defined by the equation $N = N_0 e^{-\lambda t}$, with N_0 and λ the initial number of radioactive atoms and the exponential decay constant, respectively. Each radionuclide has a characteristic physic half-life $T_{1/2}$, which is defined as the time it takes for half of the total amount of radioactive nucleus to decay; hence, $T_{1/2} = ln(2)\lambda$. The exponential decay constant λ is a property of the nucleus. The activity $A = \lambda N$ of a radionucleus is defined as the number of radioactive transformations per second; and trivially satisfies $A = -dN/dt$. The activity is given in Becquerel ($Bq = 1\,decay/s$). The old-school unit of activity was the Curie (Ci), which corresponds to the activity of $1\,g$ of ${}^{226}Ra$. The current definition is $1\,Ci = 37\,GBq$. On the other hand, the specific activity α is the activity per amount of radionuclide (Bq/g).

$$\alpha = \frac{\lambda N_A}{M} \tag{3.1}$$

M is the molar mass (g/mol) of the element and $N_A = 6.02214076 \times 10^{23}\ mol^{-1}$ is the Avogadro's number (Amedeo Avogadro, 1776–1856). The counting rate C in a detector is another important definition $C = \varepsilon A$, with ε the efficiency of the detector. A radioactivity with carrier is that in which the sample includes stable isotopes. A theoretical specific activity for a radionuclide is that from a carrier-free sample. The mean life of a radionuclide is defined as $\tau = \lambda^{-1}$. The biological half-life $T^B_{1/2}$ for a tissue is defined as the time it takes for the radionuclide concentration to be $1/2$ of its maximum concentration in the particular tissue. The effective half-life is then defined as $T^E_{1/2} = T_{1/2}T^B_{1/2} / \left(T_{1/2} + T^B_{1/2}\right)$.

3.4 IMPORTANT RADIOISOTOPES FOR BRAIN IMAGING

The energy of x-ray photons is similar to that of photons emitted from a nuclear γ decay. Therefore, an x-ray transmission imaging was performed in 1961 using a nuclear γ decay (^{131}I source, emission of $\beta^- \rightarrow \gamma$, Oldendorf, 1961) instead of the traditional method of electron-acceleration in a vacuum tube. Other radionuclides were subsequently evaluated for transmission imaging; e.g., ^{241}Am (Kuhl et al., 1966). Oldendorf's work might have encouraged Kuhl and Edwards (1963) to evaluate an ^{131}I emission imaging modality using a Masonite phantom, which can be considered as the first demonstration of SPECT image reconstruction. Although many previously contributed to stablish emission modalities using animals, the application of the β^- decaying ^{85}Kr to study regional cerebral blood flow (rCBF) in the brain is considered the first SPECT study in a human (Ingvar and Lassen, 1961, 1962). The method was based on an intra-arterial (carotid) injection of the γ emitting radionuclide bolus in a patient undergoing tumor resection, and measurements were obtained from an open craniotomy. Measurements of rCBF without the need of a craniotomy were performed later in patients using ^{133}Xe and depth-focusing collimators (Harper et al., 1964). Measurements of rCBF in cats using ^{131}I were feasible at that time using a theoretical framework (Sokoloff, 1964). However, most current SPECT-based methods to measure rCBF use a synthetic radionuclide (technetium, ^{99m}Tc) bound to either exametazime (HMPAO) or ethyl cysteinate dimer (ECD). The half-life and specific activity of ^{99m}Tc is $T_{1/2} = 361.2\,min$ and $\alpha = 30 - 50\,MBq/ng$, respectively. The maximum activity of ^{99m}Tc is obtained after 23 hours.

At this point, it is convenient to classify radiopharmaceuticals (a preparation of a radionuclide bonded or unbonded to a functionalizing molecule that complies with pharmacopoeia regulations) for SPECT into three major groups: **a)** diffusible, **b)** non-diffusible, and **c)** static. A diffusible radionuclide moves through the circulatory systems without it being metabolized. Hence, its concentration in brain tissues depends on perfusion (rate of artery delivery) and the artery-to-tissue concentration gradients. Quantitative estimates of rCBF can be obtained by combining mathematical models and two types of measurements: **1)** the arterial/venous radiotracer concentrations (input function) and **2)** the brain uptake/clearance from SPECT data. Concentrations in arteries/venous can be obtained by either direct blood samples or indirect methods (e.g., measuring γ emission directly from the lung). ^{85}Kr and ^{133}Xe are example of diffusible radiotracers. Because of its very rapid transit time and inert gas properties, ^{133}Xe has remained as an important radionuclide to study rCBF, mainly in clinical settings. Non-diffusible radionuclides (e.g., ^{131}I RISA/hippurate) are not good at crossing the brain-blood-barrier (BBB); hence, they are mainly used for angiography studies. However, mathematical models can also be used to reconstruct the CBF and cerebral blood volume (CBV) in specific brain areas. Two early proposed and still useful models, the Kety-Schmidt method for nitrous oxide and the modified Stewart-Hamilton dye-dilution model, are discussed in Meier and Zierler (1954). After an intravenous (i.v.) injection, static radionuclides (e.g., ^{123}I HIPDM/IMP, ^{99m}Tc HMPAO/ECD) are extracted by brain tissues on their first pass through the circulatory system. Note that ^{123}I has $T_{1/2} = 788\,min$ and $\alpha \cong 71.4\,MBq\,/\,ng$. Depending on how they are trapped, these radionuclides remain in the brain tissues for long periods (hours). Therefore, SPECT data might reflect rCBF around the time of the i.v. injection. These quantitative rCBF estimations are based on tracer microsphere models. Furthermore, the ^{123}I and ^{99m}Tc radionuclides have been combined with a series of molecular targets to study neurotransmission, vascular response, and metabolism in the brain. These are some of these available SPECT probes to that end: **a)** to study **receptors** (NMDAR – ^{99m}Tc-NCAM; serotonin 5-HT$_4$ – [^{123}I]SB207710; GABAA – [^{123}I]IMZ; Nicotinic ACh α4β2 – 5-[^{123}I]IA; Adenosine 2A – [^{123}I]MNI-420; Dopamine D2/D3 – [^{123}I]IBZM, [^{123}I]Epidepride; Somatostatin – [^{99m}Tc]Depreotide, ^{99m}Tc-HYNIC-TOC; Vasoactive intestinal peptide – [^{99m}Tc]TP3954) and **b)** to study **transporters** (Serotonin – [^{123}I]*p*ZIENT, Radioiodinated ADAM; Dopamine [^{123}I]FP-CIT, [^{123}I]β-CIT-FP, ^{99m}Tc-TRODAT-1; Norepinephrine – [^{123}I]INER, [^{123}I]MIBG, Vesicular acetylcholine – [^{123}I]-IBVM). A non-diffusible ^{99m}Tc-DTPA-DG (glucose transporter and hexokinases) has been useful to study BBB leaks. A more detailed list of radionuclides for SPECT can be found here https://ncbi.nlm.nih.gov/books/NBK23064/.

β^+ decaying radionuclides for PET were mostly developed after the introduction of the cyclotron. The idea of accelerating particles between two D-shaped magnets and used them to bombard atomic nucleus, which constitutes the naissance of the cyclotron, was initially proposed by Lawrence et al. (1932). However, the wartime need for nuclear reactors pushed the massive use of the cyclotron up to the 60s–70s. The four most important radionuclides in PET imaging modality are listed in Table 3.1. Historically, radionuclides are developed based on three important properties: **a)** it can be produced nearby research-laboratories/hospitals by means of cyclotrons, **b)** it exhibits a β^+ decay for most of the events, **c)** the stable related isotope is present in brain tissues (with exception of ^{18}F), and **d)** it has a favorable chemistry to produce the radiopharmaceuticals of interest.

TABLE 3.1
The most important radionuclides in PET imaging

Radionuclide	^{18}F	^{11}C	^{15}O	^{13}N
Half-Life $T_{1/2}$ (min)	110	20.4	2.07	9.97
Specific Activity α $(MBq\,/\,ng)$	3522.2	31029	227630	53686
Main Energy (MeV)	0.6	1.0	1.7	1.2
β^+ **Decay** (%)	97	100	100	100

Note: The specific activity is calculated using the theoretical limit for the isotope (carrier-free sample, Eq. 3.1). ^{18}F decays also by EC (3 %).

In this subsection, a quick overview of the development of these radionuclides is presented. Metabolism of oxygen was measured in four subjects that have inhaled $^{15}O_2$ (Dollery and West, 1960), which might be the first application of PET principles in humans. The $^{15}O_2$-based PET technique advanced considerably after this initial work. In 1980, PET-based measurements of both oxygen metabolism and CBF in a human were already possible (Frackowiak et al., 1980). Around the same period, Reivich et al. (1979) introduced a method to measure local cerebral glucose utilization using [^{18}F]FDG, known as the fluorodeoxyglucose method. Today, the standard [^{18}F]FDG-PET is the most widely used modality in basic neuroscience research and clinical practice. Recently, ^{11}C has become one of the most attractive radionuclides for PET. Both [^{11}C] CO_2 and [^{11}C]CH_3I have been widely used as a radioactive precursors to target molecules for ^{11}C addition using: **1)** thiol groups (Langstrom & Lundqvist, 1976), **2)** phenols (Ehrin et al, 1985), **3)** amides (Hashimoto et al., 1989), **4)** amines (Mathis et al., 2003), and, more recently, **5)** fatty acids (Amor-Coarasa et. al., 2015). The longer lived ^{18}F has also been used as a radioactive precursor to target organic molecules. ^{18}F can be inserted into biomolecules by changing the C-H (or C-OH) bond to a C-F one. Table 3.2 summarizes the most useful ^{18}F and ^{11}C probes in PET imaging to study important molecular/cellular mechanisms underlying brain functions. These mechanisms are lately discussed in this book.

Finally, the application of ^{13}N to study brain functions and dysfunctions has busted after its FDA approval in 2000 for PET imaging. In the [^{13}N]NH_3 form, it has been recently used to study ammonia-related mechanisms in the brain (e.g., the glutamine synthetase, GS). Radiotracers for SPECT and PET studies are mostly administered in two ways: **a)** by i.v. injection and **b)** by inhalation (gases). Injections could be either a bolus or a constant infusion.

3.5 RADIONUCLIDE PRODUCTION: PRINCIPLES IN CYCLOTRONS AND GENERATORS

When producing radioisotopes for medical imaging, the most frequently used incident particles are protiums (^1_1H) and deuterons (^2_1H). A type of nuclear reaction occurs by projecting these accelerated particles on stable nuclei or complex targets (3.2), which is the working principle of a cyclotron. Let us define the nuclear reaction $^{A_1}E_1(P_i,mP_j)^{A_2}E_2$, in which a stable isotope $^{A_1}E_1$ with atomic mass (A_1) is bombarded with a high-energy particle beam $P_i = \{^1_1H, ^2_1H\}$, producing an radionuclide $^{A_2}E_2$ with atomic mass A_2 and a radiation of m energetic particles P_j. For most nuclear reactions used in PET and SPECT modalities (Table 3.3), the radiating energetic particles are either neutrons (n) or α particles ($^4_2H_e^{+2}$).

$$^{A_1}E_1 + P_i \rightarrow {}^{A_2}E_2 + nP_j \tag{3.2}$$

Cyclotrons are typically available at research centers, local radiopharmacies, and research hospitals. The radionuclide generators are alternatives to a cyclotron for the on-site generation of some short-lived radionuclides. Parent radionuclides used in generators are produced in either cyclotrons or nuclear reactors. These parents have

TABLE 3.2
PET Radionuclides for brain imaging studies of neurotransmission, vascular response, and metabolisms

PET	^{18}F	^{11}C
Enzymes and Peptides		
GLUT/HK	[^{18}F]FDG, 6-[^{18}F]FDF	[^{11}C]D-Glucose
ACS	[^{18}F]FAC	[^{11}C]Acetate
MCI	[^{18}F]10, [^{18}F]Flurpiridaz, [^{18}F]MC1-27, [^{18}F]RP1004, [^{18}F]RP1003	**Not Found**
AChE	[^{18}F]FEP-4MA	[^{11}C]PMP, [^{11}C]MDDP, [^{11}C]MP4A(AMP)
PKM_2	**Not Found**	[^{11}C]DASA-23
AA	**Not Found**	1-[^{11}C]AA
iNOS	[^{18}F]iNOS-9	**Not Found**
Transporters		
DAT	[^{18}F]FE-PE2I, [^{18}F]PR17.MZ, [^{18}F]FECNT, [^{18}F]FE@CIT, [^{18}F]FBFNT, [^{18}F]FP-CIT, [^{18}F]PR04.MZ	[^{11}C]PE2I, [^{11}C]LBT-999, [^{11}C]PR04.MZ, [^{11}C]MP
SERT	[^{18}F]ACF, [^{18}F]FE*m*ZIENT, [^{18}F]FE*p*ZIENT, [^{18}F]1, [^{18}F]FEMcN	[^{11}C]MADAM, [^{11}C]HOMADAM, [^{11}C]DASB, [^{11}C]AFE, [^{11}C]AFM, [^{11}C]AFA, [^{11}C]*m*ZIENT, [^{11}C]*p*ZIENT, [^{11}C](+)McN5652
NET	(*S*,*S*)-[^{18}F]FMeNER-D_2, (*S*,*S*)-[^{18}F]FRB-D_4, (*S*,*S*)-[^{18}F]FRB,	[^{11}C]MENET-1, (*S*,*S*)-[^{11}C]MRB, [^{11}C]mHED, [^{11}C]NE, [^{11}C]-*p*-hydroxy-PG, [^{11}C]PHEN, [^{11}C]EPI
Gly-T_1	[^{18}F]CFPyPB	[^{11}C]CFpyPB, [^{11}C]RO5013853, [^{11}C]GSK931145, [^{11}C]SA1-3,
VMAT	[^{18}F]FP-(+)-DTBZ, [^{18}F]AV-133	[^{11}C]MTBZ, [^{11}C]mHED, [^{11}C]NE, [^{11}C]DTBZ, [^{11}C]EPI, [^{11}C]-*p*-hydroxy-PG, [^{11}C]PHEN
$SLC1A_5$	**Not Found**	L-[5-^{11}C]-Gln
Receptors		
D_2/D_3	[^{18}F]-5-OH-FPPAT, [^{18}F]Fallypride, [^{18}F]DMFP, [^{18}F]FESP, [^{18}F]5	[^{11}C]RAC, [^{11}C]NPA, [^{11}C]ZYY-339, [^{11}C]FLB 457, [^{11}C]-5-OH-DPAT, 2-Cl-[^{11}C]-(-)-NPA, [^{11}C]MNPA, [^{11}C]PPHT, [^{11}C](+)-PHNO, [^{11}C]NMSP
D_1	**Not Found**	[^{11}C]NNC 112, [^{11}C]SCH 23390
D_4	[^{18}F]3d, [^{18}F]FMTP	[^{11}C]7
5-HT_{1A}	[^{18}F]FCWAY, [^{18}F]F13714, [^{18}F]F15599, [^{18}F]FPWAY, *p*-[^{18}F]MPPF, [^{18}F]6FPWAY, [^{18}F]MeFWAY	[*carbonyl*-^{11}C](D/J)WAY, [^{11}C]CUMI-101, [^{11}C]MPT, (*R*)-[^{11}C]RWAY, [*carbonyl*-^{11}C]WAY 100635, [^{11}C]MMP

TABLE 3.2 (Continued)

PET	^{18}F	^{11}C
5-HT_{1B}	**Not Found**	[^{11}C]P943, [^{11}C]AZ10419369
5-$HT_{2A/2}$	[^{18}F]Altanserin/(d2), [^{18}F]FESP	[^{11}C]MDL 1000907, [^{11}C]CIMBI-36/5, [^{11}C]NMSP
5-HT_4	**Not Found**	[^{11}C]SB-207145
5-$HT_{6/7}$	[^{18}F]-2FP3	[^{11}C]-GSK215083
$\alpha_4\beta_2$	[^{18}F](-)-6c, [^{18}F]ZW-104, [^{18}F]NCFHEB, [^{18}F]FPhEP, [^{18}F]Nifene, 2/6-[^{18}F]FA, [^{18}F]Nifrolene, [^{18}F] Nifrolidine, [^{18}F]AZAN	**Not Found**
α_7	[^{18}F]NS10743, [^{18}F]ASEM	[^{11}C]A-844606, [^{11}C]A-752274, [^{11}C]A-582941, [^{11}C]CHIBA-1001, [^{11}C]A-833834
M_2	[^{18}F]FP-TZTP	**Not Found**
NMDA	[^{18}F]GE-179	[^{11}C]AcL703,717
$GABA_A$	[^{18}F]FFMZ	[^{11}C]FMZ, [^{11}C]Ro15-4513
$mGluR_1$	[^{18}F]FPIT, [^{18}F]MK-1312, [^{18}F]FITM, [^{18}F]FPTQ	[^{11}C]LY2428703, [^{11}C]JNJ-16567083 [^{11}C]MMTP(1), [^{11}C]ITMM(6), [^{11}C]ITDM(4)
$mGluR_5$	[^{18}F]SP 203, [^{18}F]FDEGPECO, [^{18}F]F-MTEB, [^{18}F]F-PEB, [^{18}F]FE-DABP688	[^{11}C]SP 203, [^{11}C]M-MPEP, [^{11}C]M-PEPy, [^{11}C]MPEP, [^{11}C]ABP688
$ADORA_1$	[^{18}F]CPFPX	[^{11}C]MPDX
$ADORA_{2A}$	[^{18}F]MNI444	[^{11}C]KW-6002, [^{11}C]TMSX, [^{11}C]SCH442416
$ADORA_3$	[^{18}F]FE@SUPPY	**Not Found**
H1	**Not Found**	[^{11}C]Doxepin
H3	[^{18}F]FMH3, [^{18}F]Fluoroproxyfan, [^{18}F]9	[^{11}C]GSK189254
α_1-AR	**Not Found**	[^{11}C]GB67
α_2-AR	**Not Found**	[^{11}C]Mirtazapine
AT_1	**Not Found**	[^{11}C]Telmisartan, [^{11}C]KR31173, [^{11}C]L-159,884
NPY	[^{18}F]Y1-973	**Not Found**
SOM	[^{18}F]-5b	**Not Found**

Note: Abbreviation: GLUT – Glucose transporters, HK – Hexokinase, ACS – Acetyl-CoA synthetase, MCI – Mitochondrial complex I, AChE – Acetylcholine esterase, PKM_2 – Pyruvate kinase M_2, AA – Arachidonic acid AA, iNOS – inducible nitric oxide synthase, DAT – Dopamine transporter, SERT – Serotonin transporter, NET – Norepinephrine transporter, Gly-T_1 – Glycine transporter type-1, VMAT – Vesicular monoamine transporter, $SLC1A_5$ – Glutamine transporter, D_x – Dopamine receptor type-x, 5-HT_x – Serotonin receptor type-x, $\alpha_4\beta_2$ – α-4 β-2 Nicotinic acetylcholine (ACh) receptor, α_7 – β-7 ACh receptor, M_2 – Muscarinic ACh type-2 receptor, NMDA – N-methyl-D-aspartate receptor, $GABA_A$ – γ-aminobutyric acid type-A receptor, $mGluR_x$ – Metabotropic glutamate receptor type-x, $ADORA_x$ – Adenosine receptor type-A_x, H_x – Histamine H_x receptor, α_x-AR – α_x – Adrenoceptors, AT_1 – Angiotensin II type-1, NPY – Neuropeptide Y receptor, SOM – Somatostatin receptor. A more extended list of radionuclides for PET can be found at https://ncbi.nlm.nih.gov/books/NBK22999. In particular, readers are encouraged to look at PET probes for L and D – type amino acids, which could be useful for future chapters. PET probes have been also developed to study several brain disorders, e.g., **a)** Aβ plaques and τ tangles, **b)** caspase, **c)** hypoxic cells, and **d)** inflammatory signals/receptors.

TABLE 3.3
Most relevant cyclotron produced radionuclides for PET and SPECT

Nuclear Reaction	Stable Nucleus ${}^{A_t}E_t$	Accelerated P_i	Radiation nP_j	Radionuclide
${}^{20}Ne\left({}^{2}_{1}H,1\alpha\right){}^{18}F$	Neon -${}^{20}Ne$	${}^{2}_{1}H$	$1\alpha = 1{}^{4}_{2}H_{e}^{+2}$	${}^{18}F_2 \to {}^{18}F$
${}^{18}O\left({}^{1}_{1}H,1n\right){}^{18}F$	Oxygen -${}^{18}O$	${}^{1}_{1}H$	$1n$	${}^{18}F \to {}^{18}F$
${}^{14}N\left({}^{2}_{1}H,1n\right){}^{15}O$	Nitrogen -${}^{14}N$	${}^{2}_{1}H$	$1n$	${}^{15}O_2 \to {}^{15}O$
${}^{15}N\left({}^{1}_{1}H,1n\right){}^{15}O$	Nitrogen -${}^{15}N$	${}^{1}_{1}H$	$1n$	${}^{15}O_2 \to {}^{15}O$
${}^{14}N\left({}^{1}_{1}H,1\alpha\right){}^{11}C$	Nitrogen -${}^{14}N$	${}^{1}_{1}H$	$1\alpha = 1{}^{4}_{2}H_{e}^{+2}$	${}^{11}CO_2 \to {}^{11}C$
${}^{16}O\left({}^{1}_{1}H,1\alpha\right){}^{13}N$	Oxygen -${}^{16}O$	${}^{1}_{1}H$	$1\alpha = 1{}^{4}_{2}H_{e}^{+2}$	${}^{13}N \to {}^{13}N$
${}^{124(*)}Xe\left({}^{1}_{1}H,2n\right){}^{123}Cs$	Xenon -${}^{124(*)}Xe$	${}^{1}_{1}H$	$2n$	$Na{}^{123}I \to {}^{123}I$

Note: Note that the superscript $(*)$ is used to indicate that an unstable isotope of ^{124}Xe is used in the production of ^{123}I. ^{124}X decays ($T_{1/2} \approx 1.8 \times 10^{22}$ years) to ^{124}Te by means of a double EC (XENON Collaboration, 2019). However, ^{124}X transforms into ^{123}Cs by bombarding it with protiums. ^{123}Cs is highly unstable, decaying to ^{123}Xe ($T_{1/2} \cong 5.9$ min). Finally, unstable radionuclide ^{123}Xe decays to ^{123}I ($T_{1/2} \cong 2.08$ h) by means of EC.

a longer half-life $T_{1/2}(p)$ in comparison to that of their respective daughter ($T_{1/2}(d)$) nuclides, usually at least 10 times longer. Inside a generator, the daughter is continuously produced by the decay of the parent, but at the same time, it also decays in a contiguous manner. The Bateman equation (3.3) characterizes the rate of decay of the daughter given the exponential decay constants of the parent (λ_p) and daughter (λ_d). The generator chemically separates the daughter from the parent radionuclides as they are different elements.

$$A_d(t) = A_p(t_0)\frac{\lambda_d}{\lambda_d - \lambda_p}\left(e^{-\lambda_p t} - e^{-\lambda_d t}\right) + A_d(t_0)e^{-\lambda_d t} \tag{3.3}$$

According to the characteristics of the parent/daughter mixture inside the generator after long running times, there are three particular cases to be discussed here. The first case for which $T_{1/2}(p) < T_{1/2}(d)$ is referred to the non-equilibrium state, as the daughter concentration builds up inside the generator due to fast parent decay. This half-life relationship is not feasible for a practical radionuclide generator. A practical relationship is achieved when the parent nuclide half-life is at least 10 times longer than that of the daughter nuclide, as is the case of the ^{99}Mo/^{99m}Tc generator. If $T_{1/2}(p)/T_{1/2}(d) \geq 10$ then a **transient** equilibrium has been reached. If $T_{1/2}(p)/T_{1/2}(d) \gg 10$ then a **secular** equilibrium has been reached. As an important note, reference to "equilibrium" here is not in the traditional meaning of the word, since the parts are radioactive and the amounts are changing constantly the "equilibrium" state refers only to the relationship between parent and nuclide isotopes, not to the thermodynamically unchanged state of the reaction. Under equilibrium, the ratio daughter to parent remains constant inside the generator. In these cases, after reaching the equilibrium, both species decay at the half-life of the parent with a constant ratio daughter-to-parent activity $A_d(t)/A_p(t) = \lambda_d/(\lambda_d - \lambda_p)$. A secular equilibrium is preferred to produce generators. In this case, the decay of the parent is negligible within the first 10 half-life cycles of the daughter. As under secular equilibrium $\lambda_d \gg \lambda_p$, their respective activities become equal at the equilibrium $A_d(t) = A_p(t)$.

Among all the currently available generators, four are of relevance for brain studies: **1)** the ^{99m}Tc-generator, **2)** the ^{68}Ga-generator, **3)** the ^{62}Cu-generator, and **4)** ^{44}Sc generator. ^{68}Ga and ^{62}Cu are radionuclides with β^+ decay; hence, they have been used in PET studies of metabolic alterations and oxidative stress in a variety of neurological conditions. The ^{68}Ga-generator uses metastable ^{68}Ge ($T_{1/2}(p) = 271\,days$) that decays to ^{68}Ga ($T_{1/2}(d) = 68\,min$) by EC. The generator reaches a secular equilibrium. ^{68}Ge is produced in a cyclotron by the nuclear reaction $^{69}Ga(^1_1H,2n)^{68}Ge$. Most commonly, the generator is made of a TiO_2 column and $1N\,HCl$ eluent. The nuclear reaction taken place in a cyclotron to produce the ^{62}Zn needed in a ^{62}Cu-generator is $^{63}Cu(^1_1H,2n)^{62}Zn$. A transient equilibrium may be reached in a ^{62}Cu-generator because of the particular half-life of ^{62}Zn ($T_{1/2}(p) = 9.3\,hours$) and ^{62}Cu ($T_{1/2}(d) = 9.7\,mins$). ^{62}Zn decays to ^{62}Cu by both EC and β^+ mechanisms. The ^{62}Cu -generator comprises a Dowex column and $2N\,HCl$ eluent. ^{44}Sc has been produced for decades from ^{44}Ti, which has a long half-life (~60 years) and is produced in the nuclear reaction $^{45}Sc(p,2n)^{44}Ti$. The $^{44}Ti\,/\,^{44}Sc$ generator constitutes a secular equilibrium system. However, the ^{99m}Tc-generator is the most widely used in research and clinical practice. The generator produced the isotope ^{99m}Tc from a β^- decaying ^{99}Mo radionuclide ($T_{1/2}(p) = 66\,hours$). Hence, the generator reaches a transient equilibrium. ^{99}Mo must be produced in a nuclear reactor by ^{235}U fission. The daughter ^{99m}Tc decays to ^{99m}Tc by isomeric transition, emitting γ-rays and low-energy electrons. The generator is made of a Al_2O_3 column to absorb MoO_4^{2-} and 0.9% saline to elute the soluble ^{99m}Tc. Thanks to the creative work of the Brookhaven Lab chemist Margaret Greene, the first ^{99m}Tc-generator was built by Tucker et al. (1958).

Researchers and clinicians working with either SPECT or PET imaging modalities are advised to continuously revise the guideline for safety handling of radionuclides at the official IAEA website (www.iaea.org/resources/safety-standardsAEA).

3.6 THE SCINTILLATOR COUNTERS FOR X AND γ RAYS

Transmitted x-ray photons have high energy ($10-120\,keV$). The emission of γ-rays is a common aspect of emission imaging modalities (SPECT and PET), and these rays have also high energy. PET radionuclides emit radiation with energy in the order of $511\,keV$. For SPECT, most cameras (collimators) have been optimized to accommodate the emissions of ^{123}I and ^{99m}Tc ($159\,keV$ and $140\,keV$, respectively). Note for comparison that the energy of photons in the visible spectrum ranges from $1.65\,eV$ (**red** lower energy) to $3.26\,eV$ (**violet** upper energy). Therefore, the conversion of a high-power photon into multiple lower-energy visible photons is required to obtain quantitative measurements of the x- and γ- radiation. This conversion is achieved using a scintillator, which is typically coupled to a sensor to measure the emitter light. At this point, it is good to differentiate two types of measurements of the emitted light: **a)** integrated-light and **b)** pulse-counting. The serial combination of a scintillator and a light sensor (or photodetector), a device known as a scintillator counter, was first proposed by Sir Samuel Curran in 1944 as part of the Manhattan Project. The use of scintillators in medical imaging was linked to the need to observe γ-rays. Initial detections of γ-rays were performed with the Geiger-Müller counter (Geiger and Müller, 1928), but it was not until the development of the scintillation detectors (Cassen et al., 1950) that sensitivity was good enough for medical imaging. The most important contribution in γ-ray detection was made by the creation of the Anger camera (Anger, 1952), also known as γ camera. The initial γ camera consisted of a thallium-doped sodium iodide NaI(Tl) flat screen, i.e., a scintillator, to where a transmitted x-ray beam was projected through a pinhole. The concept was proven by imaging the metastasis of a thyroid tumor at elbow with ^{131}I. Soon after the first SPECT study from an open craniotomy in a human (Ingvar and Lassen, 1961, 1962), Lassen et al. (1963) used the same scintillator counter placed over the intact scalp to observe the γ emissions of ^{85}Kr.

The quantum mechanisms underlying the production of the visible photons vary for different scintillators. Discussing these mechanisms is beyond the scope of this book. However, a transition of a primary electron from a full valence-band to an empty conduction-band while absorbing a high-energy photon constitutes a fundamental principle. By means of Coulomb scattering, a primary electron in the conduction band can generate many excited secondary electrons (*conversion process*). By interactions with other electrons and the lattice, these secondary electrons dissipate their kinetic energy (*transfer process*). Finally, many of these low-energy secondary electrons emit scintillation photons with wavelengths in the range of the visible light by the recombination with holes in certain emission centers (*luminescence process*). Therefore, the overall efficiency of a scintillator is determined by the efficiency of these three processes. After the introduction of the NaI(Tl) in the 40s, a few dozens of scintillators have been developed for radiation measurement/counting applications. To be a good scintillator, certain characteristics must be met: **1)** linearity over the energy range of the incident radiation, **2)** high stopping radiation power, **3)** high efficiency in producing visible photons from the high-energy radiation, **4)** a reduced dead-time, **5)** transparency to the emitted scintillation light, and **6)** good match with photodetectors according to their sensitivity range. Based on these characteristics, there are seven main physical parameters to be considered while looking for appropriate scintillation materials for CT and SPECT/PET: **1)** density ρ $(g\,/\,cm^3)$, **2)** effective atomic number Z_{eff}, **3)** light yield f $(Ph\,/\,keV)$, **4)** decay time τ, **5)** emitted scintillation light wavelength λ_e nm, **6)** refractive index n, and **7)** energy resolution $\Delta E\,/\,E$.

For both x and γ -ray detection, scintillators are required to have high effective atomic number ($Z_{eff} > 50$) and light yield f. The following equation provides a good approximation for $Z_{eff} \approx \sum_k (m_k Z_k^4)^{1/4}$, with m_k and Z_k the total mass and atomic number associated with the *k-th* element, respectively. There are three main types of x and γ -rays interaction with matter: photoelectric absorption, Compton scattering and pair-creation. As its probability depend on ρZ_{eff}^4, the photoelectric absorption is the type of interaction of main interest in CT and SPECT/PET medical imaging. Hence, material with high density and effective atomic number are ideal for increasing the stopping power of the x and γ radiation, a very important component to ensure efficiency in energy conversion. A good energy resolution helps separation of non-wanted events (e.g., Compton scattering and pair-creation), which leads to higher sensitivity and spatial resolution. Detector thickness is substantially reduced with the use of very dense scintillators, a critical parameter to ensure minimal parallaxial error in compact detector arrays.

For CT modality, transmitted *x*-ray photons must be integrated over a certain period. This type of integration-detector requires a low afterglow (memory). For example, for the two most widely used crystalline scintillators in CT, "*cadmium tungstate*" ($CdWO_4$) and "*thallium-activated cesium iodide*" (CsI(TI)), the afterglow at $3ms$ is 0.05% and 0.5%, respectively. It is also low for the new generation of ceramic scintillators in CT imaging: **a)** 5 % for yttrium gadolinumoxyde (YGO) – $(Y,Gd)_2O_3$ when doped/activated (europium "Eu", praseodymium "Pr", terbium "Tb"), **b)** $< 0.1\%$ for gadolinium silicate (GOS) – Gd_2O_2S when doped/activated (Pr, cerium "Ce", fluorine "F"), and **c)** $< 0.1\%$ gadolinium gallium garnet (GGG) – $Gd_3Ga_5O_{12}$ when doped/activated (chromium "Cr", Ce). Lanthanum hafnate ($La_2Hf_2O_7$) doped with titanium "Ti" is another ceramic material proposed for scintillation in CT imaging. Heat dissipation is an important aspect in *x*-ray imaging, which leads sometimes to abrupt changes in temperature in the detectors. Hence, scintillators for *x*-ray detection must have an ideal output temperature coefficient of $|c_T| < 0.1\% / {}^oC$. Imaging speed is critical for the CT modality as it requires multiple projections per second. Hence, scintillators with decay time τ for high-sampling rate ($\sim 10kHz$) are desirable. A limited sampling rate is a major disadvantage of YGO scintillators. Table 3.4 summarizes the physical characteristics of most used scintillators in the CT imaging modality.

Pulse-counting detectors are typically used for γ-ray measurements in emission modalities. Pulse-counting detectors are required to have a fast decay time and high-energy resolution. The dead-time is reduced for scintillators with short decay time. This property is crucial to increase image quality as event counting is increased, and the precision for coincidence detection in PET is enhanced. By increasing the image quality using appropriate scintillators, patient radionuclide dose can be reduced. Most of the SPECT scanner still use NaI(Tl) and CsI(Tl) scintillators. There are three other attractive scintillators for the SPECT imaging modality: **i)** Ce-doped Lanthanum Chloride ($LaCl_3$(Ce)), **ii)** Ce-doped Lanthanum Bromide ($LaBr_3$(Ce)), and **iii)** Ce-doped yttrium orthoaluminate ($YAIO_3$(Ce)). Until the 80s, the bismuth germanate (BGO, $Bi_4Ge_3O_{12}$) was the most used scintillator for the PET imaging modality. Based on phoswich technology, new scintillators with appropriate characteristics for PET imaging were developed at the beginning of this century: **a)** LSO – Lutetium oxyorthosilicate doped with cerium (Lu_2SiO_5:Ce), **b)** GSO – Gadolinium oxyorthosilicate doped with cerium (Gd_2SiO_5:Ce), and **c)** YSO – Yttrium oxyorthosilicate doped with cerium (Y_2SiO_5:Ce). The discovery of the ultra-fast ($0.6ns$) barium fluoride (BaF_2) motivated the design of the time-of-flight (TOF) systems for PET imaging. Unfortunately, BaF_2 emission is not strong ($2Ph/keV$) and in the far-ultraviolet range ($220nm$). Non-hygroscopicity is another important property while designing scintillator counters, in particular for PET. The most used scintillators in SPECT/PET imaging modalities are listed in Table 3.4.

TABLE 3.4
The most useful scintillator in transmission/emission medical imaging

Scintillator	ρ (g/cm^3)	f (Ph/keV)	τ (ns)	(nm)	n	Further Info
CT						
$CdWO_4$	7.9	27.3	14500	475	2.25	$c_T = -0.3\% / °C$
CsI(Tl)	4.52	54	1000	545	1.78	$c_T = 0.02\% / °C$
YGO	5.9	42	10^6	610	–	$c_T = 0.04\% / °C$
GOS:Pr,Ce,F	7.34	50	4000	520	–	$c_T = -0.6\% / °C$
GGG:Cr,Ce	7.09	39	150000	730	–	–
$La_2Hf_2O_7$:Ti	7.9	13	10000	475	–	–
SPECT						
NaI(Tl)	3.67	38	230	410	1.85	hygroscopic
CsI(Tl)	*see data above*					
$LaCl_3$(Ce)	3.85	49	28	350	~1.9	hygroscopic
$LaBr_3$(Ce)	5.08	63	16	380	~1.9	hygroscopic
$YAlO_3$(Ce)	5.50	21	30	350	–	–
PET						
BGO	7.13	8–10	300	480	2.15	Not hygroscopic
LSO:Ce	7.4	29	40	420	1.82	Not hygroscopic
GSO:Ce	6.71	10	60	440	1.91	Not hygroscopic
YSO:Ce	4.53	46	70	420	1.8	Not hygroscopic
LuAP:Ce	8.34	10	17	365		Not hygroscopic

3.7 THE COLLIMATION AND COINCIDENCE PRINCIPLES

CT and SPECT/PET imaging modalities have evolved over the years to allow multi-slice recording at high speed and low radiation doses. To that end, two important imaging components have been upgraded with new technological advances: **1)** the collimation for CT/SPECT and **2)** the coincident detection for PET. In this section, the most important aspects of these two components are discussed. Multi-slice transmission/emission images are obtained from γ photon data collected using two-dimensional (2D) arrays of scintillators. To provide a mechanical support to these arrays and minimize crosstalk between the emitted scintillation light for different pixels, scintillator arrays are packaged with a reflective material matrix and cut in independent blocks. The scintillator arrays are connected to light sensors. Electric signals from the latter are then magnified with low-noise preamplifiers and stored in a computer via an analog-to-digital converter. Different types of light sensor are used to measure the visible light from scintillators generated by the absorption of x and γ high-energy photons. At this level, quantum efficiency (photo-electron/scintillation-photon) is the most important property. Quantum efficiency is maximized by matching the spectral response of the photo-electron converter to that of the scintillator emission spectrum.

Due to its high-amplification gain ($> 10^6$), the photomultiplier tube (PMT, Figure 3.1) has been the predominant light sensor in modalities based on pulse-counting (SPECT and PET). It comprises a photocathode light-to-electron converter

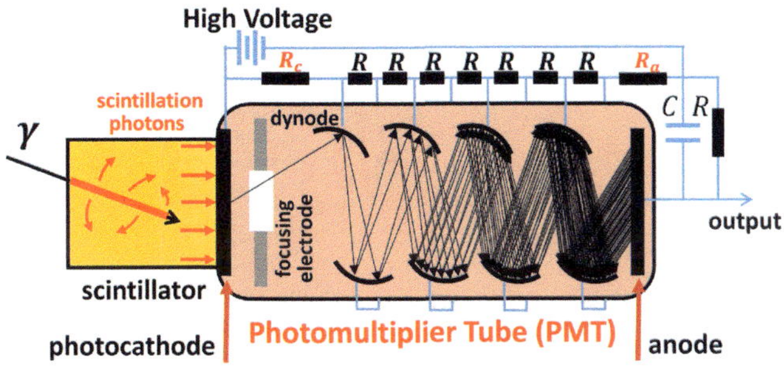

FIGURE 3.1 Working principle of the scintillator counters based on photomultiplier tube (PMT).

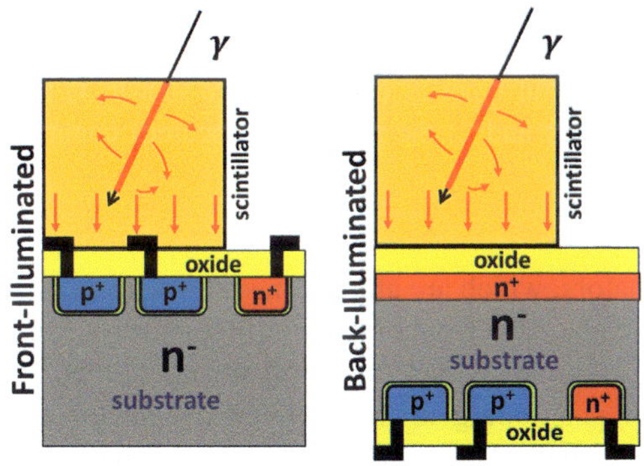

FIGURE 3.2 Illustration of the front and back – illuminated p-n photodiodes (PDs).

in contact with the scintillator, followed by electron-multiplying dynodes. In each dynode, electrons are multiplied by the application of high voltages. Voltage divisors are used to linearly increase the voltage drop at each subsequent dynode. Modern CT scanners use integrating scintillators optically coupled mostly to PD arrays. A PD comprises electron-rich regions (N^+) connected to a positive cathode and hole-rich regions (P^+) connected to a negative anode. At the interface between these regions, a depletion area is created. An electric current is produced inside a PD by the photo-electric effect. An incident photon is absorbed at the junction's depletion region, creating an electron-hole pair. As the result of a built-in electric field, electrons move toward the cathode and holes toward the anode. Two PD configurations are illustrated in Figure 3.2 in the front and back-illuminated formats.

A large portion of the *x*-rays passing the head undergoes either coherent (Rayleigh) or incoherent (Compton) scattering, which results in scattered *x*-rays emerging from the head at different angles. To separate these scattered rays from those transmitted along a straight line from a *x*-ray source to a detector, CT scanners use collimated detectors (Figure 3.3). The characteristics of these collimated detectors depend on the CT scanner generation. In general, they are formed by anti-scatter grids (ASG)

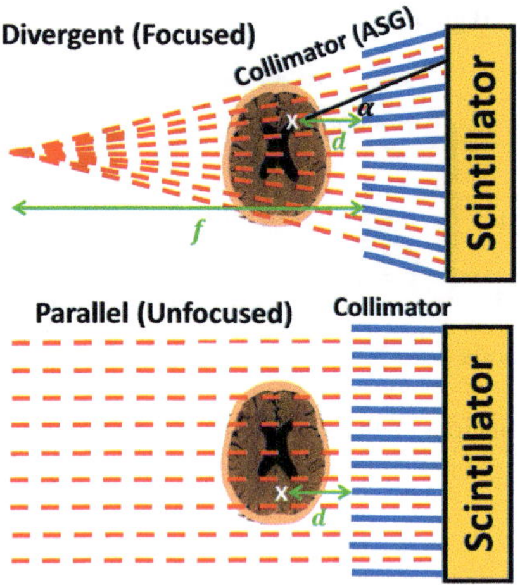

FIGURE 3.3 The most useful types of collimators are shown: **1)** the divergent (focused) mostly for CT imaging (top) and 2) the parallel (unfocused) commonly used for SPECT (bottom).

functioning as collimators, which are aligned in series with the scintillators and light sensors. The ASGs are placed in front of the scintillator arrays in two basic segmented organizations (linear, crossed). In the crossed organization, there is a one-to-one pixel representation. Event positioning in the collimated detector array is achieved by using the Anger logic. These ASGs must be made of elements with high Z number (e.g., molybdenum ^{42}Mo, tungsten ^{74}W or lead ^{82}Pb). As x-rays originate from a point source, ASG used in CT imaging are divergent (focused). Compensating filters are used together with these collimators to shape and focus the x-ray beam. For most scanners, a single x-ray vacuum tube is used. The x-ray beam can be of three different types: **1)** pencil, **2)** fan and **3)** cone. Multiple x-ray tubes can be used.

In a CT scanner, collimated detectors are arranged using two main formats: **1)** slip ring and **2)** stationary ring. For scanners with slip ring both the detector array and the x-ray vacuum tube are mounted on a gantry that rotates at hundreds *rpm* and simultaneously translates to provide data for a 3D image of μ inside the head. For scanners with a stationary ring, only the x-ray vacuum tube rotates/translate attached to the gantry while the ring remains fixed to the scanner (Figure 3.4).

The gantry provides the mechanical and electrical infrastructure for their proper function. Details about the most useful mathematical methods for image reconstruction in CT are provided later in this chapter. Figure 3.5 shows a digitalized 3D reconstruction of the skull of a monkey as a particular example of application of the CT modality. To increase imaging speed, shorter integration periods are used by selecting faster scintillators (Table 3.4). The detector-based spectral CT (SDCT) is a revolutionary multicolor x-ray imaging modality based on a dual layer of two scintillators with different characteristics. The technique decomposes contributions to μ from the low/high energy components of the x-ray. With the introduction of direct conversion

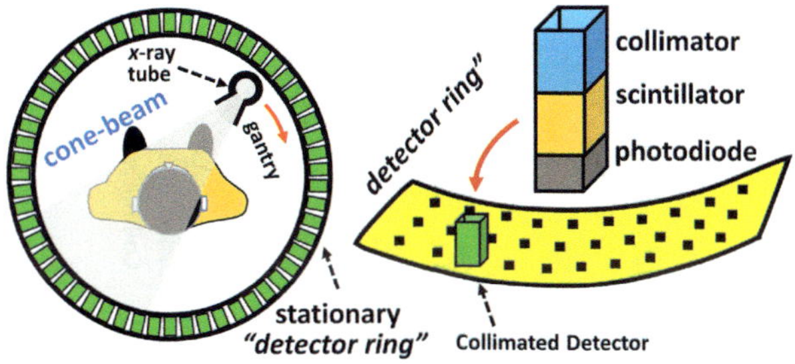

FIGURE 3.4 Illustration of the main principles in CT scanners.

Note: The stationary ring used in modern generation of scanners is shown. A segment of the ring is magnified with a photodiode-type of collimated detector. This array of collimated detectors is used in scanners with a cone-beam x-ray configuration. The x-ray source rotates around the subject and, to cover a volumetric region, it also translates along the longitudinal axis.

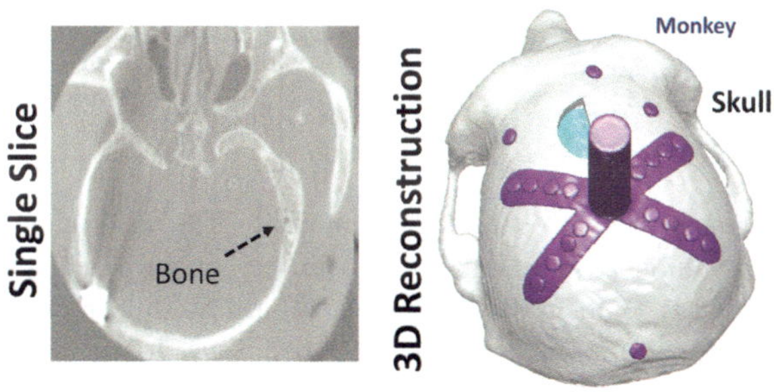

FIGURE 3.5 A single-slice image (left) of the attenuation coefficient obtained from a monkey head obtained at Vanderbilt University (courtesy of Dr. Shall). Using several slices, the 3D segmentation of the skull is achieved (right).

materials (DiCo) in CT scanners, *x*-ray photons are separated according to their individual energy, which has allowed the reconstruction of the μ spectral profiles. Combinations of these modern techniques with Compton scattering imaging are promising.

The pinhole in the original γ camera constitutes the first utilization of a collimator for SPECT imaging (Anger, 1952). The use of these collimators in SPECT imaging became popular in the early 60s. For example, Glass and Harper (1963) proposed to use depth-focusing collimators to observe the γ emitting ^{133}Xe through the intact skull of patients. Harper et al. (1964) use it a year later to measure rCBF from an intact skull in patients using ^{133}Xe. The principles underlying the use of collimator in modern SPECT scanner is similar to those for CT scanner. The geometry of the collimator array constitutes the main difference between the CT and SPECT collimation. SPECT collect emitted γ photon using head with planar organization of collimated detectors (Figure 3.6).

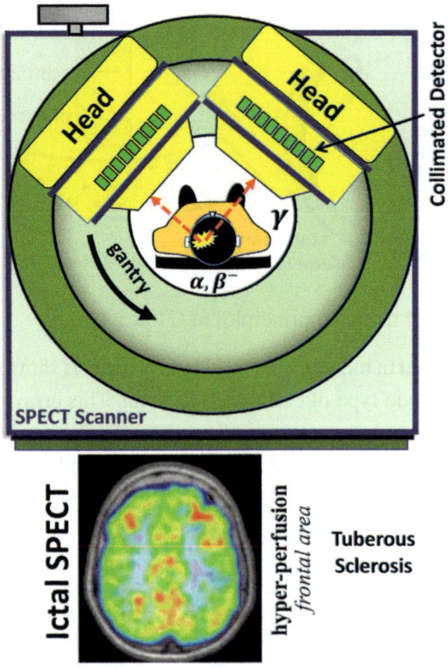

FIGURE 3.6 Imaging principles of the SPECT modality. The application of this imaging modalities in epilepsy surgical planning is shown in the bottom panel. (Reproduced with permission from Bernal B, Guillen MR, Zamora K, Altman N. Utilizing rs-fMRI's Mean to Localize Seizure Focus. *Open Journal of Radiology*, 2015, 5, 92–103).

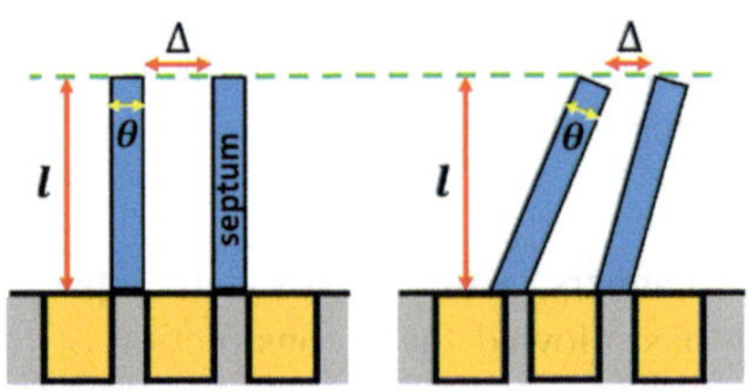

FIGURE 3.7 Typical parameters for these two collimators are also illustrated.

The hexagonal array (honeycomb) is one of the most widely used. Parallel-hole collimators are still used in some SPECT scanners. SPECT imaging reconstruction methods are based on γ photons entering perpendicular to the head. γ photons with an oblique direction must be absorbed by the collimators. Hence, parallel (unfocused) collimators are the most frequently used in SPECT scanners (Figure 3.7). SPECT scanners could collect γ photons using a single head, a dual head format or even multiple heads (Figure 3.6). These SPECT heads are attached to a gantry that rotates around the subject.

Collimated detectors are characterized by their specific-sensitivity S_C and resolution R_C. The specific-sensitivity and resolution for divergent hexagonal collimators are:

$$S_C(d,\alpha) = \sqrt{3}\Delta^4 f^n (cos(\alpha))^2 / \left(8\pi l_e^2 (\Delta+\theta)^2 (f-d)^n\right) \tag{3.4}$$

$$R_C(d,\alpha) = \Delta(l+d)(f+l/2)/\left(l_e cos(\alpha)(f+l)\right) \tag{3.5}$$

With $n = 1$ for fan-beam and $n = 2$ for cone-beam collimators. Parameter $l_e = l - 2/\mu_C$ represents the adjusted physical length of the collimator, with μ_C the collimator attenuation coefficient. The overall resolution of the collimated detector is:

$$R_S(d,\alpha) = \sqrt{\left(\frac{R_i}{m(d)}\right) + R_C(d,\alpha)^2} \tag{3.6}$$

With R_i being the intrinsic spatial resolution of the detector and $m(d) = (f+l)/(f-d)$. The specific-sensitivity and resolution for parallel hexagonal collimators are:

$$S_C(d) = \sqrt{3}\Delta^4 / \left(8\pi l_e^2 (\Delta+\theta)^2\right) \tag{3.7}$$

$$R_C(d) = \Delta(l+d)/l_e \tag{3.8}$$

$$R_S(d) = \sqrt{R_i^2 + R_C(d)^2} \tag{3.9}$$

For both cases, volume sensitivity is the average of the specific sensitivity over the entire field of view. The specific-sensitivity reflects the ratio of γ photons reaching the scintillator from all arriving to the associated collimator. In SPECT, the specific-sensitivity depends on the density and atomic number of the material used to make the collimators (e.g., lead, tungsten, platinum, or gold). However, it is generally low ($\sim 10^{-4} - 10^{-2}$). The spatial resolution is in the sub-centimeter range. As the linear momentum must be conserved, the two γ photons resulting from matter-antimatter annihilation in PET are emitted in opposite directions. Hence, it is possible to define a line-of-response (LOR) connecting the two impacted PET detectors for each decaying event. The detection of a pair of coincident γ photons constitutes indeed an electronic collimation. Therefore, all γ photons arriving within a temporal window Δt to a pair of PET detectors are counted. For this reason, the specific-sensitivity for PET is higher than that for SPECT. Using a fixed rings with grid of detectors (i.e., scintillators and PMTs), all events in opposite direction can be determined (Figure 3.8).

PET imaging has been possible with the development of fast scintillators and sensitive/smaller PMTs. Because of the coincidence detection principle, PET imaging can be used to study changes in brain chemical composition within the *pm* range. Most common image reconstruction methods in the PET imaging modality assume a

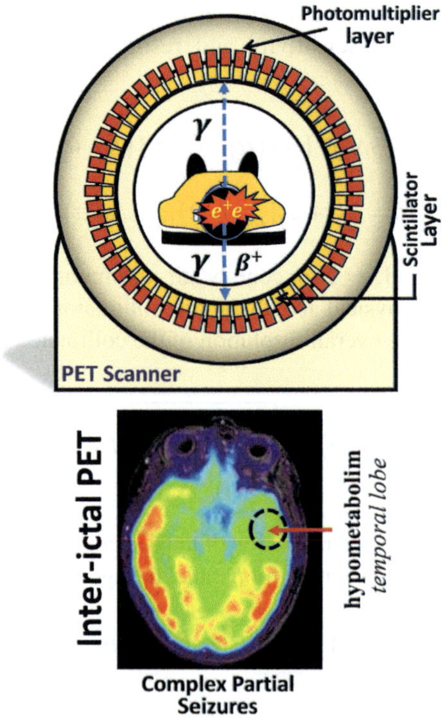

FIGURE 3.8 Imaging principles of the PET modality. The application of this imaging modalities in epilepsy surgical planning is shown in the bottom panel. (Reproduced with permission from Bernal B, Guillen MR, Zamora K, Altman N. Utilizing rs-fMRI's Mean to Localize Seizure Focus. *Open Journal of Radiology*, 2015, 5, 92–103).

uniform probability for the location of the event within the field of view (FOV) and along the LOR. However, the sensitivity and SNR can be improved by incorporating information about the specific arrival times of the paired γ photons to the opposite detectors in the LOR, a technique named time-of-flight (TOF) PET. Assuming an object of dimension D, the gain in the sensitivity and signal-to-noise ratio (SNR) is $D / \Delta x$ and $\sqrt{D / \Delta x}$ if a difference in time Δt for the photon arrivals at both detectors can be accurately observed. Note that $\Delta x = c\Delta t / 2$, with $c = 299792458\, m / s$ representing the speed of light in vacuum. For most used PET scanners based on BGO scintillators, the timing resolution is $\Delta t \approx 600\, ps$ for a $\Delta x \approx 89\, mm$. Hence, the use of the TOF-PET methodology carries no great weight. However, with the development of faster ($\Delta t \approx 400\, ps$) scintillators for PET (e.g., LSO:Ce, GSO:Ce, YSO:Ce and LuAP:Ce), significant improvements in sensitivity and SNR can be achieved. Improvements are also possible with the use of advanced technologies for optoelectronic. For example, scanners for CT and SPECT/PET imaging can measure scintillation light using other types of photodetectors; e.g., a charge-coupled device (CCD) camera, a silicon drift detector (SDD), a photovoltaic detector array (PDA), or a complementary metal oxide semiconductor (CMOS). In modern times, SPECT/PET scanners are integrated with technology for CT imaging in a single piece of equipment. These advanced hybrid devices facilitate the interpretation of SPECT and PET functional imaging by their co-registration with anatomical brain images obtained from a CT modality on the same subject. CT and SPECT/PET scanners are also endowed with sophisticated software/hardware for motion correction. The bottom panels in Figures 3.6 and 3.8 show the application of these two imaging modalities in epilepsy surgical planning.

3.8 RECONSTRUCTION TECHNIQUES

The methodology proposed by Hounsfield in 1971 to reconstruct radiographic planar images from transmitted x-ray data constitutes a cornerstone in reconstructive tomography (i.e., CT and SPECT/PET). However, his proposal was preceded by a substantial work by several other researchers in the field: **a)** the conversional focal-plane tomography for x-rays by Bocage and Takahashi, followed by its use in radioisotope emission imaging modalities from late in the 60s, e.g., the tomoscanner (Miraldi and Di Chiro, 1970) and **b)** early variations of the back-projection method in the apparatus for γ-ray imaging developed for transmission by Oldendorf and for emission by Kuhl and Edwards. However, the mathematical technique for image reconstruction developed by Allan McLeod Cormack (1924–1998) (Cormack 1963; 1964) for x-ray was crucial in this endeavor. For this work, he shared the 1979 Nobel Prize for Physiology or Medicine with Hounsfield. There are two mathematical frameworks underlying the Hounsfield and Cormack proposal for CT imaging: **1)** the Radon transform proposed by Johann Radon (1887–1956) in 1917 for solving gravitational equation (see analytical methods) and **2)** the algebraic reconstruction technique formulated by Stefan Kaczmarz (1895–1939) in 1937 (see iterative methods). These frameworks have been also used in reconstructive tomography for SPECT and PET. In reconstructive tomography, all tissues traveled by the beans are reconstructed. Under this construct, it is mathematically feasible to reconstruct laminar distribution of either the attenuation coefficient $\mu(x,y)$ (transmission) or the radioisotope density $\rho(x,y)$ (emission) if enough projections are taken. For simplicity, these two magnitudes are represented in this book by the generic function $f(x,y)$. Reconstruction of $f(x,y)$ in the Cartesian coordinate system (x,y) from many projected transmission/emission data is considerably complex from a computational perspective, requiring the use of computers. Therefore, this type of tomographic imaging is commonly referred to as computer-assisted tomography (CAT). Methods for 3D reconstructions are not presented in this book. Similarly, image reconstruction from projections acquired with a pinhole collimator are not discussed. Algorithms based on linograms, Chebyshev polynomials, and nonlinear back-projection are also not reviewed here. Reconstructive methods are classified into two major groups: **a)** analytic and **b)** Iterative ('brute-force'). Exact mathematical solutions are used in analytic methods; hence, they are faster. The filtered back-projection (FBP) is one of the most widely used analytical methods in modern CT scanners. Iterative methods are computationally costly as they use 'brute-force' to reconstruct an image. The Hounsfield's initial CT scanner at EMI used an iterative method to reconstruct the image.

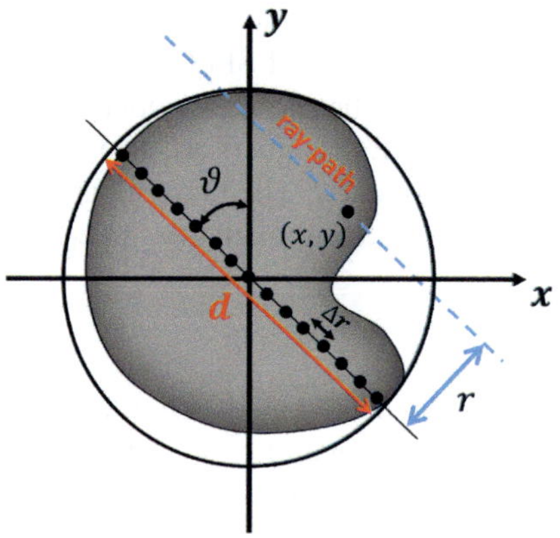

FIGURE 3.9 Parameter definition for reconstructive tomography.

A ray-path in reconstructive tomography is defined by a pair of coordinates (r, ϑ), named sinogram, where r represents the distance from the origin to the ray and ϑ its angle (Figure 3.9). The ray-sum is defined as $p(r,\vartheta) = \int_{r,\vartheta} f(x,y)dl$, with dl representing the differential path-length along the ray. The transmitted beam intensity for mono-energetic photons is $I = I_0 e^{\int_{r,\vartheta} \mu(x,y)ds}$. By definition, $p(r,\vartheta)$ satisfies the following relationship for x-rays.

$$p(r,\vartheta) = -ln\left(\frac{I}{I_0}\right) \quad \textbf{Transmission}$$

Under the no-attenuation assumption for radionuclide emission, the following relation hold for γ-rays:

$$p(r,\vartheta) = I \quad \textbf{Emission}$$

By definition, $r = xcos(\vartheta) + ysin(\vartheta)$. A complete set of ray-sums at a given angle is called a *profile* (Figure 3.10). Let the FOV be confined by a circular domain of diameter d. Then, it is appropriate to define M profiles (ϑ_m with $m = \{0,\cdots,M-1\}$) angularly spaced ($\Delta\vartheta = \pi / M$) from 0° to 180°, each consisting of $N = d / \Delta r$ ray-sums (r_n with $n = \{0,\cdots,N-1\}$) at intervals Δr. Note that $r_n = -d/2 + n\Delta r$. In practice, $f(x,y)$ must be calculated at a finite number of pixels (x,y) at equally spaced intervals from NM ray-sums $(p(r_n, \vartheta_m))$.

In this context, the first proposed method for reconstruction was the back-projection (Oldendorf's experiment and Kuhl's apparatus). It is also referred to as the summation method (3.10). The back-projected value of $f(x,y)$ is the

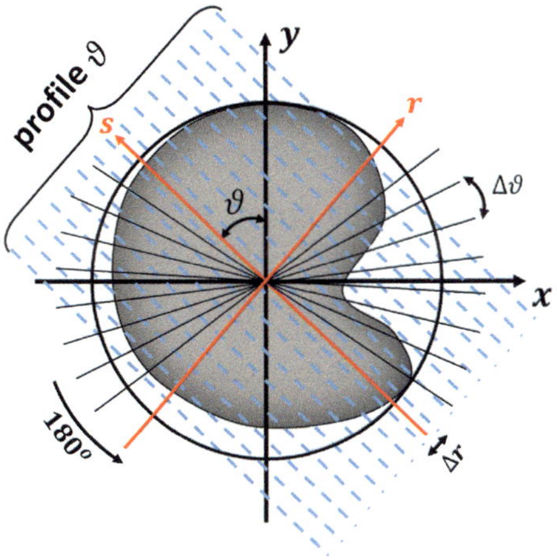

FIGURE 3.10 Profiles and associated coordinate transformations.

summation of all ray-sums passing through pixel (x, y), which are selected by the factor $xcos(\vartheta_m) + ysin(\vartheta_m)$.

$$\hat{f}(x,y) = \sum_{m=0}^{M-1} p\left(xcos\left(\vartheta_m\right) + ysin\left(\vartheta_m\right), \vartheta_m\right)\Delta\vartheta \tag{3.10}$$

Analytic methods use the spatial Fourier representation of $f(x, y)$ (3.11) to obtain exact formulas for $p(r, \vartheta)$.

$$f(x,y) = \int_{-\infty}^{\infty}\int_{-\infty}^{\infty} F\left(k_x, k_y\right) e^{2\pi i\left(k_x x + k_y y\right)} dk_x dk_y \tag{3.11}$$

The wave numbers k_x and k_y represent the spatial frequencies in each Cartesian coordinate. Image reconstruction with analytical methods require a band-limiting approximation ($k_{x,y} < k_c$) with k_c a critical frequency, implying: **1)** spacing in the $\hat{x}$ and $\hat{y}$ directions must be $\Delta r = 1/2k_c$, **2)** profiles must be sampled at the interval Δr, and **3)** a discrete Fourier transform (DFT) is used instead of the Fourier transforms. The two most popular analytical methods originate from theoretical representations of the Fourier coefficients $F(k_x, k_y)$: **a)** the 2D Fourier reconstruction (2D-FR) and **b)** the filtered back-projection (FBP). In general, the Fourier coefficients satisfy the following equation:

$$F\left(k_x, k_y\right) = \int_{-\infty}^{\infty}\int_{-\infty}^{\infty} f(x,y) e^{-2\pi i\left(k_x x + k_y y\right)} \text{dx dy} \tag{3.12}$$

The 2D Fourier reconstruction (2D-FR) – By performing a rotation of the Cartesian coordinate system an angle $\vartheta = tan^{-1}(k_y / k_x)$ (Figure 3.10) and defining $k = \sqrt{(k_x^2 + k_y^2)}$, equation (3.12) can be written as a function of $p(r, \vartheta)$: $F(k_x, k_y) = \int_{-\infty}^{\infty} p(r, \vartheta) e^{-2\pi ikr} dr$. In other words, $F(k_x, k_y)$ is exactly equal to the Fourier coefficients $P(k, \vartheta)$ of function $p(r, \vartheta)$ with respect to r for any given profile angle ϑ. The method works as follow: **1)** calculate the inverse DFT of $p(r, \vartheta_m)$ with respect to r for each profile ϑ_m, $P(k_n, \vartheta_m) = \sum_{n'=0}^{N-1} p(r_{n'}, \vartheta_m) e^{-2\pi ik_n\left(-\frac{d}{2}+n'\Delta t\right)}$, **2)** estimate $F(k_x^{(n)}, k_y^{(m)})$ on the spatial-frequency rectangular grid ($n = 0, \cdots N-1$ and $m = 0, \cdots, M-1$) by interpolating the Fourier coefficients $P(k_n, \vartheta_m)$, and **3)** use the direct DFT to estimate $f(x_n, y_m)$ from Fourier coefficient $F(k_x^{(n)}, k_y^{(m)})$, $f(x_n, y_m) = \frac{1}{NM} \sum_{n'=0}^{N-1} \sum_{m'}^{M-1} F(k_x^{(n')}, k_y^{(m')}) e^{2\pi i (k_x^{(n')} x_n + k_y^{(m')} y_m)}$.

The filtered back-projection (FBP) – There are three common filtering approaches to perform FBP: **1)** Fourier, **2)** Radon and **3)** convolution.

Fourier filtering is based on the introduction of the polar coordinate system to solve (3.11). It is demonstrated that:

$$f(x, y) = \int_0^{\pi} p^*\left(x cos(\vartheta) + y sin(\vartheta), \vartheta\right) d\vartheta \tag{3.13}$$

$$p^*(r, \vartheta) = \int_{-\infty}^{\infty} |k| P(k, \vartheta) e^{2\pi ikr} dk \tag{3.14}$$

The method works as follow: **1)** calculate $P(k_n, \vartheta_m)$ for each profile ϑ_m, **2)** take the inverse DFT after scaling it by $|k_{n,m}|$, $p^*(r_n, \vartheta_m) = \frac{1}{N} \sum_{n'=0}^{N-1} |k_{n',m}| P(k_{n'}, \vartheta_m) e^{2\pi ik_{n',m} r_n}$, **3)** back-project onto image plane, $f(x_n, y_m) = \sum_{m'=0}^{M-1} p^*(x_n cos(\vartheta_m) + y_m sin(\vartheta_m), \vartheta_m) \Delta\vartheta$, and **4)** repeat 1-3 for all profiles. p^* is imposed to be zero outside the domain of reconstruction. Interpolations, or other approximation methods, are also required here to switch from function domains.

Radon filtering is based on the Fourier transform $-1/2\pi^2 r^2$ for function $|k|$, the well-known convolution theorem and its integration by parts, which are used to demonstrate that:

$$p^*(r, \vartheta) = \frac{1}{2\pi^2} \int_{-\infty}^{\infty} \frac{\partial p(r', \vartheta) / \partial r'}{r - r'} dr' \tag{3.15}$$

The method works as follow: **1)** calculate $p^*(r_n, \vartheta_m)$ numerically from $p(r_n, \vartheta_m)$ using equation (3.15) – the filtering, **2)** back-project onto image plane $f(x_n, y_m)$ as the step 3 in the Fourier-filtering, and **3)** repeat 1–2 for all profiles.

Convolution filtering is similar to the Radon filtering, but uses a cut-off version of the Fourier transform for $|k|$, $\int_{-k_c}^{k_c} |k| e^{2\pi ikr} dk = \frac{k_c}{\pi r} sin(2\pi k_c r) - \left(\frac{sin(\pi k_c r)}{\pi r}\right)^2$.

The filtering is based on the assumption that $P(k, \vartheta) = \begin{cases} P(k_n, \vartheta_m) & k \le k_c \\ 0 & k > k_c \end{cases}$. It is demonstrated that:

$$p^*(r,\vartheta) = k_c p(r,\vartheta) - \int_{-\infty}^{\infty} p(r',\vartheta)\left(\frac{sin\left(\pi k_c (r-r')\right)}{\pi(r-r')}\right)^2 dr' \qquad (3.16)$$

The method works as follow: **1)** calculate $p^*(r_n, \vartheta_m)$ by integrating numerically (3.16) for each profile ϑ_m, **2)** back-project onto image plane $f(x_n, y_m)$ as the step 3 in the Fourier-filtering, and **3)** repeat 1–2 for all profiles.

Iterative methods approximate the object by an array of pixels uniformly distributed and the rays are considered strips: $p_j = \sum_{i=1}^{N_{Pixels}} w_{ij} f_i$, with $N_{Pixels} = \pi N^2 / 4$ inside the previously defined circumference of radius d. The weighting factors w_{ij} represent the contribution of the *i-th* pixel to the *j-th* ray-sum p_j. These factors are usually pre-calculated for later use during the image reconstruction. There are several approaches to calculate the w_{ij} factors: **1)** the physical interception between the *i-th* pixel and the *j-th* ray, **2)** a binary assignation (1,0) according to whether the center of the pixel lies within the ray, and **3)** the distance between the center of the pixel and two adjacent rays, either linearly or non-linearly modeled. Certain constraints could be imposed on to $f(x, y)$ to reduce the computational burden: e.g., $0 \le f(x, y) \le f_{max}$ and $f(x, y) = 0$ for pixels contained within rays that have zero ray-sum. The introduction of the discretization scheme leads to the following algebraic equation (3.17) with N_{Obs} observation and N_{Pixels} unknown. P is a large vector comprising all ray-sums from multiple profiles arranged with a specific order. The vector $F = vec(\{f_{n,m}\})$ comprises the unknown functions f evaluated at the pixels $f_{n,m} = f(x_n, y_m)$. The matrix W is constructed according to these definitions.

$$P = WF \qquad (3.17)$$

From an optimization viewpoint, the reconstructive tomography is formalized as $\hat{F} = min\left|P - WF\right|^2$. However, the algebraic equation (3.17) does not have a trivial solution because of one of the following reasons: **a)** W might have incomplete rank (e.g., $N_{Obs} < N_{Pixels}$), **b)** W might be ill-conditioned, resulting on the amplification of intrinsic noise/artifacts in the data, and **c)** W might be very large to be computationally inverted. Problems (**a**) and (**b**) can be solved by introducing constraints on function f. The use of a regularization term $\lambda^2 \left|F - \bar{F}\right|_\Sigma^2$ has been proposed, with $\bar{F}$ either given a priori ($\bar{F} = F_A$) or estimated from F (e.g., average $\bar{F} = |F|$). The norm $|\cdot|_\Sigma$ defines the type of regularization (e.g., smoothness, minimum norm). To deal with these problems, variants of iterative methods for large algebraic systems of equations have also been proposed: $f_i^{(l)} = f_i^{(l-1)} + \sum_{j=1}^{M} \Delta f_{ij}^{(l)}$. Corrections to each iteration are made by comparing the predicted/calculated ray-sum p_j^c and the measured one p_j. The correction factor $\Delta p_j = p_j - p_j^c$ is used in two general approaches to calculate $\Delta f_{ij}^{(l)}$:

Additive correction. Each pixel is corrected in proportion to its own weight w_{ij}, $\Delta f_{ij}^{(l)} = \dfrac{w_{ij}\Delta p_j}{\sum_{i=1}^{M}\left(w_{ij}\right)^2}$.

Multiplicative correction. Each pixel is corrected in proportion to its own value f_i^a, $\Delta f_{ij}^{(l)} = \dfrac{f_i^{(l-1)}\Delta p_j}{p_j^c}$.

Iterative methods are also classified according to the correction sequence. Three of the most classical methods are listed henceforth.

The iterative least-squares technique (ILST): Projections are calculated at the beginning of the iteration. Then, corrections are made to all pixels simultaneously.

The simultaneous iterative reconstruction technique (SIRT): The function at each pixel is corrected simultaneously for all rays passing through it. Then, these corrections are incorporated before selecting any other pixels.

The algebraic reconstruction technique (ART): A set of ray-sums are calculated and corresponding corrections applied to all pixels. Then, these updated values are used to calculate the next projection.

The Shepp-Logan phantom is the standard test to evaluate image reconstruction algorithms (Figure 3.11). It represents different tissue compartments in the human head using 10 ellipsoids.

PET and SPECT image reconstruction requires attenuation correction. There have been several methods and approaches proposed for attenuation correction (Lee et al., 2016). However, this topic is outside the scope of this book. Please consult NIST for the mass attenuation coefficients of various biological tissues (including brain

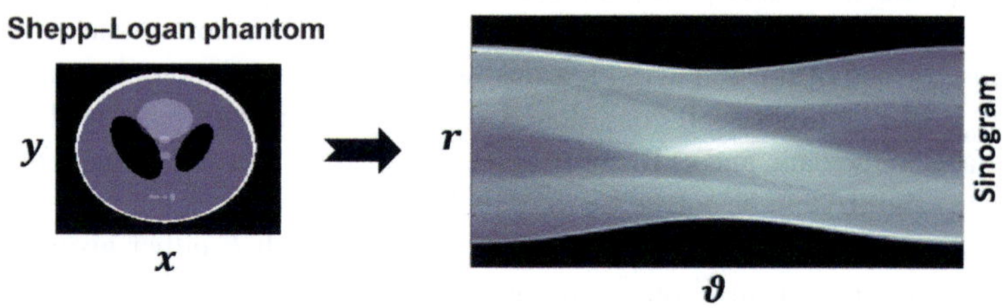

FIGURE 3.11 Original Shepp-Logan phantom and its respective sonogram. (Adapted using CT reconstruction package, MathWorks, Version 1.3.0.0 by Mark Bangert.)

and bone) for X-Ray https://nist.gov/pml/x-ray-mass-attenuation-coefficients. As for PET, its resolution is also affected by different factors (e.g., co-linearity, positron range, etc.).

3.9 PROBLEMS AND EXERCISES

Problem 3.1 – Given that in a generator $parent(p)\xrightarrow{\lambda_p} daughter(d)\xrightarrow{\lambda_d} product$, demonstrate the Bateman's equation (3.3).

Problem 3.2 – A ^{99m}Tc-generator is provided to a hospital with an initial activity of ^{99}Mo of $30.8\,Gbq$. If the activity of ^{99m}Tc was negligible at the beginning, calculate its activity of after three full days of operation. What type of equilibrium is reached?

Problem 3.3 – Calculate the energy of a photon emitted by a coherent laser at a wavelength $= 550\,nm$. Define the eV unit and its relation to the energy unit "Joule".

Problem 3.4 – Describe the following nuclear reaction: $^{7}Li(^{1}_{1}H,1\alpha)^{4}He$ in terms of targeted and produced nuclei, accelerated particle, and radiation.

Problem 3.5 – Calculate the specific-sensitivity S_C of a parallel hexagonal collimator with the following parameters: $0.25\,cm$ hole diameter, $2.5\,cm$ hole length, and $0.5\,cm$ septum thickness.

Problem 3.6 – Prove that the minimal thickness of a collimator septum with linear attenuation coefficient μ_C for a septal penetration less than 5 % is given by the following equation: $\theta \le 6\Delta/(\mu_C l - 3)$.

REFERENCES

Amor-Coarasa A, Kelly JM, Babich JW. Synthesis of [11C]palmitic acid for PET imaging using a single molecular sieve 13X cartridge for reagent trapping, radiolabeling and selective purification. *Nucl Med Biol* 42(8), 685–690, 2015.

Anger HO. Use of a gamma-ray pinhole camera for in vivo studies. *Nature* 170, 200–201, 1952.

Bocage EM. Patent No. 536,464, 1921, Paris, France, cited by Massiot J. *History of Tomography, Medica Mundi* 19(3), 106–115, 1974.

Cassen B, Curtis L, Reed CW. A sensitive directional gamma-ray detector. *Nucleonics* 8, 78–81, 1950.

Cormack DM. Representation of a function by its line integrals, with some radiological applications. *J Appl Phys* 34, 2722–2727, 1963.

Cormack DM. Representation of a function by its line integrals, with some radiological applications. II. *J Appl Phys* 35, 2908–2913, 1964.

Doi K, Morita K, Sakuma S, Takahashi M. Shinji Takahashi, M.D. (1912–1985): Pioneer in early development toward CT and IMRT. *Radiol Phys Technol* 5(1), 1–4, 2012.

Dollery CT, West JB. Metabolism of Oxygen-15. *Nature* 4743, 1121–1121, 1960.

Ehrin E, Farde L, De Paulis T, Eriksson L, Greitz T, Johnström P, Litton JE, Lars J, Nilsson G, Sedvall G, Stone-Elander S, Ögren SO. Preparation of 11C labelled raclopride, a new potent dopamine receptor antagonist: Preliminary PET studies of cerebral dopamine receptors in the monkey. *Int J Appl Radiat Isot* 36(4), 269–273, 1985.

Frackowiak RSJ, Lenzi G-L, Jones T, Heather JD. Quantitative measurement of regional cerebral blood flow and oxygen metabolism in man using 15O and positron emission tomography: Theory, procedure and normal values. *J Comput Assist Tomogr* 4(6), 727–736, 1980.

Geiger H, Müller W. The electron counting tube. *Phys Z S* 29, 829–841, 1928.

Glass HI, Harper AM. Measurement of regional blood flow in cerebral cortex of man through intact skull. *Brit Med J* 1, 593–593, 1963.

Harper AM, Glass HI, Steven JL, Granat AH. The measurement of local blood flow in the cerebral cortex from the clearance of xenon133. *J Neurol Neurosurg Psychiat* 27(3), 255–258, 1964.

Hashimoto K, Inoue O, Suzuki K, Yamasaki T, Kojima M. Synthesis and evaluation of 11C-PK 11195 for in vivo study of peripheral-type benzodiazepine receptors using position emission tomography. *Ann Nucl Med* 3(2), 63–71, 1989.

Ingvar DH, Lassen NA. Quantitative determination of regional cerebral blood flow in man. *Lancet* 2, 806–807, 1961.

Ingvar DH, Lassen NA. Regional blood flow of the cerebral cortex determined by Krypton85. *Acta Physiol Scand* 54, 325–338, 1962.

Kuhl DE, Edwards QR. Image separation radioisotope scanning. *Radiology* 80, 653–661, 1963.

Kuhl DE, Hale J, Eatonw L. Transmission scanning: A useful adjunct to conventional emission scanning for accurately keying isotope deposition to radiographic anatomy. *Radiology* 87, 278–284, 1966.

Langstrom B, Lundqvist H. The preparation of 11C-methyl iodine and its use in the synthesis of 11C-methyl-L-methionine. *Int J Appl Radiat and Isot* 27(7), 357–363, 1976.

Lassen NA, Høedt-Rasmussen K, Sørensen SC, Skinhøj E, Cronquist S, Bodforss B, Ingvar DH. Regional cerebral blood flow in man determined by a radioactive inert gas ($Krypton^{85}$). *Neurology (Minneap.)* 13, 719–727, 1963.

Lawrence EO, Livingston MS. Production of high speed light ions without the use of high voltage. *Phys Rev* 40, 19–35, 1932.

Lee TC, Alessio AM, Miyaoka RM, Kinahan PE. Morphology supporting function: Attenuation correction for SPECT/CT, PET/CT, and PET/MR imaging. *Q J Nucl Med Mol Imaging* 60(1), 25–39, 2016.

Mathis CA, Wang Y, Holt DP, Huang GF, Debnath ML, Klunk WE. Synthesis and evaluation of 11C-labelled 6-substituted 2-arylbenzothiazoles as amyloid imaging agents. *J Med Chem* 46(13), 2740–2754, 2003.

Meier P, Zierler KL. On the theory of the indicator-dilution method for measurement of blood flow and volume. *J Appl Physiol* 6, 731–744, 1954.

Miraldi F, Di Chiro G. Tomographic techniques in radioisotope imaging with a proposal of a new device: The tomoscanner. *Radiology* 94, 513–520, 1970.

Oldendorf WH. Isolated flying spot detection of radiodensity dis-continuities-displaying the internal structural pattern of a complex object. *IRE Trans Biomed Electron* 8(1), 68–72, 1961.

Reivich M, Kuhl DE, Wolf A, Greenberg J, Phelps M, Ido T, Casella V, Fowler J, Hoffman E, Alavi A, Som P, Sokoloff L. The [18F] fluorodeoxyglucose method for the measurement of local cerebral glucose utilization in man. *Circ Res* 44, 127–137, 1979.

Sokoloff L. Cerebral blood flow measured with radioisotopes. In Dynamic Clinical Studies with Radioisotopes, US Atomic Energy Commission report, June 1964.

Tucker WD, Greene MW, Weiss AJ, Murrenhoff A. Methods of preparation of some carrier-free radioisotopes involving sorption on alumina. *Trans Am Nucl Soc* 1, 160–161, 1958.

XENON Collaboration. Observation of two-neutrino double electron capture in 124Xe with XENON1T. *Nature* 568, 532–535, 2019.

4 Functional Near-Infrared Spectroscopy (fNIRS)

4.1 OVERVIEW

The visible light spectrum ($380-700\,nm$) have been known since Isaac Newton (1642–1727) discovered the sunlight composition using a prism in 1671. In 1800, radiation with wavelengths in the near-infrared range was accidentally discovered by Frederick W Herschel (1738–1822) while trying to associate a temperature value to each visible color. Using a prism, he observed changes in temperature beyond the red portion of the sunlight spectrum ($625-700\,nm$), which was not visible to humans. He named them the "calorific rays". Many technological developments in modern times are based on light in the near-infrared. One area of continuous expansion is medical imaging. It is known that light in a particular *optical window* ($680-900\,nm$) can penetrate deep into the biological tissues. An optical technique based on this property and on re-emitted light measurements via reflectance, named functional near-infrared spectroscopy (fNIRS), has proven to be very useful to image the functioning brain. Wavelengths used in fNIRS are mostly the near-infrared I (NIR-I, $700-1000\,nm$) range (Golovynskyi et al., 2018), although it could also include part of the red visible spectrum. Henceforth, the most important milestones in the development of the fNIRS for brain imaging are listed.

At Beckman Instruments, Wilbur I Kaye (1922–2006) expanded in the mid-50s the classical DU spectrophotometer to measure the near-infrared spectral components. In 1972, Takuo Aoyagi and Susumu Nakajima invented a method, known today as pulse oximetry, to compute arterial oxygen saturation using pulsatile variations in the tissue optical density in the red and NIR-I range (Severinghaus and Honda, 1987). In 1975, the method, which did not require calibration, was tested in patients. Motivated by the changes in absorption spectra of cytochromes reported earlier by Chance (1954) and in the spectrophotometric properties of hemoglobin derivatives, Jöbsis (1977) designed an experimental paradigm to measure relative changes in tissue oxygen saturation with trans-illumination of the cat intact cranium. In 1993, four independent groups were able to experimentally support the usefulness of the fNIRS technique to study the human brain (Chance et al., 1993; Hoshi and Tamura, 1993; Kato et al., 1993; Villringer et al., 1993). Then, an avalanche of work using/improving the fNIRS technique in humans took place in the 90s (e.g., Gratton et al., 1995; Maki et al., 1995; Hirth et al., 1996; Chance et al., 1997). In contrast to x and

DOI: 10.1201/9780367901509-4

γ –rays, visible/near-infrared photons entering biological tissues quickly disperse, forming a diffuse photon density that propagate inside the head. Therefore, mathematical methods used in photon transmission (CT) and emission (SPECT/PET) modalities for image reconstruction were not valid in fNIRS. In this chapter, the theoretical frameworks used in fNIRS are discussed. Also, the main technical aspects of this imaging technique are highlighted.

4.2 ABSORPTION AND SCATTERING BY BIOLOGICAL TISSUE – THE HEMOGLOBIN

The hemoglobin is an intrinsic chromophore in its two forms. Light with wavelengths $\leq 680\,nm$ are strongly absorbed inside the head by the oxy-hemoglobin (HbO_2) and the deoxy-hemoglobin (dHb). Light abortion is also significant for wavelengths $\geq 900\,nm$, but mostly by water (H_2O). Due to the low values of the extinction coefficients for these three chemical species, as well as for lipids, photons with wavelengths within the optical window (680 – 900 nm) can propagate large distances (i.e., up to a few centimeters) inside the head (Figure 4.1).

The fNIRS technique consists of measuring with an array of external sensors near-infrared photons transmitted through the head from a set of weak (~ mW) light sources (Figure 4.2). The diffusely re-emitted photons through the brain volume correspondent to a "banana shape" (yellow). The sources and sensors are placed on the scalp following particular distributions. The fNIRS raw data is defined as time-varying topographic maps of the transmitted light on the scalp, which are used to estimate the concentration of HbO_2 and dHb inside the brain (Figure 4.3). These maps are obtained by interpolating data from a set of source-sensor pairs (i.e., optodes) on the scalp at each time instant.

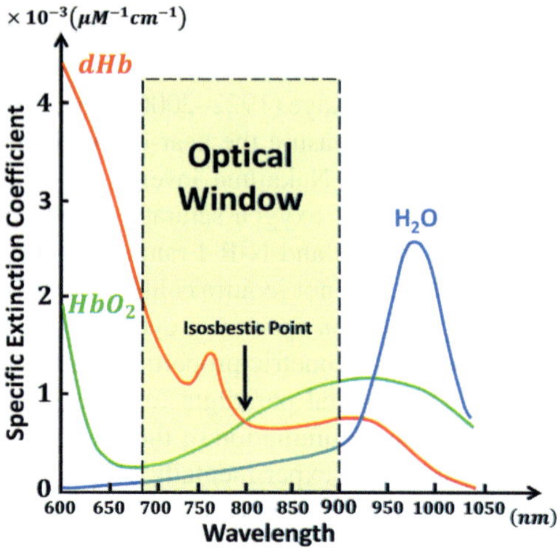

FIGURE 4.1 The optical window for fNIRS is illustrated based on the extinction coefficients for oxy-hemoglobin (HbO_2), deoxy-hemoglobin (dHb), and water (H_2O).

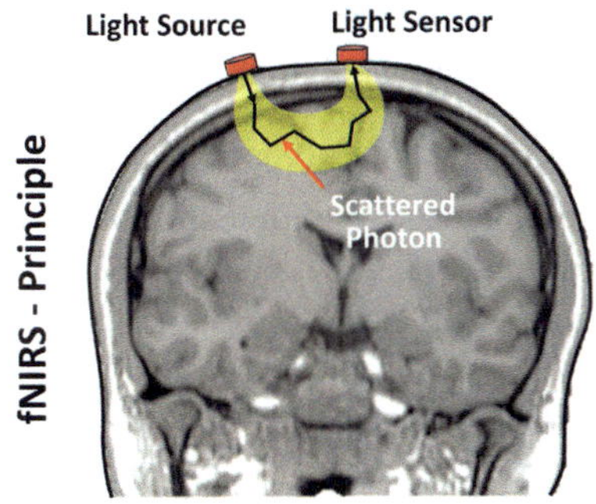

FIGURE 4.2 A schematic representation of the fNIRS principle.

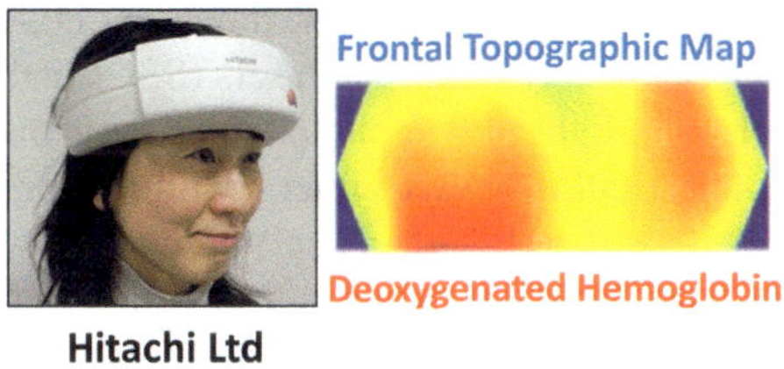

FIGURE 4.3 A portable fNIRS model provided by the Advanced Research Laboratory (Hitachi Ltd) to Ryuta Kawashima's lab (Tohoku University) in the early 2000s for the evaluation of brain activity in motorists while driving.

Generally, near-infrared photons traveling inside the head are absorbed and scattered by biological tissues forming it. These two physical phenomena are well-described by the absorption l_a and scattering l_s lengths, which represent the photon mean free-path between consecutive absorption and scattering events, respectively. The absorption and scattering probability per unit length is characterized by the absorption $\mu_a = 1/l_a$ and scattering $\mu_s = 1/l_s$ coefficients, respectively. Scattering events can happen in arbitrary directions. However, biological tissues have sometimes preferential scattering directions given by their cellular/molecular compositions. Therefore, the scattering probability in biological tissues is better represented by the reduced scattering coefficient $\mu_s' = (1-g)\mu_s$, where g is the anisotropy factor. For most biological tissues (Cheong et al., 1990), the reduced scattering coefficient μ_s' $(0.5-10\,mm^{-1})$ is much larger than the absorption coefficient μ_a $(0.005-0.02\,mm^{-1})$. Consequently, photons experience multiple scattering events before they are either absorbed or transmitted through a boundary. Hence, their distribution inside the head is approximated by a diffusion equation (Patterson et al., 1989).

The foundation of the fNIRS modality is in the differentiate absorption spectra within the red/near-infrared range of two forms of the hemoglobin: **a)** oxygenated (HbO_2) and **b)** deoxygenated (dHb). The absorption spectra of the HbO_2, with its two strong absorption bands, was first reported by Felix Hoppe-Seyler (1825–1895) in 1862. Two years later, a mathematician George G Stokes (1819–1903) demonstrated that these two bands are replaced by a single diffuse band after treating blood with a reducing agent, named later the Stokes' solution. The two bands recovered by shaking the solution with air. Today, the full spectra of these two species are well-known in a wide range of optical wavelengths. Within the biological optical window, the HbO_2 and dHb extinction coefficients are different for each wavelength, with exception of an isosbestic point around $800\,nm$. As discussed later, this property is crucial to estimate dynamic changes of HbO_2 and dHb in the brain from fNIRS data. From absolute measurements of the concentration of these two species, the blood oxygen saturation is calculated by the following expression $SO_2 = HbO_2 / tHb$, with the total hemoglobin defined as $tHb = HbO_2 + dHb$.

4.3 LIGHT SOURCE AND SENSORS IN fNIRS

fNIRS systems can be built with different types of light sources: **1)** light emitting diodes (LED), **2)** continues-wave/pulsed laser diodes (LD), **3)** broadband light source with filtering components, **4)** solid state lasers (SSL), and **5)** high-power fiber lasers. LEDs emit incoherent light based on spontaneous emission with a typical $25-50\,nm$ bandwidth. LEDs could produce either continuous or intensity-modulated ($\sim MHz$) light. Coherent light (bandwidth $< 1\,nm$) is ensured with LDs, which are based on stimulated emission. LDs can be used to produce a continues-wave and a short ($10^1 - 10^2\ ps$)/ultra-short ($10^1 - 10^2\ fs$) pulse at a frequency of up to $100\,MHz$. Broadband light sources are useful for multi-spectral approaches. In this case, a continuous absorption spectrum is reconstructed using sources emitting light in a continuous portion of the wavelengths. SSLs have usually a broad wavelength range with the NIR-I range (e.g., Ti:Sapphire). Time-multiplexing of the source on multiple positions is possible using SSL due to their high power and wavelength tenability. Recently, fNIRS with high-power fiber lasers have been evaluated. These lasers exploit modulational-instability initiated supercontinuum (SC) mechanism. Measurements of the transmitted light are performed with three main types of photodetectors (i.e., the sensors): **1)** classical/MCP photomultipliers tubes (PMT), **2)** single/avalanche photodiodes (PD/APD), and **3)** cooled charge-coupled device (CCD) arrays. Photo-detection takes place by the photoelectric effect, under which a photon generates free charge carriers detected as an electric signal. The most used type of photodetector, the PMT, is based on an external photoelectric effect. PD/APD and CCD arrays are based on the internal photoelectric effect. The fundamental working principles of the PMT and p-n PD were discussed in Chapter 3. There are three major sources of noises in photodetector devices: **a)** shot, **b)** dark current and **c)** thermal. In general, sensors of fNIRS are directly placed on the scalp; however, optical fibers are also used to transmit the observed light to the fNIRS system. In most cases, several optodes are used to collect transmitted light simultaneously from multiple locations on the scalp, a principle called overlapping measurements. Although its spatial resolution is low, fNIRS has been extensively used in research and clinical practice because of: **1)** better temporal resolutions ($\sim 10-100\,ms$), **2)** low-cost, and **3)** compatibility with other neuroimaging modalities.

4.4 RECORDING MODES: CONTINUOUS-WAVE, FREQUENCY/TIME DOMAINS

Depending on the temporal profile of the incident light, fNIRS systems can be classified into three main categories (Figure 4.4): **a)** continuous-intensity, **b)** intensity-modulated and **c)** time-resolved. The mathematical methods used in these three types of fNIRS systems for physiological parameters (i.e., hemodynamics/oxidative-metabolism) estimation and image reconstruction are different. These methods are discussed later in this chapter.

fNIRS systems based on a continuous-intensity mode use multiple discrete/broadband wavelengths. In this type of fNIRS systems, the differences ΔI between the incident I_0 and observed I light are used to characterize the temporal evolution of physiological parameters in the tissue under a logarithmic scale, it will appear as the I / I_0 ratio. They are also known as fNIRS systems with a continuous-wave (**CW**) method. The CW method is implemented in most of the commercial fNIRS systems. Unfortunately, absolute changes in the optical properties (μ_a and μ_s') of brain tissues cannot be fully characterized with the CW method, which constitutes its main disadvantage. Hence, obtaining absolute values for the HbO_2 and dHb is not feasible. However, fNIRS systems based on the CW method are relatively cheaper. Due to its operational simplicity, miniaturized and wireless versions of CW-fNIRS systems have been commercially available for some time. Most commercially available systems are based on the use of two wavelengths. However, there are some that use a third wavelength. The number of sources and sensors varies with the fNIRS

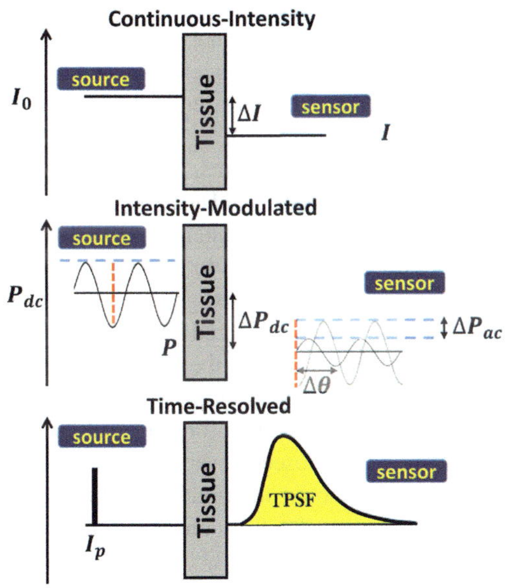

FIGURE 4.4 Types of temporal profiles of light sources in fNIRS systems.

system, ranging from a few to several dozens. In an excellent review, Scholkmann et al. (2014) discussed the current state of instrumentation and methodology of the continuous-intensity mode to create fNIRS contrasts. An optimization problem is required to select the optimal wavelengths set for a best image reconstruction of the HbO_2 and dHb. The modified Lambert-Beer's law is used to model CW-fNIRS data. To improve sensitivity via phase-locked detection techniques, light sources in CW-fNIRS are sometime modulated with a low frequency signal (~ kHz). Depth sensitivity in CW-fNIRS measurements increases by separating the source-sensor pair. LEDs and continues-wave LDs are the most used light sources in CW-fNIRS systems, with PMT and PD/APD as preferable light sensors.

Under an intensity-modulated mode, the amplitudes of the sources are modulated at a radio frequency (RF, ~ $0.1-1.0\,GHz$), and differences in the dc power off-set ΔP_{dc}, ac power modulation ΔP_{ac} and ac phase $\Delta\theta$ are used to estimate the time course of optical parameters in the brain tissue. Similar types of light sources, but with amplitude controllability, are used in intensity-modulated fNIRS systems. This mode is also referred as the fNIRS frequency domain (**FD**) method (Franceschini et al., 1997). With the FD-fNIRS, absolute values of μ_a and μ_s' are measured inside the brain. An intensity-modulated light source creates a diffuse photon density wave (DPDW) that propagates inside the head, which is considered a turbid medium. Based on the typical optical parameters for brain tissues, the DPDW has a wavelength of $20-30\,cm$ and an attenuation length of about $0.6\,cm$. Single/multi-distance methods used in FD-fNIRS are based on analytical solutions for the DPDW. Instrumentation for FD-fNIRS measurements continue to develop over the years, with the most recent promising advances in the area of: **a)** homodyne/heterodyne detection, **b)** frequency division multiplexing (FDM), **c)** vertical-cavity surface-emitting lasers (VCSEL), **d)** direct sampling multi-wavelength, **e)** RF network analyzers, and **f)** electronics (e.g., CMOS-based integrated circuitry, frequency-sweeping boards, and hybrid FPGA-assisted demodulation). Readers interested in more details are encouraged to read a review by Fantini and Sassaroli (2020).

Compared to CW and FD –fNIRS systems, fNIRS systems based on a time-resolved mode, also called the time domain (**TD**) method, are bulky and very expensive. For that reason, there is only one dual-channel TD-fNIRS system commercialized by Hamamatsu Photonics. Benaron et al. (1995) performed the first evaluation of a TD-fNIRS tomographic system to monitor oxygenation and image hemorrhage in neonatal brains. This pioneering work triggered the technological development of multi-channel TD-fNIRS instruments during the last few years of the XX century (Torricelli et al., 2014). TD-fNIRS systems employ a pulsed light source with duration of a few tens of picoseconds (ps). Therefore, these systems require the use of LDs and light sensors with a sub-nanosecond (ns) temporal resolution. These light sensors are used to measure the photon distribution of time-of-flight (DTOF) in a diffusive medium caused by a light pulse from an LD. In some literature, the DTOF is also called temporal point spread function (TPSF). As a result of tissue scattering, the DTOP constitutes a delayed/broadened version of the light pulse. As for the other two recording modes, the measured light is also attenuated due to tissue absorption and diffusion-out. In the classical approach, sensors, and LDs in TD-fNIRS are in the range of 10 – 40 mm apart. Recently, a null-distance approach has been proposed to reduce a significant loss of photons (Torricelli et al., 2005). The depth sensitivity in the TD-fNIRS is improved by exploiting the temporal information. Pulsed DL and SSL are the most used light sources in TD-fNIRS. The use of high-power fiber lasers using supercontinuum (SC) generation has been evaluated in recent years. TD-fNIRS measurements on the scalp consist of time series of DTOFs, which are typically acquired at least with two wavelengths. DTOF time series are in the range of $0.1-10\,ns$. Therefore, detection techniques must have sufficient sensitivity with a temporal resolution of $1-250\,ps$. Initial measurements were made with a modified optical time domain reflectometer (OTDR). However, more efficient detection techniques have been developed over the years: **a)** non-linear optical-gating (Kerr effect, parametric amplification, non-linear up-conversion), **b)** the streak camera ($1-10\,ps$) with bi-dimensional design for multi-wavelength, **c)** the time-correlated single photon counting (TCSPC) technique, **d)** the time-gated intensified charge-coupled device (ICCD) camera, consisting of a photocathode, an MCP PMT, and a phosphor screen, and **e)** pseudo-random-bit sequence light sources combined with a cross-correlation method to retrieve the impulse response. The TCSPC techniques are the most widely used.

A list with commercially available fNRIS system can be found in the society for functional Near-Infrared Spectroscopy website (https://fnirs.org/resources/instruments/).

4.5 THE OPTICAL PROPERTIES OF THE HUMAN HEAD

The tissue transparency window in NIR-I has been the key element in the development of the fNIRS. Recently, two additional optical windows have been identified for brain imaging techniques: NIR-II ($\sim 1000-1350\,nm$) and NIR-III ($\sim 1550-1870\,nm$). NIR-III is also known as short-wave IR (SWIR). However, current fNIRS methods are based mostly on the optical properties of the head for the NIR-I window; hence, it will be the focus of this section. In infinite homogeneous media, the NIR-I diffuse photon density propagates as a spherical wave. The spherical symmetry is disrupted by objects, a phenomenon called photon scattering (or light diffraction). Due to microscopic discontinuities in the refractive index *n*, surfaces limiting two different biological tissues also create deformation of the spherical symmetry by photon reflections and refractions. To reach the brain, photons emitted by a light source on the scalp must cross a series of tissue layers (Figure 4.5), which have very different optical properties. To be observed, dispersed photons coming from the brain have also to cross these layers once again to reach the light sensors placed on the scalp.

There are three main tissue layers to be considered in fNIRS: **a)** the skin (i.e., scalp), **b)** the bone (i.e., skull) and **c)** the CSF. For this reason, optical properties (μ_a, μ_s' and n) of these tissues have been in the past characterized for humans. These tissue layers have different thicknesses. The scalp is formed by an outer sub-layer, called epidermis, and three internal sub-layers (dermis, hypodermis, and galea). Depending on gender and age, the thickness of the scalp for adults ranges from $1.8-4.7\,mm$. The skull is formed by the diploë, which is limited by the internal and external plates. The thickness of the skull depends not only on gender/age, but

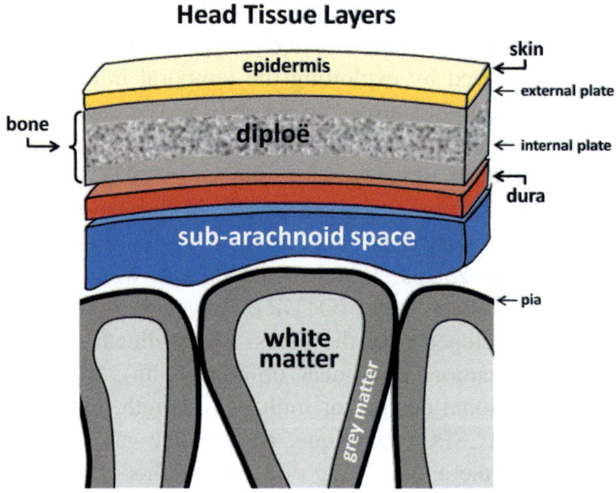

FIGURE 4.5 A schematic of the most relevant tissue layers for fNIRS measurements from the human head.

also on the rostral to caudal position in an individual. In general, the skull thickness ranges from $6.5-7.1\,mm$. The sub-arachnoid space is filled by cerebrospinal fluid (CSF) and includes major blood vessels and cisterns. The space is supported by the arachnoid trabeculae/membrane. The thickness of the sub-arachnoid space is approximately 2 *mm*. A dura mater of $\sim 270\,\mu m$ in thickness is located between the internal plate of the bone and the sub-arachnoid space.

Based in observations from multiple studies, a theoretical equation has been proposed to describe the wavelength dependence of scattering in the brain, skull, and scalp of humans (Jacques, 2013), with separate contributions of Rayleigh and Mie scatterings. Under this model, the wavelength λ is normalized by a reference wavelength $\lambda_{ref} = 500\,nm$.

$$\mu_s'(\lambda) = a\left(\underbrace{f_{Ray}\left(\frac{\lambda}{\lambda_{ref}}\right)^{-4}}_{\text{Rayleigh scattering}} + \underbrace{\left(1-f_{Ray}\right)\left(\frac{\lambda}{\lambda_{ref}}\right)^{-b_{Mie}}}_{\text{Mie scattering}}\right)$$

The scaling factor a equals $\mu_s'(\lambda_{ref})$. In the context of biomedical optics, Mie scattering is referred to scattering by typically large spherical particles. On the other hand, Rayleigh scattering is referred to scattering by particles with mass density fluctuations smaller than the wavelength of light. Parameters a, f_{Ray}, and b_{Mie} in this theoretical model were fitted to data from *N* different studies (Table 4.1). The theoretical curves are shown in Figure 4.6.

TABLE 4.1
The mean ± SD of model parameters a, f_{Ray}, and b_{Mie} in the theoretical expression for $\mu_s'(\lambda)$

Tissue	$a(cm^{-1})$	f_{Ray}	b_{Mie}	N
Scalp	48.0 ± 10.6	0.409 ± 0.178	0.702 ± 0.351	7
Skull	15.3 ± 7.9	0.022 ± 0.032	0.326 ± 0.298	2
Brain	27.4 ± 10.5	0.315 ± 0.368	1.087 ± 1.386	8

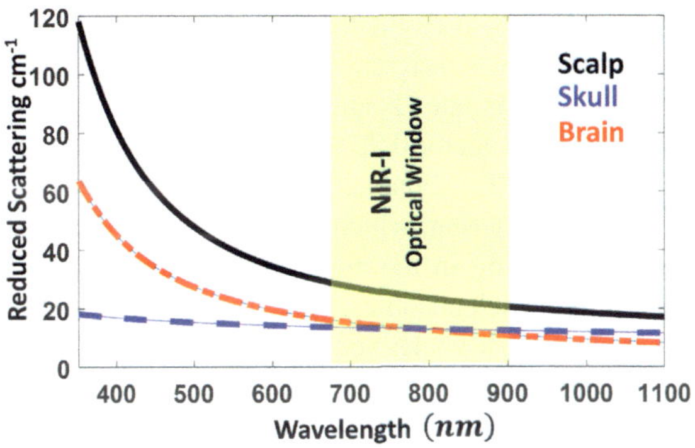

FIGURE 4.6 Theoretical curves of the reduced scattering coefficients for human scalp, skull, and brain.

For the human cranial bone, a more accurate theoretical estimation of $\mu_s'(\lambda) = 1533.02\lambda^{-0.65}$ within the NIR-I range was provided by Bashkatov et al. (2006). The scattering length in the CSF is not smaller than all system dimensions and absorption lengths in the head. Therefore, the diffusion approximation (discussed below) is not accurate. Using simulations in a three-dimensional head model based on the Monte Carlo multi-layered (MCML) method (see details later in this chapter), Custo et al. (2006) demonstrated that the reduced scattering coefficient of CSF increases up to the inverse of its typical thickness, or also referred as the line-of-sight distance; hence, under the diffusion approximation it can be assumed that $\mu_s' \leq 0.5\,mm^{-1}$ for the sub-arachnoid layer in humans within the NIR-I range. Light waveform distortions due to the presence of the dura mater has been disregarded in most fNIRS studies.

The following methods are used to experimentally measure μ_s and g from tissue samples. Biased estimation of these parameters could result from **1)** sample desiccation and **2)** tissue heterogeneity. Typically, light transmission measurements T from a collimated bean through a very thin tissue sample is used to estimate the **scattering coefficient** $\mu_s = -ln(T)/L$ by means of the Lambert's law under a uniform attenuation assumption. Light transmission $T = \Phi_t / \Phi_i$ is defined from the incident Φ_i and transmitted/collimated (observed) Φ_t radiant power (Watt, W). For this estimation to be accurate the thickness of the sample must be one mean-free-path $L \sim 1/\mu_s \approx 100\,\mu m$.

Observations of deflected photons could affect the estimation of μ_s. The estimation of the **anisotropy factor** g is more complicated. The anisotropy factor can be mathematically expressed in terms of the phase function $p(cos(\theta))$, i.e., $g = \int_{4\pi} p(cos(\theta))cos(\theta)d\Omega$, with $d\Omega$ the differential solid angle. The phase function represents the probability that an individual beam is scattered in any particular direction θ. Several parametric models for the phase function have been proposed in the past (next section). Goniophotometers have been used in the past to measurement the probability of photon scattering by a thin tissue sample at different angular θ direction (Firbank et al., 1993). Additional hemispherical lenses on the tissue-air interface are used to reduce the effect of refraction. g could be also calculated from the combined use of diffuse light and collimated transmission measurements to estimate $\mu_s^{'}$ and μ_s, respectively (i.e., $g = 1 - \mu_s^{'} / \mu_s$). Recent proposals are based on the use of a confocal microscope to measure the attenuation of backscattered reflectance at different tissue depths (Jacques, 2013). Calculation using an optical system with collimated bean, a dual-pinhole collimator and a turntable goniometer reveal an anisotropy factor $0.92 \le g \le 0.95$ for the human skull within the NIR-I range (Firbank et al., 1993). Filamentous proteins in the scalp sub-layers are the major source of photon scattering, which is mostly well-described with the Mie's scattering model. Light scattering in the epidermis occurs majorly due to the large concentration of keratins and melanosomes. Collagen constitutes the major light dispersive source in the dermis. Cell nuclei and membranes cause particulate scattering. Light scattering in the scalp is also due to the existence of micro vessels with different thicknesses and particular depth distributions. The epidermis is densely populated with capillaries, while terminal arterioles and post-capillary venules can be mainly found in the dermis. Based on multiple observations, van Gemert et al. (1989) proposed that the epidermal and dermal g factors for humans may be considered identical, and mathematically represented as $g(\lambda) = 0.62 + 0.29 \times 10^{-3}\ \lambda(nm)$ for visible light. Observations by Jacques et al. (1987) at $632.8\,nm$ (HeNe laser) suggest this formula is valid also for the NIR-I range. Within the NIR-I range, the scattering coefficients and anisotropy factors have differences for gray and white matter (Figure 4.7).

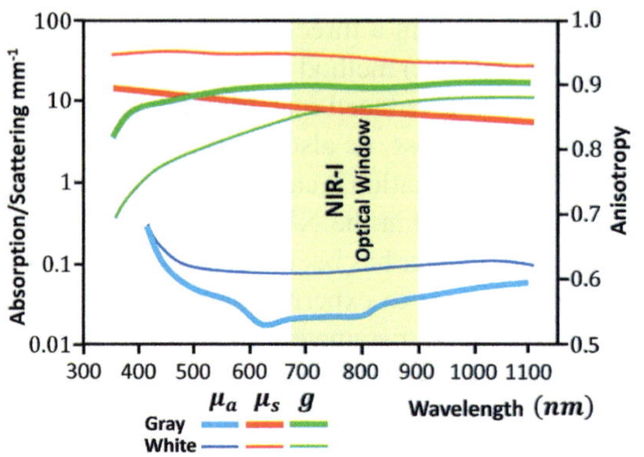

FIGURE 4.7 Experimental values of the absorption/scattering coefficients and the anisotropy factor for grey and white matter (Yaroslavsky et al., 2002).

The **refractive index** is another important optical parameter in fNIRS. This complex index can be modeled as $n = n' + in''$, with a real component n' describing energy storage and an imaginary component n'' linked to energy dissipation. The imaginary component is related to the absorption coefficient, $n''(\lambda) = \lambda\mu_a(\lambda)/4\pi$. The real component determines the speed of light in a medium and scales as a function of its water content (i.e., the water volume fraction r_W): $n' = n'_d - (n'_d - n'_w)r_W$ (Jacques, 2013). The real component for the refractive index of water is $n'_w = 1.33$. For most biological tissues the dry mass density is $\sim 1.53\,g/cm^3$, for an approximated refractive index of $n'_d \approx 1.51$. Water content in the skin increases drastically with depth, from 20% for the surface of the epidermis up to 70 % for the deep layers of the dermis. These water content profiles agree with measurements of skin n' over the visible and NIR-I range (Lister et al., 2012), revealing that the refractive index is approximately $1.41 \le n \le 1.49$ for the epidermal tissues and $1.36 \le n \le 1.41$ for the dermal tissues. The mass density for skull is a little larger $1.615 - 1.922\,g/cm^3$. Therefore, the refractive index of human cranial bone is $1.55 \le n' \le 1.56$, depending on the content of natural hydroxyapatite. The refractive index for gray and white matter are 1.39 and 1.41, respectively. The CSF has a refractory index similar to that of water. Measurements of the refractive index can be performed using the MRI technique (Biswas and Luu, 2011).

The **absorption coefficient** of a medium is defined as the fraction of absorbed incident light per incremental path-length dl: $\mu_a = -\frac{1}{T}\frac{dT}{dl}$. Hence, the fraction of transmitted light along a path-length L is $T = e^{-\int_0^L \mu_a(l)dl}$. For N absorbing substances in the medium, the absorption coefficient can be written as $\mu_a = \sum_{k=1}^{N} \alpha_k \rho_k$, with α_k and ρ_k representing the attenuation cross-section and density of the *k-th* substance, respectively. Through the Avogadro constant, these parameters related linearly to the molar absorption (extinction) coefficient ε_k $(cm^{-1}M^{-1})$ and concentration c_k (M) for the *k-th* substance: $\alpha_k = \varepsilon_k ln(10)/N_A$ and $\rho_k = c_k N_A$. On the other hand, the medium absorbance A along the path-length L is defined as the sum of the absorbance of all substances A_k; i.e., $A = \sum_{k=1}^{N} A_k = \sum_{k=1}^{N} \varepsilon_k \int_0^L c_k(l)dl$. The Beer-Lambert's law can be written in the following general form:

$$log_{10}(T) = log_{10}\left(\frac{\Phi_t}{\Phi_i}\right) = -\sum_{k=1}^{N} \varepsilon_k \int_0^L c_k(l)dl$$

Note that $\mu_a = ln(10)\sum_{k=1}^{N} \varepsilon_k c_k$ and $T = 10^{-A}$.

Biological tissues are composed of multiple substances, called chromophores or optical species. From an optical viewpoint, hemoglobin, water, melanin, yellow pigment, fat, and adipose tissue are among the most important for fNIRS. Jacques (2013) proposed a formula for the absorption coefficient μ_a in any generic biological tissue, with the following parameters: s – hemoglobin oxygen saturation in the arterio-venous vasculature ($SatO_2$), r_B – average blood volume fraction (BVF), r_M – melanosome volume fraction, r_f – fat volume fraction, bilirubin (concentration c_{bi}

and extinction coefficient ε_{bi}), and β-carotene (concentration $c_{\beta c}$ and extinction coefficient $\varepsilon_{\beta c}$). μ_{a,HbO_2}, $\mu_{a,dHb}$, $\mu_{a,W}$, $\mu_{a,F}$ and $\mu_{a,M}$ represent the absorption coefficients for HbO_2, dHb, water, fat, and melanosome, respectively.

$$\mu_a = r_B s\mu_{a,HbO} + r_B(1-s)\mu_{a,dHb} + r_W\mu_{a,W} + r_f\mu_{a,F} + r_M\mu_{a,M} + \underbrace{2.3\varepsilon_{bi}c_{bi} + 2.3\varepsilon_{\beta c}c_{\beta c}}_{Yellow\ Pigments}$$

For NIR-I, $\varepsilon_{bi} \approx 1.5\times10^2\, cm^1M^{-1}$ and $\varepsilon_{\beta c} \approx 8\times10^2\, cm^1M^{-1}$ (Jacques, 2013). These extinction coefficients seem to be wavelength-independent in the NIR-I range. The specific extinction coefficient of the hemoglobin varies not only with wavelength but also with its oxygen saturation. The wavelength-dependency of the extinction coefficients for the fully-oxygenated hemoglobin (HbO_2) and fully-deoxygenated hemoglobin (dHb) are shown in Figure 4.1 – A. Note that at the isosbestic point ($\sim 806\,nm$), $\varepsilon_{HbO_2} = \varepsilon_{dHb} \approx 818\,cm^{-1}M^{-1}$. In adult humans, the hemoglobin molecule is formed by the globin (two α and two β polypeptide chains) and four heme groups, named tetramer. A γ chain substitute the β chain in infants. These metalloproteins are embedded into the red blood cells (erythrocytes). Oxygen can bind to these four iron-containing heme pigments to be transported bloodstream from the lung alveoli to the brain capillaries for delivery on demand.

Hemoglobin exists in either a taut form (T) with low affinity to oxygen or a relaxed form (R) with high affinity to oxygen. The switch between these forms at the tissue-capillary interface is determined by three main factors: pH, CO_2 and 2,3-bisphosphoglyceric acid (2,3-BPG). Low pH and high CO_2/2,3-BPG favor the T-form; hence, oxygen release to the tissue. The tissue partial pressure of oxygen (PaO_2) influences also the favorable form. High PaO_2 switches the hemoglobin to an R-form, typically occurring in the alveoli, while a low PaO_2 favors the T-form in the brain capillaries. Once an oxygen molecule binds a single iron-heme, the hemoglobin suffers a conformational shift facilitating oxygen link to the other three iron-binding sites. Therefore, the hemoglobin could be in partially oxygenated states. To model light absorption by the hemoglobin, the contribution of the two pure species (HbO_2 and dHb) is weighted by the hemoglobin oxygen saturation s. Figure 4.8A shows a typical hemoglobin oxygen saturation curve as a function of PaO_2. The mass concentration of hemoglobin in blood varies with age and gender, ranging from $110\,g/L$ to $172\,g/L$. The molecular weight of the hemoglobin is $MW_{Hb} = 64458\,g/mol$. Assuming a mean hemoglobin mass concentration ($110\,g/L$), then, the concentration of hemoglobin in tissue is $c_t \approx 2.187\,r_B\, mM$. Measurements of the cerebral BVF distribution in human cerebral cortices indicates that $r_B \approx 0.031\pm0.003$, with a 37 % of it located in arteries. However, r_B varies significantly among brain tissues.

The proposed theoretical model can be used to represent absorption phenomena in the most relevant tissue layers for fNIRS measurements (Table 4.2).

The dependency of $\mu_{a,W}$, $\mu_{a,F}$ and $\mu_{a,M}$ in the NIR-I range are shown in Figure 4.8B. For melanin, a theoretical formula has been proposed in the

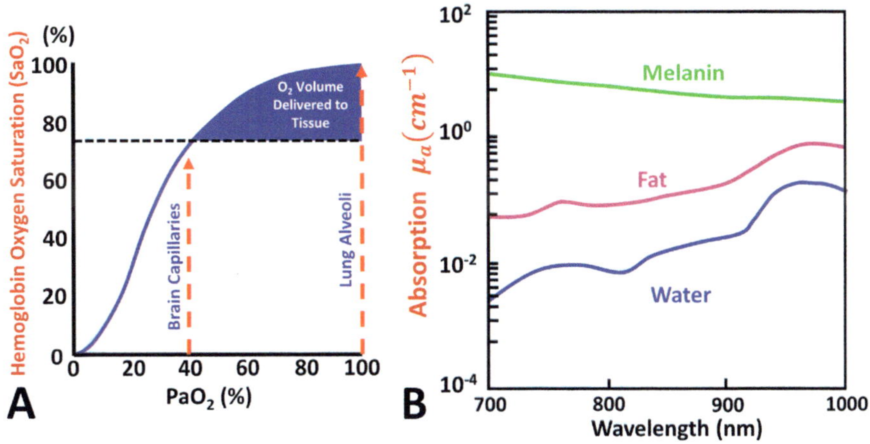

FIGURE 4.8 A. The hemoglobin oxygen saturation curve. B. Theoretical values of the absorption coefficients for water, fat, and melanin as a function of the wavelength in the NIR-I (Jacques, 2013).

TABLE 4.2
Approximate values of parameters in the theoretical model for $\mu_a(\lambda)$ in different head tissue layers

Tissue	r_B(%)	s(%)	r_W(%)	r_f(%)	r_M(%)
Epidermis	0	0	na	na	2.5
Dermis	0.2	39.0	65.0	0	0
Skull	1.36	na	na	na	0
CSF	0	0	99	0	na

Note: Acronym: na – non available.

past: $\mu_{a,M}(\lambda) = 519\ cm^{-1}\left(\dfrac{\lambda}{500\,nm}\right)^{-3.5}$. Melanin and yellow pigments are not commonly found inside the brain. Also, absorption effects due to water and fat, although dependent on the wavelength, can be considered constant over time. Hence, the contribution of HbO_2 and dHB species to the total light absorption coefficient μ_a inside the brain depends on the hemoglobin oxygen saturation $SatO_2$; i.e., $\mu_a = r_B s \mu_{a,HbO_2} + r_B(1-s)\mu_{a,dHb}$. This constitutes the most important property underlying fNIRS. Changes in oxygen consumption (metabolism) and vascular BVF (perfusion) occur during cellular activation in the brain, particularly in neurons. These changes will cause temporal variations in s and r_B, respectively. Therefore, estimations of $\mu_a(t)$ from fNIRS data can be used to infer temporal changes in these two physiological variables.

4.6 LIGHT PROPAGATION: MULTIPLE APPROXIMATIONS

In 1872, Ludwig Boltzmann (1844–1906) proposed a general transport equation to describe the statistical behavior of systems in non-equilibrium states. For photon migration inside a medium, the transport equation takes a particular form (4.1).

$$\left(\frac{1}{v(\vec{r})}\frac{\partial}{\partial t}+\hat{s}\cdot\nabla+\mu_{tr}(\vec{r})\right)L(\vec{r},\hat{s},t)=\mu_s(\vec{r})\int_S L(\vec{r},\hat{s}',t)\Lambda(\hat{s}\cdot\hat{s}',\vec{r})d\hat{s}'+q(\vec{r},\hat{s},t) \quad (4.1)$$

In the context of this book, $L(\vec{r},\hat{s}',t)$ $(Wsr^{-1}m^{-2})$ describes the photon radiance at position $\vec{r}$ and time t with velocity in unit direction $\hat{s}$. However, a similar equation can be formulated for the number of photons per unit volume. Surface S is a unitary sphere centered at $\vec{r}$. The steradian (sr) is the unit of solid angle. $\mu_{tr}(\vec{r})$ is the transport coefficients (m^{-1}) at position $\vec{r}$. By definition, $\mu_{tr}(\vec{r})=\mu_a(\vec{r})+\mu_s(\vec{r})$, with $\mu_a(\vec{r})$ and $\mu_s(\vec{r})$ the respective absorption and scattering coefficient at $\vec{r}$, which were referred earlier in this chapter. The speed of light in the medium satisfies $v(\vec{r})=c/n(\vec{r})$, with $n(\vec{r})$ being the refractive index of the material at point $\vec{r}$. The term $q(\vec{r},\hat{s},t)$ $(Wsr^{-1}m^{-3})$ reflects the spectral radiance in wavelength of a light source at $\vec{r}$ injecting photons per unit of solid angle and volume at time t in the direction $\hat{s}$. The light is refracted if it strikes a particle with an index of refraction different from that of its environment. The refraction angle is a function of: **a)** the size/shape of the particle, **b)** the light wavelength, and **c)** the incidence angle. Hence, each particle has a scattering profile, called the phase function. The phase function differs from particle to particle. However, an average phase function, describing bulk-related scattering phenomena, is used in most applications. The normalized phase function $\Lambda(\hat{s}\cdot\hat{s}',\vec{r})$ describes the probability that a photon at position $\vec{r}$ moving in the direction $\hat{s}$ being scatter to a new direction $\hat{s}'$. For most applications, $\Lambda(\hat{s}\cdot\hat{s}',\vec{r})$ is assumed to be independent of $\vec{r}$ within a given tissue compartment, i.e., $\Lambda(\hat{s}\cdot\hat{s}')$. Normalization implies $\int_S \Lambda(\hat{s}\cdot\hat{s}')\,d\hat{s}'=1$. The phase function for scattering $\Lambda(\hat{s}\cdot\hat{s}')$ represents the albedo times the probability density function that photons are scattered from $\hat{s}$ to $\hat{s}'$. Under the assumption that light scatterers are randomly distributed (i.e., non-spatially correlated structures), the phase function for scattering $\Lambda(\hat{s}\cdot\hat{s}')$ depends only on the angle θ between $\hat{s}$ and $\hat{s}'$; i.e., $\Lambda(\hat{s}\cdot\hat{s}')=p(cos(\theta))$, with the abovementioned anisotropy factor defined as $g=\langle\hat{s}\cdot\hat{s}'\rangle$. Note that $\langle x\rangle$ is the expected value of magnitude x. Scattering phenomena could be also divided into: **a)** isotropic, **b)** forward and **c)** backward. Under isotropic scattering, $p(cos(\theta))=c$, with $c=1/4\pi$ a specific constant for normalization. For non-isotropic phase functions, the anisotropic factor g (i.e., the average cosine of the phase function) is introduced to characterize the degree of anisotropy.

For slightly forward scattering, the Eddington phase function is used $p_E(cos(\theta))=c(1+3g_E cos(\theta))$. Note that $g_E=0$ implies isotropic scattering. In this context, $g_E=1$ and $g_E=-1$ represent forward and backward scattering respectively. The more general delta-Eddington approximation is used for some biological tissues: $p_{\delta E}(cos(\theta))=c(2f\delta(1-cos(\theta))+(1-f)(1+3g_E cos(\theta)))$. f represents the fraction of light scattered into the forward peak. As $f\rightarrow 1$, $p_{\delta E}(cos(\theta))$ approaches to the delta function $\delta(\cdot)$. The Eddington approximation is obtained for $f=0$.

However, the forward scattering phase function in biological tissues has been modeled with the Henyey-Greenstein (HG) parametric model for Mie scattering, with (i.e., the anisotropy factor, $0 \leq g_{HG} \leq 1$).

$$p_{HG}\left(cos(\theta)\right) = c\left(1 - g_{HG}^2\right) / \left(\left(1 + g_{HG}^2 - 2g_{HG}cos(\theta)\right)^{3/2}\right)$$

To include isotropic scattering and a proportion of backward scattering, Jacques et al. (1987) introduced a more general HG model for the phase function, which includes two parameters: **1)** the proportion of isotropic/backward scattering b and **2)** the HG anisotropy factor g_{HG}.

$$p_{mHG}\left(cos(\theta)\right) = c\left(b + (1-b)\left(1 - g_{HG}^2\right) / \left(\left(1 + g_{HG}^2 - 2g_{HG}cos(\theta)\right)^{3/2}\right)\right)$$

According to this model, isotropic scattering is obtained with $b = 1$, while for purely forward scattering $b = 0$.

It can be mathematically demonstrated that:

$$g_E = \frac{g_{HG}\left(1 - g_{HG}\right)}{\frac{1}{1-b} - g_{HG}^2}$$

The anisotropy factors g_E and g_{HG} in these two approximations are related to the original average cosine factor g by the following relationships: $g = f + (1-f)g_E$ and $g = (1-b)g_{HG}$.

In general, the source term $q(\vec{r}, \hat{s}, t)$ must be specified at specific locations $\vec{r}_s$ on the tissue-air interface Ω (i.e., scalp) where the sources are. $q(\vec{r}, \hat{s}, t)$ is zero elsewhere. For example, a collimated source incident at $\vec{r}_s$ in the direction $\hat{v}$ is mathematically represented as: $q(\vec{r}, \hat{s}, t) = A(t)\delta(\vec{r} - \vec{r}_s)\delta(\hat{s} - \hat{v})$. As for the boundary condition at Ω, it is commonly assumed that photons can only travel in an inward direction at the source locations $\vec{r}_s$, i.e., $L(\vec{r}, \hat{s}, t) = 0$ for any $\vec{r} \neq \vec{r}_s$, independent of the direction $\hat{s}$, and $L(\vec{r}, \hat{s}, t) = 0$ for any direction satisfying $\hat{s} \cdot \hat{n}_s < 0$, with $\hat{n}_s$ the normal to the boundary at $\vec{r}_s$. Under the referred boundary conditions, the most accurate description of light propagation in diffusive media is obtained by solving equation (4.1) by means of the MCML method (Wang et al., 1995), which typically requires long computational times.

The MCML method, which is based on a stochastic modeling, was initially proposed by Metropolis and Ulam (1949). For light transport problems, the method consists of recording histories of N photons as they move inside a medium with scattering and absorption effects. There are several free MCML code available on the web (e.g., https://omlc.org/software/index.html). Simulated photons inside the medium experience three main phenomena: **1)** absorption, **2)** scattering, and **3)** reflection by boundaries. The accuracy of the method is proportional to $1/\sqrt{N}$. Typically, the photons are directed into the medium from a collimated light source. For diffused light, the initial photon traveling directions are randomly selected. For sources with a finite incident area, the inflowing location is selected based on the beam spatial profile. Once launched, the kth photon travels a distance Δs in the specified direction $\hat{s}$. At this location, the photon is either absorbed, scattered, reflected (back into medium) or transmitted (out of the medium). If the photon is absorbed, the position is recorded. If the photon is scattered, a new direction is set for the photon to move in the next cycle. If the photon reaches the boundary of the medium, a transmission/reflection effect is recorded. In case of transmission, a new photon is launched. This process is repeated an N number of times. Figure 4.9 shows the MCML method in its two variants: **1)** fixed step (red line) and 2) variable step (blue-line). A black line is used to indicate common steps in these two variants. For the fixed step variant, Δs is always fixed and determined by the optical properties of the medium $\Delta s \ll 1/\mu_{tr}$. For optimal results, it has been recommended to use $\Delta s \approx 0.1/\mu_{tr}$. To determine if the photon is absorbed or scattered, a random number ζ uniformly distributed in the interval $[0,1]$ is generated. If $\zeta < P_a$ the photon is absorbed, recorded and a new photon is launched. If $P_a \le \zeta < P_a + P_s$, the photon is instead scatted. $P_a \approx \mu_a \Delta s$ and $P_s \approx \mu_s \Delta s$ are the probability for the photon to be absorbed and scattered, respectively. Based on the normalized phase function $\Lambda(\hat{s}, \hat{s}', \vec{r})$, a new moving direction $\hat{s}'$ is set for the scattered photon. If the photon is neither absorbed nor scattered, it will continue its free path in the next iteration along the same direction. For the variable step variant, $\Delta s = -ln(\xi)$ is calculated in each iteration using a random number ξ, uniformly distributed in the interval $[0,1]$. A random number ζ uniformly distributed in the interval $[0,1]$ is also generated to determine if the photon is scattered. If $\zeta < \mu_a / \mu_{tr}$, the photon is scattered. Otherwise, the photon is absorbed and recorded. The variable step variant was proposed to substantially reduce N keeping the same accuracy. To reduce the MCML variance, several techniques have been proposed (e.g., implicit capture, roulette, and converse). Hendricks and Booth (1983) reviewed several variance-reducing techniques. The Monte Carlo eXtreme (MCX) method, which is a more sophisticated version of MCML simulator (http://mcx.space/wiki/index.cgi?MCX), is preferable for many fNIRS applications.

However, the approximations discussed below have been used to reduce complexity and computational costs.

The P_N approximation: It is convenient to expand functions $L(\vec{r}, \hat{s}, t)$ and $q(\vec{r}, \hat{s}, t)$ in terms of the spherical harmonics $Y_{l,m}$ (D.2) for unit direction $\hat{s} = (sin(\theta)cos(\varphi), sin(\theta)sin(\varphi), cos(\theta))$. The following notation is used henceforth $Y_{l,m}(\hat{s}) = Y_{l,m}(\theta, \varphi)$.

$$L(\vec{r}, \hat{s}, t) = \sum_{l=0}^{\infty}\sum_{m=-l}^{l}\sqrt{\frac{4\pi}{2l+1}}\ell_{l,m}(\vec{r},t)Y_{l,m}(\hat{s}) \tag{4.2}$$

$$q(\vec{r}, \hat{s}, t) = \sum_{l=0}^{\infty}\sum_{m=-l}^{l}\sqrt{\frac{4\pi}{2l+1}}q_{l,m}(\vec{r},t)Y_{l,m}(\hat{s}) \tag{4.3}$$

Based on the addition theorem (D.4), the normalized phase function $(\Theta_0 = 1)$ is rewritten as:

$$\Lambda(\hat{s}\cdot\hat{s}') = \sum_{l=0}^{\infty}\sum_{m=-l}^{l}\Theta_l Y_{l,m}^{*}(\hat{s}')Y_{l,m}(\hat{s}) \tag{4.4}$$

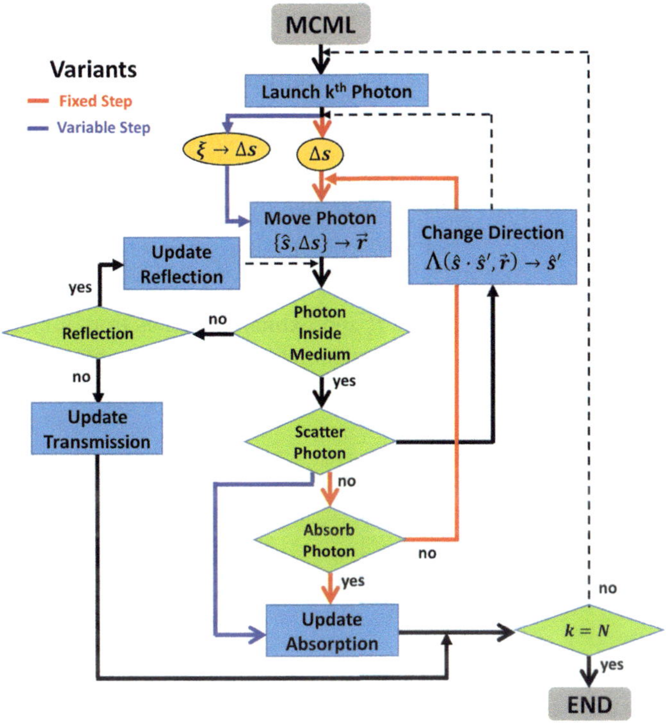

FIGURE 4.9 The flow diagram of the two variants of the MCML method: fixed (red) and variable (blue).

Note that $\Theta_l = \langle P_l(\hat{s} \cdot \hat{s}') \rangle$, with P_l the Legendre polynomials of degree l. Note that theoretically $g = \Theta_1$. Substituting these expansions in equation (4.1) and using both the orthonormality property (D.5) and the recurrent relationships (D.6-D.8) for the spherical harmonics, the following infinite set of coupled equations is obtained.

$$
\begin{aligned}
&\left(\frac{1}{v(\vec{r})}\frac{\partial}{\partial t} + \mu_{tr}(\vec{r})\right)\ell_{l,m}(\vec{r},t) \\
&+\frac{1}{2l+1}\left[\frac{\partial}{\partial z}\left(\sqrt{(l+1-m)(l+1+m)}\,\ell_{l+1,m}(\vec{r},t) + \sqrt{(l-m)(l+m)}\,\ell_{l-1,m}(\vec{r},t)\right)\right. \\
&-\frac{1}{2}\left(\frac{\partial}{\partial x} - i\frac{\partial}{\partial y}\right)\left(\sqrt{(l+m)(l+m-1)}\,\ell_{l-1,m-1}(\vec{r},t) - \sqrt{(l-m+2)(l-m+1)}\,\ell_{l+1,m-1}(\vec{r},t)\right) \\
&-\frac{1}{2}\left(\frac{\partial}{\partial x} + i\frac{\partial}{\partial y}\right)\left(-\sqrt{(l-m)(l-m-1)}\,\ell_{l-1,m+1}(\vec{r},t) + \sqrt{(l+m+1)(l+m+2)}\,\ell_{l+1,m+1}(\vec{r},t)\right) \\
&\qquad\qquad = \mu_s(\vec{r})\Theta_l\ell_{l,m}(\vec{r},t) + q_{l,m}(\vec{r},t)
\end{aligned}
$$

The P_N approximation is obtained by assuming $\ell_{l,m}(\vec{r},t)=0$ for $l>N$. Therefore, there are $(N+1)^2$ coupled first-order partial differential equations (PDEs) that are required to be solved in order to describe the diffusion of photons inside the medium from source q.

The P_1 approximation: Two important physical magnitudes are defined from function $\phi(\vec{r},\hat{s}',t)$: **a)** photon density $\phi(\vec{r},t)=\int_S L(\vec{r},\hat{s},t)\,d\hat{s}$ (Wm^{-2}) and **b)** photon current flux $\vec{J}(\vec{r},t)=\int_S \hat{s}L(\vec{r},\hat{s},t)\,d\hat{s}$ (Wm^{-2}). Note that by definition $\phi(\vec{r},t)=\ell_{0,0}(\vec{r},t)$ and $\vec{J}(\vec{r},t)=\frac{1}{\sqrt{2}}\left(\ell_{1,-1}(\vec{r},t)-\ell_{1,1}(\vec{r},t),\frac{1}{i}\ell_{1,-1}(\vec{r},t)+\frac{1}{i}\ell_{1,1}(\vec{r},t),\sqrt{2}\ell_{1,0}(\vec{r},t)\right)'$. The symbol $\vec{x}'$ denotes transpose of vector $\vec{x}$. The P_1 approximation results from imposing $N=1$. In such a case, it is demonstrated that equation (4.1) transforms into a simplified set of PDEs.

$$\left(\frac{1}{v(\vec{r})}\frac{\partial}{\partial t}+\mu_{tr}(\vec{r})\right)\phi(\vec{r},t)+\nabla\cdot\vec{J}(\vec{r},t)=\mu_s(\vec{r})\phi(\vec{r},t)+q_{0,0}(\vec{r},t) \tag{4.5}$$

$$\left(\frac{1}{v(\vec{r})}\frac{\partial}{\partial t}+\mu_{tr}(\vec{r})\right)\vec{J}(\vec{r},t)+\frac{1}{3}\nabla\phi(\vec{r},t)=\Theta_1\mu_s(\vec{r})\vec{J}(\vec{r},t)+\vec{q}_1(\vec{r},t) \tag{4.6}$$

Where $\vec{q}_1(\vec{r},t)=\int_S \hat{s}\,q(\vec{r},\hat{s},t)\,d\hat{s}$ (Wm^{-3}).

Defining the diffusion coefficient $D(\vec{r})=v(\vec{r})/3(\mu_a(\vec{r})+\mu_s'(\vec{r}))$ and recalling that the reduced scattering coefficient $\mu_s'(\vec{r})=(1-\Theta_1)\mu_s(\vec{r})=(1-g)\mu_s(\vec{r})$, equations (4.5) and (4.6) are rather simplified for the particular case of an isotropic light source $q_0(\vec{r},t)=q_{0,0}(\vec{r},t)$ (Wm^{-3}).

$$\left(\frac{1}{v(\vec{r})}\frac{\partial}{\partial t}+\mu_a(\vec{r})\right)\phi(\vec{r},t)+\nabla\cdot\vec{J}(\vec{r},t)=q_0(\vec{r},t) \tag{4.7}$$

$$\left(\frac{1}{v(\vec{r})}\frac{\partial}{\partial t}+\frac{v(\vec{r})}{3D(\vec{r})}\right)\vec{J}(\vec{r},t)+\frac{1}{3}\nabla\phi(\vec{r},t)=\vec{q}_1(\vec{r},t) \tag{4.8}$$

Considering the particular case of biological tissues with approximately constant refractive index $v(\vec{r})\approx v=c/n$, these two equations can be combined to obtain the *telegraph formula* for each medium:

$$\left(\frac{3D(\vec{r})}{v}\frac{\partial}{\partial t}+v\right)\left(q_0(\vec{r},t)-\mu_a(\vec{r})\phi(\vec{r},t)-\frac{1}{v}\frac{\partial\phi(\vec{r},t)}{\partial t}\right)$$
$$-\frac{v\vec{J}(\vec{r},t)\cdot\nabla D(\vec{r})}{D(\vec{r})}+D(\vec{r})\nabla^2\phi(\vec{r},t)=3D(\vec{r})\nabla\cdot\vec{q}_1(\vec{r},t)$$

fNIRS systems are typically built using source-sensor separation larger than the mean reduced scattering length $\langle 1/\mu_s' \rangle$. In this case, the P_1 approximation is adequate for media with single scattering albedo μ_s/μ_{tr} close to unity and a quasi-isotropic phase function. Taking into consideration the following property $\int_S \hat{s}\hat{s}\, d\hat{s} = \frac{3}{4\pi} I$, the radiance can be written under the P_1 approximation as $L(\vec{r}, \hat{s}, t) = \phi(\vec{r},t)/4\pi + 3\vec{J}(\vec{r},t)\cdot\hat{s}/4\pi$. Note that symbol $\vec{a}_1\vec{a}_2$ represents the dyadic product between vectors $\vec{a}_1$ and $\vec{a}_2$. I is the second-order unit tensor. Similarly, the spectral radiance is represented in terms of its monopolar $\vec{q}_0$ (isotropic) and dipolar $\vec{q}_1$ components: $q(\vec{r}, \hat{s}, t) = \vec{q}_0(\vec{r},t)/4\pi + 3\vec{q}_1(\vec{r},t)\cdot\hat{s}/4\pi$.

The diffusion approximation: In the presence of only isotropic sources, $\vec{q}_1(\vec{r},t) \approx 0$. Therefore, equation (4.8) takes the form: $\vec{J}(\vec{r},\omega) = -\frac{D(\vec{r})}{(i\omega/\mu_{tr}(\vec{r}) + v)}\nabla\phi(\vec{r},\omega)$ in the frequency domain. As mentioned before, absorption is 10^2 weaker that scattering in biological tissues (i.e., $\mu_s' \gg \mu_a$). Taking into account that $v \gg i\omega/\mu_s'(\vec{r})$, it is demonstrated that $v\vec{J}(\vec{r},t) = -D(\vec{r})\nabla\phi(\vec{r},t)$ with $D(\vec{r}) = v/3\mu_s'(\vec{r})$. This assumption leads to further simplification (4.9):

$$-\nabla\cdot\left(D(\vec{r})\nabla\phi(\vec{r},t)\right) + v\mu_a(\vec{r})\phi(\vec{r},t) + \partial\phi(\vec{r},t)/\partial t = vq_0(\vec{r},t) \tag{4.9}$$

In some cases, it has been suggested that $\vec{J}$ decays exponentially over time with time constant $v\lambda$. Under such approximation, the following relation is satisfied $(v - 3D(\vec{r})\lambda)\vec{J}(\vec{r},t) = -D(\vec{r})\nabla\phi(\vec{r},t)$ in the optical medium, which is equivalent to $v\vec{J}(\vec{r},t) = -\hat{D}(\vec{r})\nabla\phi(\vec{r},t)$ with an effective diffusion coefficient $\hat{D}(\vec{r}) = v/3(\mu_s'(\vec{r}) - \lambda)$.

In the frequency domain, the diffusion equation (4.9) transforms into the Helmholtz equation (4.10), with the complex wavenumber $k(\vec{r},\omega)^2 = (v\mu_a(\vec{r}) + i\omega)/D(\vec{r})$.

$$-\frac{\nabla\cdot\left(D(\vec{r})\nabla\phi(\vec{r},\omega)\right)}{D(\vec{r})} + k(\vec{r},\omega)^2\,\phi(\vec{r},\omega) = \frac{v}{D(\vec{r})}q_0(\vec{r},\omega) \tag{4.10}$$

Under the diffusion approximation, it is always used $q_0(\vec{r},\{t,\omega\}) = A(\{t,\omega\})\delta(\vec{r} - \vec{r}_s)$ for collimated light sources. For an anisotropic light source, the term $vq_0(\vec{r},\{t,\omega\})$ on the right side of equations (4.9) and (4.10) can be substituted by $vq_0(\vec{r},\{t,\omega\}) - 3\nabla\cdot\left(D(\vec{r})\vec{q}_1(\vec{r},\{t,\omega\})\right)$ (Arridge, 1999). Note that the time and frequency domain are both represented in this formalism.

The solutions of equations (4.9) and (4.10) for an infinite and homogeneous diffusive medium due to a point light source at $\vec{r}_s$ depends on the type of domain used. For the time domain (TD), the light is assumed to be an ultra-short pulse of total spectral intensity I_p (Ws) at time t_0 (i.e., $q_0(\vec{r},t) = I_p\delta(t - t_0)\delta(\vec{r} - \vec{r}_s)$). It can be assumed that $t_0 = 0$. For the frequency domain (FD), the light comprises a dc term with constant power P_{dc} (W) plus an ac term with an oscillatory power P_{ac} (W) at frequency ω. Hence, the source amplitude is $q_0(\vec{r},t) = (P_{dc} + P_{ac}exp(i\omega t))\delta(\vec{r} - \vec{r}_s)$.

The solution of equation (4.9) in the TD represents the DTOF (or TPSF):

$$\phi(\vec{r},t) = \frac{vI_p}{(4\pi Dt)^{3/2}} exp\left(-\frac{|\vec{r}-\vec{r}_s|^2}{4Dt} - \mu_a vt\right) \qquad \text{TD}$$

The solution of equation (4.10) in the TD represents the DPDW:

$$\phi(\vec{r},t) = \frac{v}{4\pi D|\vec{r}-\vec{r}_s|}\left(P_{dc} exp\left(-\frac{|\vec{r}-\vec{r}_s|}{\delta}\right) + P_{ac} exp\left(-\left(k(\omega)|\vec{r}-\vec{r}_s| - i\omega t\right)\right)\right) \qquad \text{FD}$$

The dc attenuation coefficient is defined as $\delta = 1/\sqrt{3\mu_a(\mu_a + \mu_s')}$. For a homogeneous medium, the complex wavenumber is written as $k(\omega) = \sqrt{(v\mu_a + i\omega)/D} = k_R(\omega) + ik_I(\omega)$. The real $k_R(\omega)$ and imaginary $k_I(\omega)$ parts of the complex wavevector can be defined as a function of the absorption relaxation time $\tau = 1/v\mu_a$:

$$k_R(\omega) = \sqrt{\frac{3}{2}\mu_a(\mu_a + \mu_s')(\sqrt{1+(\omega\tau)^2} + 1)^{1/2}}$$

$$k_I(\omega) = \sqrt{\frac{3}{2}\mu_a(\mu_a + \mu_s')(\sqrt{1+(\omega\tau)^2} - 1)^{1/2}}$$

4.7 THE BOUNDARY CONDITIONS IN THE DIFFUSION APPROACH – THE REFRACTION AND REFLECTION

Equations (4.9) and (4.10) constitute a second-order partial differential equation in space. Therefore, in order to solve it, a boundary condition at the surface limiting the head and free-space (air) must be specified, which should also include the effect of the light sources. Photons that impact an interface S with mismatching refraction indexes n and n' can be either transmitted or reflected. In general, the radiance $L_B(\vec{r},t)$ reflected back from the interface given the incident radiance $L(\vec{r}, \hat{s}, t)$ will satisfies the following equation: $L_B(\vec{r},t) = \int_{S,\hat{s}\cdot\hat{n}_s>0} R_F(\hat{s})L(\vec{r},\hat{s},t)\hat{s}\cdot\hat{n}_s d\hat{s}$, with $\hat{n}_s$ the normal to the boundary at the position $\vec{r}$. Function $R_F(\hat{s})$ is the Fresnel reflection coefficient for light impinging the interface in the direction $\hat{s}$. Defining θ as the angle of incident radiance and the interface normal (i.e., $cos(\theta) = \hat{s}\cdot\hat{n}_s$), the Fresnel reflection coefficient can be represented as:

$$R_F(\theta) = \begin{cases} \frac{1}{2}\left(\frac{ncos(\theta') - n'cos(\theta)}{ncos(\theta') + n'cos(\theta)}\right)^2 + \frac{1}{2}\left(\frac{ncos(\theta) - n'cos(\theta')}{ncos(\theta) + n'cos(\theta')}\right)^2 & 0 \le \theta \le \theta_c \\ 1 & \theta_c \le \theta \le \pi/2 \end{cases}$$

The refracted angle θ' must satisfy the Snell's Law: $nsin(\theta) = n'\sin(\theta')$. The critical angle θ_c is given by $nsin(\theta_c) = n'$. The reflected radiance $L_B(\vec{r},t)$ can also be written as: $L_B(\vec{r},t) = R_\phi \phi(\vec{r},t)/4 - R_j \vec{J}(\vec{r},t)\cdot\hat{n}_s/2$, with the auxiliar parameters R_ϕ and R_j defined as $R_\phi = 2\int_0^{\pi/2} sin(\theta)cos(\theta)R_F(\theta)d\theta$ and $R_j = 3\int_0^{\pi/2} sin(\theta)(cos(\theta))^2 R_F(\theta)d\theta$, respectively. The back reflected radiance $L_B(\vec{r},t)$ should also satisfy the equation $L_B(\vec{r},t) = \int_{S,\,\hat{s}\cdot\hat{n}_s<0} L(\vec{r},\hat{s},t)\hat{s}\cdot(-\hat{n}_s)d\hat{s} = \phi(\vec{r},t)/4 - \vec{J}(\vec{r},t)\cdot\hat{n}_s/2$. Therefore, it can be demonstrated that $\phi(\vec{r},t) = -2\vec{J}(\vec{r},t)\cdot\hat{n}_s(1+R_j)/(1-R_\phi)$.

Three types of boundary conditions have been proposed in the past at the scalp-air interface: **1)** the partial-current boundary condition, **2)** the extrapolated boundary condition, and **3)** the zero-boundary condition. Note that for this particular case, $n = n_{scalp}$ and $n' = n_{air}$.

Under the partial-current boundary condition, it is assumed that the total photon flux pointing inward $\vec{J}_{in}(\vec{r},t) = \int_{S,\,\hat{s}\cdot\hat{n}_s<0} \hat{s}L(\vec{r},\hat{s},t)d\hat{s}$ at any position $\vec{r}$ on the scalp S, except at the source locations, is zero, i.e., $\vec{J}_{in}(\vec{r},t) = 0$. As defined above, $\hat{n}_s(\vec{r})$ is the normal to the scalp at location $\vec{r}$. At the light source locations $\vec{r}_s$, $\vec{J}_{in}(\vec{r}_s,t) \neq 0$. In order to incorporate "inner" reflections by the scalp, a more general equation is used: $\vec{J}_{in}(\vec{r},t) = \int_{S,\,\hat{s}\cdot\hat{n}_s>0} \hat{s}R_F(\hat{s})L(\vec{r},\hat{s},t)d\hat{s}$ with $R_F(\hat{s})$ defining the direction-dependent Fresnel reflection coefficient. Note that $L_B(\vec{r},t) = \vec{J}_{in}(\vec{r},t)\cdot\hat{n}_s(\vec{r})$. The partial-current boundary condition leads to the Robin mathematical expression (4.11) for the photon density in the time $\phi(\vec{r},t)$ and frequency $\phi(\vec{r},\omega)$ domains at the air-scalp interface, which represent a Dirichlet-Neuman boundary condition. This boundary condition can be applied to any numerical solution of the diffusion equations (4.9) and (4.10).

$$\phi(\vec{r},\{t,\omega\}) + \frac{2D(\vec{r})A}{v}\nabla\phi(\vec{r},\{t,\omega\})\cdot\hat{n}(\vec{r}) = 0 \tag{4.11}$$

With parameter $A = (1+R_{eff})/(1-R_{eff})$ to be determined from the effective refractive index R_{eff}, which represents the fraction of photons internally and diffusely reflected at the boundary. Note that if light reflection by the inner scalp is not considered $R_{eff} = 0$, then $A = 1$. Two methods to estimate R_{eff} were discussed in Godavarty et al. (2002):

- An empirical method based on a relationship between the effective refractive index R_{eff} and the relative refractive index of the scalp with respect to air ($n_r = n_{scalp}/n_{air}$). This relationship is only valid for a tissue surrounded by air.

$$R_{eff} = -1.440\,n_r^{-2} + 0.710\,n_r^{-1} + 0.668 + 0.064\,n_r$$

- A theoretical method based on Fresnel's reflection theory, which is valid for any unpolarized light incident at an angle of incidence θ upon the boundary between two contiguous media with a mismatch in their refraction indexes $n = n_{scalp}$ and $n' = n_{air}$.

$$R_{eff} = \frac{R_j + R_\phi}{2 + R_j - R_\phi}$$

$$L_B(\vec{r},t) = R_{eff}\left(\frac{\phi(\vec{r},t)}{4} - \frac{\vec{J}(\vec{r},t)\cdot\hat{n}_s}{2}\right)$$

Although the partial-current boundary condition is the most widely accepted, the extrapolated boundary condition has been sometimes used to obtain more accurate analytical solutions for turbid media. It stablishes that $\phi(\vec{r}\,',t) = 0$ at the linear extrapolation distance into the air $\vec{r}\,' = \vec{r} + r_b\hat{n}(\vec{r})$ for any $\vec{r}$ at the scalp-air interface. In the absent of absorption and reflection by the scalp, the asymptotic photon density is linear, and a better agreement with the photon transport equation is achieved by using $r_b = 0.7104/\mu_s'$. This distance, which is calculated using the Milne problem, is known as the extrapolated end-point. A correction to parameter r_b is necessary if absorption and reflections at the scalp-air interface are also included: $r_b = 2(1+R_{eff})/(3(\mu_a + \mu_s')(1-R_{eff}))$. This theoretical expression was obtained using the Fresnel reflection theory (Haskell et al., 1994). For the case of neither absorption $\mu_a = 0$ nor reflection $R_{eff} = 0$, a distance of $r_b = 2/3\mu_s'$ has been used (Boas, 1996).

The zero-boundary condition has been proposed for simplicity, although it constitutes a violation of the diffusion approximation. It stablishes that $\phi(\vec{r},t) = 0$ at the physical boundary (scalp). Trivially, the latter is a particular case of the extrapolated boundary condition with case $r_b = 0$.

If equations (4.9) and (4.10) are required to be solved in a complex medium, as the head, formed by concentric tissue compartments with different but homogeneous optical properties (e.g., scalp, skull, CSF, and brain), then two conditions are required at any interior interface. Let $\phi(\vec{r},t)$ and $\vec{J}(\vec{r},t)$ be the respective photon density and flux of the tissue compartment with refraction index n. Similarly, let us define the same magnitudes $\phi'(\vec{r},t)$ and $\vec{J}'(\vec{r},t)$ for a contiguous medium with refraction index n'. At any time t and position $\vec{r}$ on the interface $S_{n,n'}$ separating the two media, the photon flow at both side of the interface, and perpendicular to it, can be defined as $j(\vec{r},t) = \vec{J}(\vec{r},t)\cdot\hat{n}(\vec{r})\big|_{S_{n,n'}}$ and $j'(\vec{r},t) = \vec{J}'(\vec{r},t)\cdot\hat{n}(\vec{r})\big|_{S_{n,n'}}$. Under the diffusion approximation, it is commonly assumed that this photon flux is continuous across $S_{n,n'}$: $j(\vec{r},t) = j'(\vec{r},t)$. For photons, a continuity equation for its density must include reflections on the interface. Let us define the relative index of refraction $n_r = n'/n$ such that $n_r > 1$. Based on Fresnel reflections, Aronson (1995) obtained a theoretical continuity equation relating $\phi(\vec{r},t)$ and $\phi'(\vec{r},t)$ on $S_{n,n'}$ (4.12).

$$\phi'(\vec{r},t) - n_r^2 \phi(\vec{r},t) = C(n_r)\, j(\vec{r},t) \tag{4.12}$$
$$C(n_r) = \left(2 - B_1(n_r) - B_2(n_r)\right) / A_2(n_r)$$

Theoretical formulas for the interface constants $B_1(n_r)$, $B_2(n_r)$ and $A_2(n_r)$ are provided in Aronson (1995, Appendix A). However, a simple numerical fitting demonstrated that:

$$C(n_r) = \begin{cases} 25.6p^3\left(1 - 1.41p + 3.95p^2\right) & 1 < n_r < 1.82 \\ 14.3\left(1 + 9.35p_1 + 49.1p_1^2 + 327p_1^3 + 1800p_1^4\right) & 1.82 \le n_r \le 3.73 \end{cases}$$

With $p = \sqrt{(n_r - 1)/(n_r + 1)}$ and $p_1 = p - 0.64$.

To consider more realistic situations with heterogeneities in the absorption and scattering coefficients, FEM are implemented in freely available tools based on meshing from MRI anatomical data (e.g., NIRFAST, Dehghani et al., 2008; TOAST, Schweiger and Arridge, 2014). However, theoretical expressions can be obtained using some approximations. For example, either the Born or the Rytov approximation can be used if the absorption coefficient in the head tissue is considered heterogeneous (i.e., $\mu_a(\vec{r})$, μ_s' is constant). For these two approximations, it is: **a)** assumed $\mu_s' \gg \mu_a(\vec{r})$ (i.e., $D = v/3\mu_s'$), and **b)** used the perturbation theory $\mu_a(\vec{r}) = \mu_{a,0} + \delta\mu_a(\vec{r})$. Under the Born approximation: $\phi(\vec{r},t) = \phi_0(\vec{r},t) + \phi_{SC}(\vec{r},t)$, with the incident field $\phi_0(\vec{r},t)$ much larger than the scattered field $\phi_{SC}(\vec{r},t)$ (i.e., $\phi_{SC}(\vec{r},t) \ll \phi_0(\vec{r},t)$. Under the Rytov approximation: $\phi(\vec{r},t) = exp(\Phi_0(\vec{r},t) + \Phi_{SC}(\vec{r},t))$, with $(exp(\Phi_{SC}(\vec{r},t)))^2 \ll o(\vec{r})$. Note that $o(\vec{r}) \approx \delta\mu_a(\vec{r})/D$ is a very small function. Equivalent approximations can be used for scattering heterogeneities by introducing perturbations to the diffusion coefficient $D(\vec{r}) = D_0 + \delta D(\vec{r})$, also under the assumption that $\mu_s'(\vec{r}) \gg \mu_a(\vec{r})$ (i.e., $D(\vec{r}) = v/3\mu_s'(\vec{r})$)).

4.8 FNIRS – OBSERVATIONS AND PARAMETERS FITTINGS

As a result of the proximity of sources and sensors in fNIRS, the scalp is considered a semi-infinite plane (scalp) at $z = 0$ separating the surrounding air and the head tissue. The head is assumed to have homogeneous optical properties (i.e., constant μ_a and μ_s'). Let us assume that variable ρ is the radial distance on a planar scalp from a central source at $z = 0$ and z is the distance into the head from the scalp. The distance between the source and sensor is defined as d. The observations in fNIRS from the source-sensor pairs are used to estimate temporal changes in these optical parameters using methods that depend on the fNIRS modality.

The method underlying the CW-fNIRS modality was inspired in the Beer-Lambert law. Initial observations of light propagation through red wine might have motivated scientists (e.g., French mathematician Pierre Bouguer, 1698–1758) to propose the theoretical foundations underlying the Beer-Lambert law. However, the final formulation of the law as a solution to the Bhatnagar–Gross–Krook operator is attributed to Johann Heinrich Lambert (1728–1777). In 1852, August Beer (1825–1863) generalized it to quantify concentrations. The Beer-Lambert law can only be applied in the case of non-scattering media. Therefore, it has been used in situations involving straightforward light transmission through a medium. The fNIRS technique requires that light penetrating the scalp from a source, returns to it at the sensor position, and for that scattering is needed. Delpy et al. (1988) proposed a modification to the Beer-Lambert law to account for multiple scattering, which is the mathematical framework for the CW-fNIRS modality. According to the modified Beer-Lambert law, the attenuation of the light satisfies:

$$I = I_0 exp\left(-\mu_a \pi d\right) + g \tag{4.13}$$

I_0 is the intensity of the incident light from the source and I is the observed light intensity at the sensor location. Parameter g is a factor to account for the geometry. Parameter π accounts for increases in the photon path-length due to multiple scattering. For the semi-infinite geometry, it depends on d, μ_a and μ_s' according to the following equation:

$$\pi\left(d,\mu_a,\mu_s'\right) = \frac{\sqrt{3\mu_s'}}{2\sqrt{\mu_a}} \frac{d\sqrt{3\mu_s'\mu_a}}{\left(d\sqrt{3\mu_s'\mu_a}+1\right)} = \frac{3\mu_s' d}{2\left(d\sqrt{3\mu_s'\mu_a}+1\right)} \tag{4.14}$$

The observation in the CW-fNIRS is given as an optical density $OD = log(I / I_0)$. To avoid the dependency with the source-detector distance d, some methods use the formula for the infinite medium $\pi(\mu_a, \mu_s') = \sqrt{3\mu_s'} / 2\sqrt{\mu_a}$.

For TD and FD-fNIRS modalities, the equations (4.9) and (4.10) are solved using analytical models based on the method of image sources (Figure 4.10) (Haskell et al., 1994; Hielscherf et al., 1995). Henceforth, cases are discussed according to the boundary condition used. fNIRS observations are based on reflectance measurements at specific points on the scalp (i.e., the sensors). The reflectance is defined as the number of photons that reach the surface per unit area per unit time at a given radial source-detector separation ρ, which is defined by the Fick's law: $R(\rho,t) = -(D/v)\nabla\phi(\rho,z,t)\cdot(-\hat{z})\big|_{z=0}$.

The DTOF solution for the extrapolated boundary condition is:

$$\phi(\rho,z,t) = \frac{vI_p}{\left(4\pi Dt\right)^{3/2}} exp\left(-\mu_a vt\right)\left(exp\left(-\left(r_0\left(z,\rho\right)\right)^2 / 4Dt\right) - exp\left(-\left(r_c\left(z,\rho\right)\right)^2 / 4Dt\right)\right) \tag{4.15}$$

Note that this is the TPSF for an ultra-short pulse of total spectral intensity I_p. The method assumes there are contributions from point sources at $z_0 = (\mu_a + \mu_s')^{-1}$ (*first term*) and $-(z_0 + 2r_b)$ (negative image, *second term*). The source image is in the non-diffuse medium (air). Note that the source is placed not at $z = 0$, but at the location of the first interaction with tissue z_0, an assumption proposed by Patterson et al. (1989). The z-dependent diagonal distances to the actual source r_0 and to its image r_c are defined as $r_0(z,\rho) = \sqrt{(z - z_0)^2 + \rho^2}$ and $r_c(z,\rho) = \sqrt{(z + z_0 + 2r_b)^2 + \rho^2}$, respectively. Note that r_b was defined in the previous section.

The DPDW solution for the extrapolated boundary condition is $\phi(\rho,z,t) = \phi_{dc}(\rho,z) + \phi_{ac}(\rho,z,t)$, with:

$$\phi_{dc}(\rho,z) = \frac{vP_{dc}}{4\pi D}\left(\frac{exp(-r_0(z,\rho)/\delta)}{r_0(z,\rho)} - \frac{exp(-r_c(z,\rho)/\delta)}{r_c(z,\rho)}\right) \tag{4.16}$$

$$\phi_{ac}(\rho,z,t) = \frac{vP_{ac}}{4\pi D} exp(i\omega t)\left(\frac{exp(-k(\omega)r_0(z,\rho))}{r_0(z,\rho)} - \frac{exp(-k(\omega)r_c(z,\rho))}{r_c(z,\rho)}\right) \tag{4.17}$$

In this case, the source comprises a dc term (contact power P_{dc}) and ac term (oscillatory power P_{ac} at frequency ω). The DPDW solution leads to theoretical expressions for the observable magnitudes at the sensor (i.e., $z = 0$, $\rho = d$) in the FD-fNIRS modality (i.e., dc power off-set ΔP_{dc}, ac power modulation ΔP_{ac} and ac phase $\Delta\theta$).

$$\Delta P_{dc}(d) = P(d) - P_{dc} \tag{4.18}$$

The reflectance at the sensor produced by the dc term in the source, which does not depend on time, is defined as $P(d) = R(d)A_s = -(D/v)\nabla\phi_{dc}(d,z)\cdot(-\hat{z})\big|_{z=0}$, with A_s the surface area covered by the sensor.

$$\Delta\theta(\omega,d) = k_I(\omega)r_0^0(d) - arctan(I_m(\omega,d)/R_e(\omega,d)) \tag{4.19}$$

With $r_0^0(d) = r_0(0,d)$

$$\Delta P_{ac}(\omega,d) = \frac{(R_e(\omega,d))^2 + (I_m(\omega,d))^2}{\vartheta(\omega,d)} \tag{4.20}$$

The parametric functions in these equations are defined as:

$$\begin{aligned} R_e(\omega,d) = &\left[1 + \left(k_R(\omega) + \frac{1}{r_0^0(d)}\right)\frac{(z_0)^2}{r_0^0(d)}\right]\frac{exp(-k_R(\omega)r_0^0(d))}{r_0^0(d)} \\ &+ \left\{\left[\left(k_R(\omega) + \frac{1}{r_0^c(d)}\right)\frac{(2r_b + z_0)z_0}{r_0^c(d)} - 1\right]cos\left(k_I(\omega)\left(r_0^c(d) - r_0^0(d)\right)\right)\right. \\ &\left. + k_I(\omega)\frac{(2r_b + z_0)z_0}{r_0^c(d)} sin\left(k_I(\omega)\left(r_0^c(d) - r_0^0(d)\right)\right)\right\}\frac{exp(-k_R(\omega)r_0^c(d))}{r_0^c(d)} \end{aligned}$$

$$I_m(\omega,d)=k_I(\omega)\frac{(z_0)^2}{r_0^0(d)}\frac{exp\left(-k_R(\omega)r_0^0(d)\right)}{r_0^0(d)} +\left\{\left[1-\left(k_R(\omega)+\frac{1}{r_0^c(d)}\right)\frac{(2r_b+z_0)z_0}{r_0^c(d)}\right]sin\left(k_I(\omega)\left(r_0^c(d)-r_0^0(d)\right)\right) +k_I(\omega)\frac{(2r_b+z_0)z_0}{r_0^c(d)}cos\left(k_I(\omega)\left(r_0^c(d)-r_0^0(d)\right)\right)\right\}\frac{exp\left(-k_R(\omega)r_0^c(d)\right)}{r_0^c(d)}$$

$$\vartheta(\omega,d)=\left[1+\left(\frac{1}{\delta}+\frac{1}{r_0^0(d)}\right)\frac{(z_0)^2}{r_0^0(d)}\right]\frac{exp\left(-r_0^0(d)/\delta\right)}{r_0^0(d)} +\left[\left(\frac{1}{\delta}+\frac{1}{r_0^c(d)}\right)\frac{(2r_b+z_0)z_0}{r_0^c(d)}-1\right]\frac{exp\left(-r_0^c(d)/\delta\right)}{r_0^c(d)}$$

With $r_0^c(d)=r_c(0,\rho)=\sqrt{(z_0+2r_b)^2+d^2}$

The solution for the zero boundary condition results from imposing $r_b=0$ in equations (4.15), (4.16–4.17) and (4.18–4.20). Trivially, the negative image (*second term*) is at $-z_0$ for this case.

Figure 4.10 provides details of the relative positions of the sources and its image for these three boundary conditions.

The partial-current boundary condition is more complicated as assumed the existence of two positive sources at $z=z_0$ and $z=-z_0$, as well as a space-varying sink $\left(-2exp\left(-\frac{|z|-z_0}{r_b}\right)/r_b\right)$ at $z=-(z_0+r_b)$.

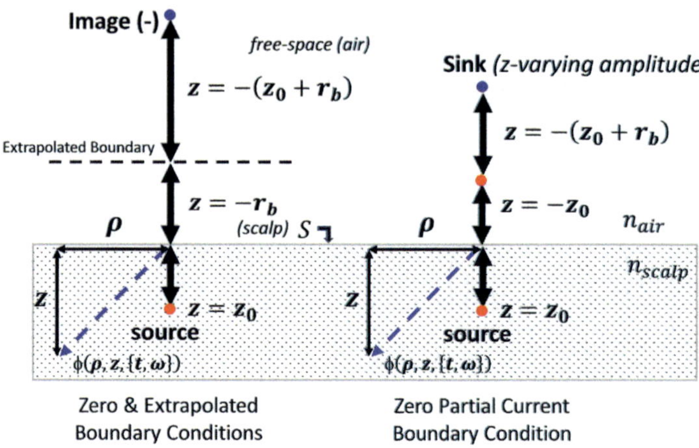

FIGURE 4.10 The boundary conditions for the scalp-air interface: the method of images.

The DTOF solution for the partial-current boundary condition is:

$$\phi(\rho,z,t) = \frac{vI_p}{(4\pi Dt)^{3/2}} exp\left(-\frac{\rho^2}{4Dt}-\mu_a vt\right)\Big(exp(-(z-z_0)^2/4Dt)-exp(-(z+z_0)^2/4Dt) - \frac{2}{r_b}\int_0^\infty exp(-l/r_b-(z+z_0+l)^2/4Dt)dl\Big) \tag{4.21}$$

The DPDW solution for the extrapolated boundary condition is $\phi(\rho,z,t)=\phi_{dc}(\rho,z)+\phi_{ac}(\rho,z,t)$, with:

$$\phi_{dc}(\rho,z) = \frac{vP_{dc}}{4\pi D}\left(\frac{exp\left(-r_0(z,\rho)/\delta\right)}{r_0(z,\rho)}+\frac{exp\left(-r_p(z,\rho)/\delta\right)}{r_p(z,\rho)} - \frac{2}{r_b}\int_0^\infty \frac{exp\left(-l/r_b-\sqrt{(z+z_0+l)^2+\rho^2}/\delta\right)}{\sqrt{(z+z_0+l)^2+\rho^2}}dl\right) \tag{4.22}$$

$$\phi_{ac}(\rho,z,t) = \frac{vP_{ac}}{4\pi D}exp(i\omega t)\left(\frac{exp\left(-k(\omega)r_0(z,\rho)\right)}{r_0(z,\rho)}+\frac{exp\left(-k(\omega)r_p(z,\rho)\right)}{r_p(z,\rho)} - \frac{2}{r_b}\int_0^\infty \frac{exp\left(-l/r_b-k(\omega)\sqrt{(z+z_0+l)^2+\rho^2}\right)}{\sqrt{(z+z_0+l)^2+\rho^2}}dl\right) \tag{4.23}$$

With $r_p(z,\rho)=\sqrt{(z+z_0)^2+\rho^2}$

After obtaining $P(d)$ by using (4.22) to estimate $R(d)$, $\Delta P_{dc}(d)$ is then calculated by the use of equation (4.18). Similarly, equations (4.19) and (4.20) can be used to estimate $\Delta\theta(\omega,d)$ and $\Delta P_{ac}(\omega,d)$, respectively. In this case, the auxiliary functions $R_e(\omega,d)$, $I_m(\omega,d)$ and $\vartheta(\omega,d)$ are evaluated using the following expressions:

$$R_e(\omega,d) = \frac{exp\left(-k(\omega)r_0^0(d)\right)}{r_0^0(d)} - \frac{1}{r_b}\int_0^\infty \frac{exp\left(-l/r_b-k_R(\omega)r_p(l,d)\right)}{r_p(l,d)}cos\left(k_I(\omega)\left(r_p(l,d)-r_0^0(d)\right)\right)dl$$

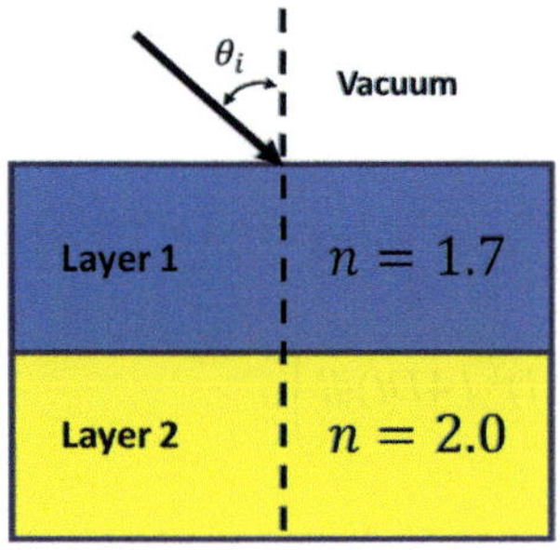

FIGURE 4.11 Light propagation across two-layer interface.

$$I_m(\omega,d)=\frac{1}{r_b}\int_0^{\infty}\frac{exp\left(-l/r_b-k_R(\omega)r_p(l,d)\right)}{r_p(l,d)}sin\left(k_I(\omega)\left(r_p(l,d)-r_0^0(d)\right)\right)dl$$

$$\vartheta(\omega,d)=\frac{exp\left(-r_0^0(d)/\delta\right)}{r_0^0(d)}-\frac{1}{r_b}\int_0^{\infty}\frac{exp\left(-l/r_b-r_p(l,d)/\rho\right)}{r_p(l,d)}dl$$

Several optimization techniques are used with these analytical models to estimate optical parameters (μ_a, μ_s') inside the head from direct observations of either OD (CW), $\phi(\rho,z,t)$ (DTOF, TD), or $\{\Delta P_{dc},\Delta P_{dc},\Delta\theta\}$ (DPDW, FD) in multiple source-sensor pairs on the scalp, which are referred as inverse methods. Different approaches are employed to develop these methods: a) non-linear Levenberg–Marquardt, b) Bayesian, c) regularization, and d) genetic algorithms. Regularization methods for diffuse optical tomography, largely adopted for processing CW data, have proved to be effective for TD-fNIRS data (Arridge, 1999; Gao et al., 2004; Selb et al., 2007). There is a limit on maximum output power for fNIRS system: Lasers (International Commission on Non-Ionizing Radiation Protection, 2000) and LEDs (International Commission on Non-Ionizing Radiation Protection, 1997). Note that, I have not discussed in this chapter more advanced optical neuroimaging techniques, such as diffuse optical spectroscopy (DOS) and diffuse optical tomography (DOT) in the near-infrared spectrum.

4.9 PROBLEMS AND EXERCISES

Problem 4.1 – Describe the scattering characteristic of media with the following g values: a) medium 1 ($g = 0$) and medium 2 ($g = 1$). .

Problem 4.2 – Given the following two-layer interface (Figure 4.11): **1)** Calculate the speed of light v and wavelengths λ in each layer, and **2)** What is the minimum θ_i required to create total internal reflection between layers 1 and 2? Note that c is the speed of light in the vacuum. The incident light has a wavelength of $750\,nm$ and a power of $3\,W$. When θ_i is 0, what is the light power reflected by layer 1?

Problem 4.3 – Use the Mie scattering calculator (https://omlc.org/calc/mie_calc.html) to determine the scattering coefficients of the following samples:

	Sample 1	**Sample 2**
Sphere diameter (μm)	0.1	1
n sphere	1.5	1.5
n medium	1.33	1.33
Wavelength (microns)	0.8	0.8
Concentration (spheres/μm^3)	100	0.01

REFERENCES

Aronson R. Boundary conditions for diffusion of light. *J Opt Soc Am A* 12, 2532–2539, 1995.

Arridge SR. Optical tomography in medical imaging. *Inverse Probl* 15, R41–R93, 1999.

Bashkatov AN, Genina EA, Kochubey VI, Tuchin VB. Optical properties of human cranial bone in the spectral range from 800 to 2000 nm. Proc. SPIE 6163 616310-1, 2006.

Benaron DA, Van Houten JP, Cheong W-F, Kermit EL, King RA. Early clinical results of time-of-flight optical tomography in a neonatal intensive care unit. In: Chance B, Alfano RR (Eds.) *Optical Tomography, Photon Migration, and Spectroscopy of Tissue and Model Media: Theory, Human Studies, and Instrumentation*. Proc SPIE 2389, pp. 582–596, 1995.

Biswas T, Luu T. In vivo MR measurement of refractive index, relative water content and T2 relaxation time of various brain lesions with clinical application to discriminate brain lesions. *Internet J Radiol* 13(1), 1–9, 2011.

Boas DA. Diffuse photon probes of structural and dynamical properties of turbid media. Theory and biomedical applications. Dissertation in Physics. 1996.

Chance B. Spectrophotometry of intracellular respiratory pigments. *Science* 120(3124), 767–775, 1954.

Chance B, Luo Q, Nioka S, Alsop DC, Detre JA. Optical investigations of physiology. A study of intrinsic and extrinsic biomedical contrast. *Proc R Soc London B Biol Sci* 352, 707–716, 1997.

Chance, B . Zhuang Z, Unah C, Alter C, Lipton L. Cognition-activated low-frequency modulation of light absorption in human brain. *PNAS* 90, 3770–3774, 1993.

Cheong W-F, Scott AP, Ashley JW. A review of the optical properties of biological tissues. *IEEE J Quantum Electron* 26(12), 2166–2185, 1990.

Custo A, Wells III WL, Barnett AH, Hillman EMC, Boas DA. Effective scattering coefficient of the cerebral spinal fluid in adult head models for diffuse optical imaging. *Appl Opt* 45(19), 4747–4755, 2006.

Dehghani H, Eames ME, Yalavarthy PK, Davis SC, Srinivasan S, Carpenter CM, Pogue BW, Paulsen KD. Near infrared optical tomography using NIRFAST: Algorithm for numerical model and image reconstruction. *Commun Numer Meth Eng* 25(6), 711–732, 2008.

Delpy DT, Cope M, van der Zee P, Arridge S, Wray S, Wyatt J. Estimation of optical path length through tissue from direct time of flight measurements. *Phys Med Biol* 33, 1433–1442, 1988.

Fantini S, Sassaroli A. Frequency-domain techniques for cerebral and functional near-infrared spectroscopy front. *Neurosci* 14(300), 1–18, 2020.

Firbank M, Hiraoka M, Essenpreis M, Delpy DT. Measurement of the optical properties of the skull in the wavelength range 650-950 nm. *Phys Med & Biol* 38, 503–510, 1993.

Franceschini MA, Moesta KT, Fantini S, Gaida G, Gratton E, Jess H, Mantulin WW, Seeber M, Schlag PM, Kaschke M. Frequency-domain techniques enhance optical mammography: Initial clinical results. *PNAS USA* 94, 6468–6473, 1997.

Gao F, Zhao H, Tanikawa Y, Yamada Y. Optical tomographic mapping of cerebral haemodynamics by means of time-domain detection: Methodology and phantom validation. *Phys Med Biol* 49, 1055–1078, 2004.

Godavarty A, Hawrysz DJ, Roy R, Sevick-Muraca EM, Eppstein MJ. Influence of the refractive index-mismatch at the boundaries measured in fluorescence enhanced frequency-domain photon migration imaging. *Opt Express* 10(15), 653–662, 2002.

Golovynskyi S, Golovynska I, Stepanova LI, Datsenko OI, Liu L, Qu J, Ohulchanskyy TY. Optical windows for head tissues in near-infrared and short-wave infrared regions: Approaching transcranial light applications. *J Biophotonics* 11(12), e201800141, 2018.

Gratton G, Corballis PM, Cho E, Faniani M, Hood DC. Shades of gray matter: Noninvasive optical images of human brain responses during visual stimulation. *Psychophysiology* 32, 505–509, 1995.

Haskell RC, Svaasand LO, Tsay T-T, Feng T-C, McAdams MS, Tromberg BJ. Boundary conditions for the diffusion equation in radiative transfer. *J Opt Soc Am A* 11 (10), 2727–2741, 1994.

Hendricks JS, Booth TE. MCVP variance reduction overview. In: Alcouffe R, Dautray R, Forster A, Ledanois G, Mercier, B (Eds.), *Methods and Applications in Neutronics, Photonics and Statistical Physics*. Springer-Verlag, New York, pp. 83–92, 1983.

Hielscherf AH, Jacques SL, Wag L, Tittel FK. The influence of boundary conditions on the accuracy of diffusion theory in time-resolved reflectance spectroscopy of biological tissues. *Phys Med & Biol* 40, 1957–1975, 1995.

Hirth C, Obrig H, Villringer K, Thiel A, Bernarding J, Mühlnickel W, Flor H, Dirnagl U, Villringer A. Non-invasive functional mapping of the human motor cortex using near-infrared spectroscopy. *NeuroReport* 7, 1977–1981, 1996.

Hoshi Y, Tamura M. Dynamic multichannel near-infrared optical imaging of human brain activity. *J Appl Physiol* 75, 1842–1846, 1993.

Jacques SL. Optical properties of biological tissues: A review. *Phys Med & Biol* 58, R37–R61, 2013.

Jacques SL, Alter CA, Prahl SA. Angular dependence of He-Ne laser light scattering by human dermis. *Lasers Life Sci* 1, 309–333, 1987.

Jöbsis FF. Noninvasive, infrared monitoring of cerebral and myocardial oxygen sufficiency and circulatory parameters. *Science* 198(4323), 1264–1267, 1977.

Kato T, Kamei A, Takashima S, Ozaki T. Human visual cortical function during photic stimulation monitoring by means of near-infrared spectroscopy. *J Cereb Blood Flow Metab* 13, 516–520, 1993.

Lister T, Wright PA, Chappell PH. Optical properties of human skin. *J Biomed Opt* 17(9), 090901, 2012.

Maki A, Yamashita Y, Ito Y, Watanabe E, Mayanagi Y, Koizumi H. Spatial and temporal analysis of human motor activity using noninvasive NIR topography. *Med Phys* 22, 1997–2005, 1995.

Metropolis N, Ulam S. The Monte Carlo method. *J Am Stat Assoc* 44, 335–341, 1949.

Patterson MS, Chance B, Wilson BC. Time resolved reflectance and transmittances for the noninvasive measurement of tissue optical properties. *Appl Opt* 28, 2331–2336, 1989.

Scholkmann F, Kleiser S, Jaakko Metz A, Zimmermann R, Mata Pavia J, Wolf U, Wolf M. A review on continuous wave functional near-infrared spectroscopy and imaging instrumentation and methodology. *Neuroimage* 85, 6–27, 2014.

Schweiger M, Arridge S. The Toast++ software suite for forward and inverse modeling in optical tomography. *J Biomed Opt* 19(4), 040801, 2014.

Selb J, Dale A, Boas DA. Linear 3D reconstruction of time-domain diffuse optical imaging differential data: Improved depth localization and lateral resolution. *Opt Express* 15, 16400–16412, 2007.

Severinghaus JW, Honda Y. History of blood gas analysis. VII. Pulse oximetry. *J Clin Monit* 3(2), 135–138, 1987.

Torricelli A, Contini D, Pifferi A, Caffini M, Re R, Zucchelli L, Spinelli L. Time domain functional NIRS imaging for human brain mapping. *NeuroImage* 85, 28–50, 2014.

Torricelli A, Pifferi A, Spinelli L, Cubeddu R. Time-resolved reflectance at null source-detector separation: improving contrast and resolution in diffuse optical imaging. *Phys Rev Lett* 95, 078101, 2005.

van Gemert MJC, Jacques SL, Sterenborg HJCM, Star WM. Skin optics. *IEEE Trans Biomed Eng* 36(12), 1146–1154, 1989.

Villringer A, Planck J, Hock C, Schleinkofer L, Dirnagl U. Near infrared spectroscopy (NIRS): A new tool to study hemodynamic changes during activation of brain function in human adults. *Neurosci Lett* 154, 101–104, 1993.

Wang L, Jacques SL, Zheng L. MCML – Monte Carlo modeling of light transport in multi-layered tissues. *Comput Meth Prog Biomed* 47, 131–146, 1995.

Yaroslavsky AN, Schulze PC, Yaroslavsky IV, Schober R, Ulrich F, Schwarzmaier H-J. Optical properties of selected native and coagulated human brain tissues in vitro in the visible and near infrared spectral range. *Phys Med & Biol* 47, 2059–2073, 2002.

5 The Neurons in Equilibrium

5.1 OVERVIEW

A dense packing of neurons and glia cells suspended in a highly conducting extracellular space (ECS) forms the brain. Local electric currents caused by the opening of transmembrane gated channels in these cells with conductance to particular ions are named "*primary volumetric current sources*" $s^{(p)}$ (Am^{-3}). In neurons, these transmembrane electric currents underlie the genesis and propagation of the impulses these cells use to communicate with each other. The transmembrane electric currents, which are largely generated by neurons, can easily propagate within the conductive brain, which are named secondary electric currents. In general, both primary and secondary electric currents inside the brain take place because of the movements of ions with different charges according to electrochemical potentials. Together with the resulting ionic electrodiffusion, osmotic water exchange might also occur between different cellular compartments in the brain. A resting membrane potential of different magnitudes in these cells is the fundamental component in such a cellular electrogenesis. The resting membrane potential represent a state of thermodynamic equilibrium. The resting membrane potential in neurons results mostly from the continues pumping of K^+ ions across their membrane, which consume energy in the form of ATP. Pumping of K^+ ions take place mostly via the Na-K ATPase, which is discussed later in this chapter. K^+ ions leak continuously tandem pore-domain channels in the membrane of neurons. The existence of large negatively charged elements (proteins/amino acids) inside brain cells, which are membrane impermeable, constitutes the other contributing mechanism to the resting membrane potential because of the Donald equilibrium. Neurons are endowed with many types of gated channels, which might open during neuronal activation. In this chapter, we will study the foundations for the neuronal electrophysiology at the equilibrium, i.e., when these channels are partially closed. Some historical moments and important scientists are also highlighted.

5.2 ELECTROCHEMICAL POTENTIALS – NERNST EQUILIBRIUM

Electric currents in the brain happen because of the electrolytic properties of its liquid constituents. An electrolyte is an ionic solution produced by dissolving particular salts in a polar solvent. If the salts are completely hydrated (ionized) and the

DOI: 10.1201/9780367901509-5

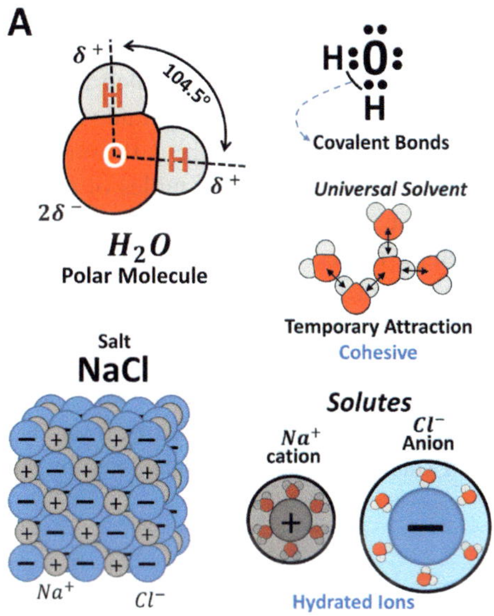

FIGURE 5.1 Schematic representations of the water polarity (top) and the hydration of Na^+ and Cl^- ions when **NaCl** is dissolved in water (bottom).

number of ions in the solution is large, the electrolytes is called "strong". About 75 % of the brain is made of water (H_2O). Therefore, electricity conduction happens in aqueous electrolytes. Cells in the brain, like neurons, separate by a membrane two strong electrolytes with different ionic compositions: **i)** – interstitial fluid in the ECS and **ii)** – the cell's cytoplasm (also named cytosol). To understand aqueous electrolyte, we need to know how they form. A water molecule is comprised of one atom of oxygen ($^{16}_{8}$O, orbitals – $2s^1, 2s^2\, 4p^2 \rightarrow$ needs 2 e^-) in covalent bonds (V-shaped) with two atoms of hydrogen ($^{1}_{1}$H, orbital – $1s^1 \rightarrow$ needs 1 e^- each) sharing electron pairs. Because of its high electronegativity, electrons in the H_2O molecule spend longer times orbiting around the oxygen atom, which creates a redistribution of differential charges (Figure 5.1): positive (δ^+) and negative ($2\delta^-$). Hence, water is a polar molecule and, in its liquid stage, is temporally attracted by weak hydrogen bonds (cohesive attraction), making it a universal polar solvent. In ionic salts, cations and anions elements are also bond together by electrostatic forces, forming highly-ordered solid "crystals". For example, Na^+ and Cl^- ions are periodically intercalated because of the opposite charge attraction to form the solid table salt (**NaCl**(s)). If an ionic salt is dissolved in water, ions easily dissociate. The negatively charge oxygen atoms of the molecule will surround cations (e.g., Na^+) and the positively charged hydrogen atoms will surround anions (e.g., Cl^-).

Ions surrounded by a shell of water molecules are hydrated (Aq). The smaller the core radius of the ions the larger the hydrated regions. Therefore, it will be useful to introduce the concept of activity a_k for the aqueous ion k, which is proporsional to the size of its hydrating shell. The activity coefficient accounts for interactions between the species in an electrolyte, which try to mask their effective charges by creating ionic atmospheres of opposite charge around each ion. The activity coefficient of ion k depends linearly on its molar concentration c_k ($M = moles / L$) in the solution (sometime referred in this book as $[k]$) scaled by dimensionless factor γ_k named activity coefficient, i.e., $a_k = c_k \gamma_k$. The magnitude c_k is defined as $c_k = n_k / V$, where n_k and V are the number of moles of the k-th ion and total volume of the solution, respectively. In general, the number of moles of a substance s (either a salt or an ion) is calculated by the equation $n_s = W_s / MW_s$, where W_s and MW_s are its weight and molecular weight, respectively. The MW for an ionic salt is calculated using the atomic mass (m_a) of each component that forms it. To illustrate it, let us calculate it for a complex molecule as magnesium sulfate heptahydrate (**$MgSO_4$-$7H_2O$**(s)): MW = ${}^{24}_{12}\mathrm{Mg}\ (m_a = 24.305) + {}^{32}_{16}\mathrm{S}\ (m_a = 32.07) + 4 \times {}^{16}_{8}\mathrm{O}(m_a = 16.00) + 7 \times (2 \times {}^{1}_{1}\mathrm{H}(m_a = 1.01) + {}^{16}_{8}\mathrm{O}(m_a = 16.00)) \cong 246.515 g / mol$. Another important term to define is ionic strength μ of a solution containing a variety of ions with different molar concentration c_k and valences z_k, which is calculated by the equation (5.1).

$$\mu = \frac{1}{2}\sum_k c_k \left(z_k\right)^2 \tag{5.1}$$

The activity coefficient for an ion k with valence z_k in an aqueous solution depends on the ionic strength of the solution as defined by the **extended** Debye-Hückle equation at 25 $°C$ (valid in the range $0 \le \mu \le 0.1 M$), named after Peter Debye (1884–1966) and Erich Hückel (1896–1980):

$$log(\gamma_k) = \frac{-0.51(z_k)^2\sqrt{\mu}}{1+\alpha_k\sqrt{\mu}/305} \tag{5.2}$$

Where α_k is the effective radius of the hydrated ion (pm). For equation (5.2), μ should be in M.

Let us define the solubility product K_{sp} for a general hydrolyzed ionic salt in equilibrium, with metallic M (valence z^+) and nonmetallic N (valence z^-) components. The stoichiometry reaction for the dissolution is: $M_p^{z^+} N_q^{z^-}(s) + H_2O \rightarrow pM^{z^+}(Aq) + qN^{z^-}(Aq)$.

$$K_{sp} = (c_{M^{z+}}\gamma_{M^{z+}})^p (c_{N^{z-}}\gamma_{N^{z-}})^q \tag{5.3}$$

To illustrate all these magnitudes, we will calculate them for two common ionic solutions: **NaCl**(s) and **$CaCl_2$**(s).

Example 1 – Let us calculate the activity coefficients for Na^+ and Cl^- ions in a solution formed by dissolving $7.25 g$ of **NaCl**(s) in $1\ L$ of water. Equation (5.4) represents the stoichiometry for this reaction. The number of **NaCl** moles in the solution will be $n_{\mathbf{NaCl}} = W_{\mathbf{NaCl}} / MW_{\mathbf{NaCl}} = 7.25 g / 58.44 g\, mol^{-1} = 0.124\, moles$. Therefore, the molar concentrations of Na^+ and Cl^- ions will be $c_{Na^+} = 124\, mM$ and $c_{Cl^-} = 124\, mM$. Hence, the ionic strength will be $\mu = 0.124 M$. As the radii of these hydrated ions are $\alpha_{Na^+} = 225\, pm$ and $\alpha_{Cl^-} = 150\, pm$, the activity coefficients are $\gamma_{Na^+} = 0.72$ and $\gamma_{Cl^-} = 0.70$. Therefore, the activity of these two ions in the solution are $a_{Na^+} = 89.28\, mM$ and $a_{Cl^-} = 86.80\, mM$. The solubility product is $K_{sp} = (a_{Na^+})(a_{Cl^-}) = 7.75 \times 10^3\, (mM)^2$.

$$NaCl(s) + H_2O \rightarrow Na^+(Aq) + Cl^-(Aq) \tag{5.4}$$

Example 2 – Given that $MW_{\mathbf{CaCl_2}} = 147.02\,gmol^{-1}$, there are $n_{\mathbf{CaCl_2}} = 2.52\times10^{-3}\,moles$ in $0.37\,g$ of **$CaCl_2$**(s). Hence, if $0.37\,g$ of this salt is dissolved in $100\,mL$ of water (also referred to as at 0.37 %), the final molar concentration will be $c_{\mathbf{CaCl_2}} = 25.2\,mM$ of **$CaCl_2$**. What will these activities be for the ions in this aqueous solution? The molar concentration of these ions will be $[Ca^+] = 25\,mM$ and $[Cl^-] = 25\,mM \times \dfrac{2\,mol(Cl^-)}{1\,mol(\mathbf{CaCl_2})} = 50\,mM$. The ionic strength will be $\mu = 0.075\,M$. The radius of the hydrated Ca^{2+} is $\alpha_{Ca^{2+}} = 300\,pm$. Hence, the activity coefficients will be $\gamma_{Ca^{2+}} = 0.78$ and $\gamma_{Cl^-} = 0.75$. The solubility product is $K_{sp} = (c_{Na^+}\gamma_{Na^+})(c_{Cl^-}\gamma_{Cl^-})^2 = 1.43\times10^3\,(mM)^3$.

These are the protocols used in my laboratory (https://nmd.fiu.edu/) to prepare the CSF and internal solution:

Artificial CSF – In many experimental situations, it is necessary to simulate the interstitial fluid (e.g., in acute slice recording or in vivo recording via a craniotomy). To that end, the cerebral spinal fluid (CSF), which has a similar chemical composition, is used. A typical artificial CSF (aCSF) solution contains ($g\,/\,L$): 7.25 **NaCl** (58.44), 0.37 **KCl** (74.55), 0.49 **$MgSO_4$-$7H_2O$** (246.52), 0.17 **NaH_2PO_4-H_2O** (137.99), 0.37 **$CaCl_2$-$2H_2O$** (147.02) and 1.85 **$NaHCO_3$** (84.01). The MW of these salts are in brackets ($g\,/\,mol$). The osmolality of the aCSF solution should be $300-315\,mOsm\,/\,kg$. The pH should be between 7.3–7.4. Bubbling with 95 %-CO_2/5 %-O_2 mixed gas, and adding if necessary HEPES buffer, is used to stabilize pH. To have an ideal artificial CSF, it is recommended to create two stock solutions, one containing only concentrated **$CaCl_2$-$2H_2O$** and the other with the following solutes (**NaCl**, **KCl**, **$MgSO_4$-$7H_2O$** and **NaH_2PO_4-H_2O**). They should be mixed on the same day as the experiments. Finally, $1\,L$ of aCSF must also contain $1.2\,g$ of glucose (Glc, **$C_6H_{12}O_6$**), which should be added and stirred for $5\,min$ the day of the experiment together with the **$NaHCO_3$**.

Internal solution – To simulate the cytosol in micropipettes for whole-cell/outside-out patch-clamp recordings, researchers use an *internal solution*. Either a Cs^+-based or a K^+-based approach is used to create the internal solution. Many factors, such as Ca^{2+} buffers (EGTA/BAPTA, HEPES), energy resources to fuel pumps (ATP, GPT) and pH adjusters (KOH, CsOH), must be carefully considered when creating the internal solution. Here, we will refer only to a particular protocol using the K^+-based approach that is composed ($g\,/\,L$) of: 2.43 **HEPES** (238.34), 27.48 **K-Gluconate** (234.27), 1.52 **KCl** (74.55), 1.03 **MgATP** (507.15), 1.12 **Na_2ATP** (551.15), 2.60 **Phosphocreatine** (sodium salt, 255.1), and 0.16 **NaGTP** (523.18). Preparation: **1)** Dissolve the amount of the corresponding chemicals into distilled water slightly less than 80% of the final volume, **2)** Titrate the solution with **KOH** until the pH becomes 7.2–7.3 (the final concentration of K^+ will depend on the total volume of **KOH**), and **3)** Keep $1\,ml$ aliquots with the final solution in the freezer ($-20\,^oC$) to be used within 3–6 months. Final osmolality should be $280-285\,mOsm\,/\,kg$; correct if needed with **ddH_2O**. The concentration of Ca^{2+} in the neuron cytosol is $\sim 80\,nM$; however, salts containing this cation is not usually incorporated because of its very small concentration.

The brain conducts electricity in the form of a superposition of fluxes of different ionic species (e.g., Na^+, K^+, Ca^{2+} and Cl^-). The net flux for the *k-th* ion depends on its electrochemical potential $\bar{\mu}_k = \mu_k + z_k F\varphi$, which has two components; the chemical potential determined by thermal motion $\mu_k = \mu_k^0(T,P) + RTln(a_k)$ and the electromotive force induced on the ion by the presence of electric potential φ. $\mu_k^0(T,P)$ represents the standard-state chemical potential (i.e., the activity is 1), which should be constant for processes occurring at constant temperature T (*isothermal*) and pressure P (*adiabatic*). If otherwise not specified, we will study only these particular types of processes. T and P are given in Kelvin (K) and Pascal (Pa), respectively. F is the Faraday constant (Chapter 2). The ideal gas constant $R = 8.31446261815324\, JK^{-1}mol^{-1}$ is defined as the product of the Avogadro's number N_A and the Boltzmann constant $k_B = 1.380649 \times 10^{-23} JK^{-1}$ (Ludwig Boltzmann, 1844–1906). Brain cells are continuously exchanging ions with the interstitial fluid while keeping equal pressure across their membrane. For a reversible cellular system with a chemical composition dynamically changing over time dn_k and subjected to an external transmembrane electrical potential $\Delta\varphi$ forcing a number of moles dn with valence z to leave the cell, it is helpful to use the differential variation of the Gibbs free energy (Josiah W Gibbs (1839 –1903): $dG = -SdT + VdP - Fz\Delta\varphi dn + \sum_k \mu_k dn_k$. Under isothermal and adiabatic assumptions, this equation simplifies to:

$$dG = -Fz\Delta\varphi dn + \sum_i \mu_i dn_i \tag{5.5}$$

Nernst equilibrium: Consider a system composed by two regions (intracellular "*i*"and extracellular "*e*") separated by a permeable membrane (Figure 5.2). The system is composed only of ion k, and it does not exchange this ion with his environment (i.e., it is closed). Let us assume that at time $t = 0$, the activities of ion k are a_k^i and a_k^e for the intra and extracellular space, respectively. At this time, an external electric potential difference is applied between the two regions: $\Delta\varphi = \varphi^e - \varphi^i$, forcing a number ions dn_k to move to the extracellular space. The system will evolve until it reaches the Nernst equilibrium (Walther H Nernst, 1864–1941). The Gibbs free energy for the entire system will be $dG = -Fz_k\Delta\varphi dn_k + (\mu_k^0 + RTln(a_k^i))dn_k^i + (\mu_k^0 + RTln(a_k^e))dn_k^e$. By definition, the standard-state chemical potential μ_k^0 for ion k should be the same in both regions and $dn_k^e = -dn_k^i = dn_k$. Hence, $dG = (-Fz_k\Delta\varphi - RTln(a_k^i) + RTln(a_k^e))dn_k$. As at the Nernst equilibrium $dG = 0$, the following relationship must hold:

$$E_k = \Delta\varphi = \frac{RT}{Fz_k} ln\left(\frac{a_k^e}{a_k^i}\right) \tag{5.6}$$

The magnitude E_k is called the equilibrium potentials for ion k. Therefore, at the thermodynamic equilibrium ion k follows a Boltzmann distribution: $a_k^e = a_k^i e^{(Fz_k E_k/RT)}$.

Table 5.1 shows typical molar concentrations of different ions in the extracellular (c^e) and intracellular (c^i) spaces of neurons. The ionic strengths (μ) for these particular media are also shown, as well as the calculated activity for all ions (Eq. 5.2).

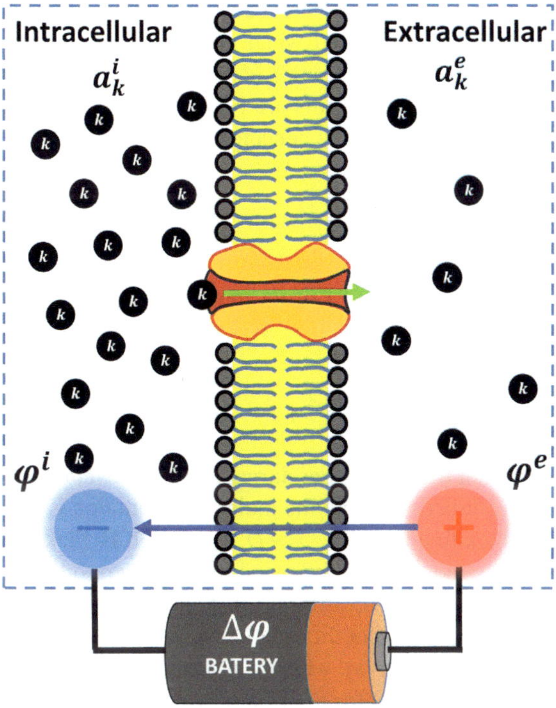

FIGURE 5.2 Physical foundations underlying the Nernst equation.

TABLE 5.1
Typical molar concentrations of ions in the extracellular and intracellular (cytosol) spaces for neurons

	Extracellular		Intracellular		Nernst Potential	
Ion	$c^e\,(mM)$	$a^e\,(mM)$	$c^i\,(mM)$	$a^i\,(mM)$	$E\,(mV)$	*approximated*
Na^+	145	103.06	10	7.60	+69.68	+71.47
K^+	5	3.46	140	104.59	−91.10	−89.06
Cl^-	110	76.24	8	5.98	−68.00	−70.00
Ca^{2+}	2	1.45	80×10^{-6}	61.7×10^{-6}	+134.50	+135.32
μ	$0.137\,M$		$0.080\,M$			

Note: The ionic strengths for these solutions were used to calculate the activities for the ions in their respective region. Note that there are large negatively charged proteins (impermeable anions) that are not considered. The Nernst equilibrium potential for each ion at $T = 310.15\,K$ is compared with that calculated from assuming an ideal solution. Note that for this calculation we also needed the radius of hydrated $K^+, \alpha_{K^+} = 150\,pm$.

Note that the differences between the Nernst equilibrium potentials calculated using equation (5.6) and those approximated when the molar concentrations are instead used $E_k^{app} = (RT / Fz_k) ln(c_k^e / c_k^i)$ are small ($<2.8\,mV$). In practice, with exception of intracellular Ca^{2+}, activity coefficients are between $0.7-0.9$; hence, we will use molar concentrations instead of activity henceforth, if not specified.

5.3 TRANSMEMBRANE IONIC FLUXES – POISSON-NERNST-PLANCK EQUATION FOR ELECTRODIFFUSION

The total flux density $\vec{J}_k$ ($mol\,m^{-2}\,s^{-1}$) of an inorganic ion k (Table 5.1) across the membrane is determined by three additive forces: **1)** a diffusional (*diff*), **2)** an electrical (*ele*), and **3)** a primary (*p*).

$$\vec{J}_k = \vec{J}_k^{(diff)} + \vec{J}_k^{(ele)} + \vec{J}_k^{(p)} \tag{5.7}$$

The diffusional force result from collisions during thermal movements which makes the ion to move toward the direction where the molar concentration is minimum, a molar flux described by the Fick first law, $\vec{J}_k^{(diff)} = -D_k \nabla c_k$ (Adolf E. Fick, 1829–1901). This spontaneous flux is a consequence of the second law of thermodynamics for isolated systems (maximal entropy). In 1905, Albert Einstein (1879–1955) proposed a linear relationship between the diffusion coefficient of an ion $D_k = u_k k_B T / ez_k$, given in m^2 / s, and its electric mobility u_k (m^2 / Vs), which was combined with the classical finding by George G. Stokes (1819–1903) for the friction of a spherical particle moving in a liquid with low Reynolds number to obtain the Einstein-Stokes equation (5.8).

$$D_k = \frac{k_B T}{6\pi\eta\alpha_k} \tag{5.8}$$

With η (Js / m^2) and α_k being the viscosity of the solvent and the radius of the hydrated ion, respectively. Table 5.2 provides the diffusion coefficient of most common monatomic ions in water.

The electrical force is linearly proportional to an electric potential gradient $\nabla\varphi$, with a linear coefficient depending on the electric conductivity. There are several approaches to calculate theoretically the electric conductivity σ_s (S / m) of a strong aqueous electrolytic solution, but all are approximated. To calculate σ_s, it is necessary first to define the molar limiting conductivity $\Lambda_{m,k}^0$ ($S\,cm^2\,mol^{-1}$, Table 5.2) of an ion k as a function of its diffusion coefficient D_k by means of the Nernst-Einstein Equation.

$$\Lambda_{m,k}^0 = z_k^2 F^2 D_k / RT \tag{5.9}$$

For non-ideal aqueous solutions, the electric conductivity is not related to $\Lambda_{m,k}^0$, but to the actual molar conductivity $\Lambda_{m,k}$ by the Kohlrausch's law (Friedrich WG Kohlrausch, 1840–1910): $\sigma_s = \sum_k \Lambda_{m,k} c_k$, with $\Lambda_{m,k}$ empirically defined as $\Lambda_{m,k} = \Lambda_{m,k}^0 - \left(K / |z_k|^{3/2}\right)\sqrt{c_k}$. The coefficient K is not trivially defined for any given electrolyte. A simplified parametrization for σ_s as a result of the Kohlrausch's law was proposed by C Anthony J Appelo (**Software II.1** – PHREEQC).

TABLE 5.2
The diffusion coefficients D_k and the molar limiting conductivity $\Lambda^0_{m,k}$ for different monatomic ions

Ion	$D_k \times 10^{-9} (m^2/s)$	$\Lambda^0_{m,k} (Scm^2/mol)$	$(\gamma_k)^{a_k}$
Na^+	1.330	50.0	0.82
K^+	1.960	73.6	0.80
Cl^-	2.030	76.2	0.80
Ca^{2+}	0.793	119.1	0.87

Note: Calculations by **Software 5.1** – PHREEQC (Version 3, www.usgs.gov/software/phreeqc-version-3) – A computer program for speciation, batch-reaction, one-dimensional transport, and inverse geochemical calculations. The diffusion coefficients are taken from the thermodynamic database "phreeqc.dat". https://doi.org/10.3133/tm6A43.

$$\sigma_s = \frac{F^2}{RT}\sum_k D_k z_k^2 (\gamma_k)^{a_k} c_k \tag{5.10}$$

With $a_k = 0.6/\sqrt{|z_k|}$ for electrolytic solutions with an ionic strength $\mu \lesssim 0.36$. For those ions (Na^+, K^+ and Cl^-) crucial in determining the value of σ_s, the parameter $(\gamma_k)^{a_k} \approx 0.8$ (Table 5.2). As we are using molar concentrations instead of activity, we will ignore this scaling factor hereafter. Based on the Ohm's law, we will determine the component in the net flux for ion k due to the electrical force:

$$\vec{J}_k^{(ele)} = -\frac{Fz_k}{RT} D_k c_k \nabla\varphi \tag{5.11}$$

Because of the mass conservation, we can define a continuity equation (5.12) relating dynamic changes in the molar concentration c_k of the ion species k in a region R to the divergence of its net flux $\vec{J}_k$:

$$\nabla \cdot \vec{J}_k + \frac{\partial c_k}{\partial t} = 0 \tag{5.12}$$

By combining the equations (5.7) and (5.11), and the Fick's first law, we can obtain the Nernst-Planck equation:

$$\frac{\partial c_k}{\partial t} = \nabla \cdot \left(D_k \nabla c_k + \frac{Fz_k}{RT} D_k c_k \nabla\varphi \right) - \nabla \cdot J_k^{(p)} \tag{5.13}$$

Note that the total ion flux density can be expressed as the sum of the primary current and a term that is a function of the gradient of the electrochemical potential, $\vec{J}_k = -\frac{D_k c_k}{RT}\nabla\bar{\mu} + J_k^{(p)}$. In the absence of primary flux $\vec{J}_k^{(p)} = 0$, the molar concentration of a non-ionic solute ($z_k = 0$) will satisfy the Fick's second law of diffusion: $\frac{\partial c_k}{\partial t} = \nabla \cdot (D_k \nabla c_k)$.

To solve equation (5.13), the electric potential in the entire region R must be calculated at each time instant. To that end, Poisson's equation for electrostatics can be used.

$$\varepsilon \nabla^2 \varphi = -\rho \tag{5.14}$$

Parameter ε represents the electric permittivity of region R (Chapter 2). Trivially, the charge concentration in R can be defined as a function of the molar concentrations of all ionic species $\rho = F\sum_k z_k c_k$. The system of equations (5.13) and (5.14) is named the Poisson-Nernst-Planck (PNP) theoretical framework. ε is usually expressed as a factor of the electric permittivity for the vacuum $\varepsilon_o = 8.854 \times 10^{-12}\, F/m$; $\varepsilon = \varepsilon_r \varepsilon_o$. The factor ε_r, named *relative electric permittivity*, depends on the frequency of the applied electromagnetic field (Chapter 2). For the brain (Gabriel et al., 1996), it is approximately $10^5 \leq \varepsilon_r \leq 10^8$ for the frequency range in electrophysiology ($0 \leq \omega \leq 10\, kHz$). Hence, the electric permittivity of the brain is small ($\varepsilon < 10^{-3}\, F/m$), which leads to an *electroneutrality* principle from the Poisson equation (5.14); i.e., $\sum_k z_k c_k \approx 0$. Electroneutrality also means that charge-relaxation processes in the brain tissue take place at very small spatiotemporal scales. There have been several approaches to solve numerically the PNP system of equations, from the use of an implicit scheme with backward Euler type discretization (Mori and Peskin, 2009) to the proposal of a Kirchhoff-Nernst-Planck (KNP) scheme (Solbrå et al., 2018). The PNP framework has also been useful to create multidomain models of electrodiffusion, a coarse-grained approach in physiology that represents molar concentrations and electric potentials for a continuous bulk of cellular compartments (Mori, 2015).

Given equation (5.10), an extracellular aqueous solution with molar concentrations (Table 5.1) and diffusion coefficients (Table 5.2) will have a relatively high electric conductivity, $\sigma_s = 13.23\, mS/cm$. However, the brain consists of a very dense suspension of cells, with polarized double-layer (membrane), in the extracellular aqueous solution, forming some sort of spherical colloid. Therefore, the electric conductivity must be corrected by the presence of this cells bulk. Assuming that the membranes of these cells are very thin compared to the distance between cells, we can use a theoretical correction for the electric conductivity proposed by Keh and Hsu (2002):

$$\sigma = \sigma_s \left(1 - \frac{3\vartheta}{2 + \vartheta}\right) \tag{5.15}$$

The parameter ϑ represents the cell volume fraction in the entire suspension. For the brain, the ECS/cell volume fraction is about 30 %; hence, $\vartheta \approx 0.7$, resulting in a final brain conductivity of $\sigma \approx 3\, mS/cm$. Measurements of the electric conductivity in the brain can be obtained by combining the four-electrode method (Schwann, 1963) and some models of it as a volume conductor. To characterize anisotropies and inhomogeneities in the electric conductivity of the cerebral cortex of rats, Goto et al. (2010) introduced a method based on the latest advances for both microelectrode fabrication, an anisotropic piecewise homogeneous volume conductor model and a

TABLE 5.3
A summary of previous measurements of the electric conductivity in the brain

Study	Animal	Conductivity (mS/cm)			
1963 – Ranck	Rabbit	2.73 – 3.62 (5 – *cycles / s*) 3.75 – 5.17 (5,000 – *cycles / s*)			
1968 – Li et al.	Cat	1.66 – 1.96 (36°C)			
		Layer	**Vertical (SD)**	**Mediolateral (SD)**	**Anteropost (SD)**
1979 – Hoeltzell & Dykes	Cat	**I**	1.7	–	–
		II/III	5.5(1.9)	2.9(1.2)	4.9(0.8)
		IV	5.0(1.2)	4.1(1.3)	2.9(0.6)
		V	–	–	–
		VI	5.5(1.6)	3.6(1.4)	3.5(1.1)
2007 – Logothetis et al.	Monkey	3.04 – 6.06			
		Layer	$\sigma_{\parallel}(SD)$	$\sigma_{\perp}(SD)$	
2010 – Goto et al.	Rat	**II/III**	3.19(0.43)	2.31(0.56)	
		IV	3.25(0.67)	2.40(0.93)	
		V	3.53(0.63)	2.28(0.47)	
		VI	2.94(0.62)	2.68(0.67)	

Note: References in this table are available in Goto et al. (2010).

nonlinear least-square optimization strategy. The authors found covariations of σ with the organization of myelin. Table 5.3 summarizes previous measurements of the brain electrical conductivity, which is to some extent in agreement with the theoretical prediction.

5.4 THE CONSTANT FIELD APPROACH AND THE GOLDMAN/GIBBS-DONNAN EQUILIBRIUM

The cellular membrane is a stable phospholipid bilayer of about $\Delta \cong 4\,nm$ in thickness, which separates the ECS and the cytosol. It is the most important cellular element in living organisms. Because of the spontaneous thermal movement of the phospholipids defining it, the membrane is represented by a *fluid mosaic model* (Figure 5.3). Phospholipids are amphiphilic molecules as they are formed by a polar hydrophilic head and a non-polar hydrophobic section (fatty acid). Membranes also comprise integral **(1)** and peripheral **(2)** proteins playing important roles in different cellular functions such as transmembrane communication (channels/transporters/pumps) and intra/inter-cellular signaling. Peripheral proteins could be either cytosolic (e.g., Phospholipases: $cPLA_2$ and PLC) or secretory (e.g., Phospholipase: $sPLA_2$). Together with cytosolic actin protein, a type of sterol (cholesterol) in the membrane helps maintain cellular shape and its overall cytoskeleton stable. Carbohydrates are also bound to some of these proteins and

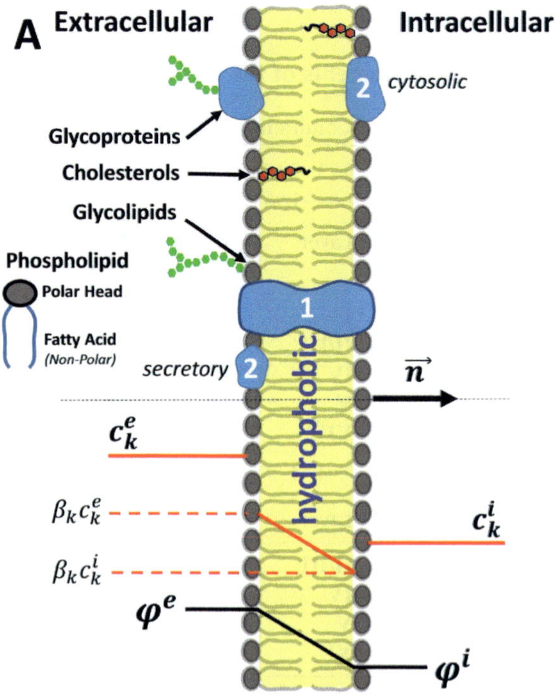

FIGURE 5.3 Schematic representations of the membrane with its main components.

lipids creating glycoproteins and glycolipids, respectively. We have been defining cellular membranes as semipermeable, which means ions can cross it due to the electrochemical force through not only channels/pores (integral proteins) but also gaps originated from lipid thermal movements. Ions are translocated against their electrochemical gradients by means of ATPase pumps and (co/counter)-transporters. The inner section of the membrane is non-polar and the intracellular/extracellular aqueous solution surrounding it is polar. In addition, transmembrane channels are selectively permeable to specific ions and their permeability magnitude varies among ion species and neuron type. Therefore, the solubility of ions inside the membrane is lower than that in the aqueous solution; hence, it is helpful to define a *partition coefficient* β_k for ion k, which describes jumps in its concentration across the membrane due to differences in solubility.

If neurons are at resting, the absence of primary currents ($\vec{J}_k^{(p)} = 0$) implies:

$$\vec{J}_k = -D_k \nabla c_k - \frac{F z_k}{RT} D_k c_k \nabla \varphi \tag{5.16}$$

Let us solve equation (5.16) inside the membrane. Under the assumptions that inside the membrane (Figure 5.3): **1)** ions flow in the direction $\vec{n}$ perpendicular to it, **2)** the electric field is constant, i.e., $\nabla \varphi = (\varphi^e - \varphi^i)\vec{n} / \Delta = V_m \vec{n} / \Delta$, and **3)** the concentration of ion k decreases linearly from $\beta_k c_k^e$ at the border with the ECS to $\beta_k c_k^i$ at the border with the cytosol, we can transform (5.16) into (5.17).

$$\int_0^{\Delta} \frac{Fz_k V_m}{RT\Delta} dn = -\int_{c_k^i}^{c_k^e} \frac{dc_k}{\left(\frac{J_k RT}{Fz_k P_k V_m} + c_k \right)} \tag{5.17}$$

The Goldman–Hodgkin–Katz (GHK) equation (5.18) constitutes the analytical solution of equation (5.17), which describes the flux across the membrane of ion k given the transmembrane difference in its concentration and a membrane resting potential V_m. The coefficient $P_k = D_k \beta_k / \Delta$ represents the permeability (m/s) of the membrane to ion k.

$$J_k = \frac{P_k F z_k V_m}{RT} \frac{\left(c_k^i - c_k^e e^{-Fz_k V_m / RT} \right)}{1 - e^{-Fz_k V_m / RT}} \tag{5.18}$$

To transform ion flux J_k into electric current density i_k (A/m^2), it is necessary to multiply equation (5.18) by the change of the ion k; i.e., $i_k = Fz_k J_k$. At the Goldman equilibrium potential V_m (David E Goldman, 1910–1998), the total electric current density crossing the membrane must be zero, i.e., $\sum_k i_k = 0$. As three monovalent ion species (Na^+, K^+, and Cl^-) contribute to the resting membrane potential, the following equation is obtained:

$$V_r = \frac{RT}{F} ln \left(\frac{P_{Na}[Na^+]_e + P_K [K^+]_e + P_{Cl}[Cl^-]_i}{P_{Na}[Na^+]_i + P_K [K^+]_i + P_{Cl}[Cl^-]_e} \right) \tag{5.19}$$

Note that if the membrane is only permeable to a particular ion, equation (5.19) leads to the Nernst potential for that ion. This is as the Nernst equation can be also deduced by imposing $J_k = 0$ in equation (5.18). Initial estimations of the ion permeability in the giant axon of the squid indicated that at rest $P_K : P_{Na} : P_{Cl} = 1:0.04:0.45$; however, at the time of the action potential (AP) peak $P_K : P_{Na} : P_{Cl} = 1:20:0.45$ (Hodgkin and Katz, 1949). Given the typical molar concentrations in neurons (Table 5.1), we obtain the following resting membrane potential $V_r = -68.9\,mV$ at $T = 310\,K$ and $V_{peak} = 53.7\,mV$. As E_{Cl}^{app} is close to the resting membrane potential (Table 5.1), the impact of Cl^- to V_r is insignificant. In addition, its permeability remains the same at the AP peak. Therefore, it is frequent to see equation (5.19) without the Cl^- component ($V_r = \frac{RT}{F} ln \left(\frac{P_{Na}[Na^+]_e + P_K [K^+]_e}{P_{Na}[Na^+]_i + P_K [K^+]_i} \right) = -68.5\,mV$) and this ion specie is not included in theoretical models for the AP genesis. Cl^- anions are very important in the homeostatic and excitation/inhibition balance. Increases in the expression of the electroneutral K^+/Cl^- cotransporter (KCC) during neuronal development implies changes in $[Cl^-]_i$. These changes can move E_{Cl} from values less negative than V_r to values more negative than it, a fact that makes γ-aminobutyric acid (GABA) neurotransmission to change qualitatively from excitatory to inhibitory.

The membrane permeability to Na^+ and K^+ cations at rest is determined by the presence of the so-called "leak channels". Two pore-domain channel family (K2P) with four transmembrane helices (4TM) is the most commonly accepted leak channels for K^+ (Goldstein et al., 2005). These 4TM/2P channels, also known as tandem pore-domain channels (TWIK, TREK, TASK, TALK, THIK, and TRESK), play a very important role in pH-based regulation of V_r and has been the target for the development of anesthetics. For Na^+, leak is linked to the sodium leak channel non-selective protein (NALCN), a member of the 24-transmembrane domain superfamily (Ren, 2011). The pore-forming α subunits in these channels consist of four homologous domains (I–IV), with each domain having six transmembrane segments (S1–S6). However, many other voltage/ligand-gated channels might also contribute to V_r in a neuron-dependent manner. For example, the resting membrane potential could shift up to 10 mV in the apical dendrites of layer 5 neocortical pyramidal cells due to the presence of a hyperpolarization-activated cyclic nucleotide-gated (HCN) channel. The HCN channel (HCN1-4) is a subfamily of the pore-loop cation channel super-family. The gating kinetics for the current I_h produced by the HCN1 channel is provided in Table 6.1.

The overall molar concentrations in the intracellular and extracellular bulks are electroneutral (Figure 5.4). However, molar concentrations (Table 5.1), which are responsible for the membrane potential, do not satisfy the electroneutrality principle in close proximity to the membranes. The superficial charge ρ^{ext} created by the flow of ions across the membrane will create an electric potential φ and a redistribution of the free ions in the electrolyte to try to screen that superficial charge according to Coulomb forces (Charles-Augustin de Coulomb, 1736–1806); i.e., $\vec{F}_k^e = -\rho_k \nabla\varphi = F z_k c_k \nabla\varphi$. If the electrolyte is considered in thermodynamic equilibrium at each time instant, the molar concentration of ion k will follows a Boltzmann distribution $c_k = \bar{c}_k e^{(-Fz_k\varphi/RT)}$, with $\bar{c}_k$ representing the mean molar concentration of the free ion k in the solution. Therefore, the electric potential in the electrolyte, assuming a weak-coupling limit ($Fz_k\varphi \ll RT \;\rightarrow\; e^{(-Fz_k\varphi/RT)} \approx 1 - Fz_k\varphi/RT$), will satisfy the Debye–Hückel equation (5.20).

$$\nabla^2\varphi = \lambda_D^{-2}\varphi - \frac{\rho^{ext}}{\epsilon} \tag{5.20}$$

There will be an excess of charge in close proximity to the membrane as a consequence of the ρ^{ext} screening. However, the difference between the ion concentration for each species (Na^+, K^+, and Cl^-) in the bulk and that nearby the membrane is negligible ($\sim 0.01\,mM$).

Equation (5.19) links differences in molar concentrations for Na^+, K^+, and Cl^- with the resting membrane potential V_m. Let us discuss the physical foundations for the establishment of these molar concentrations. We will demonstrate that the molar concentrations at the equilibrium are due to two main causes: **i)** the existence of impermeable negatively charged macromolecules in the neuronal cytosol and **ii)** the

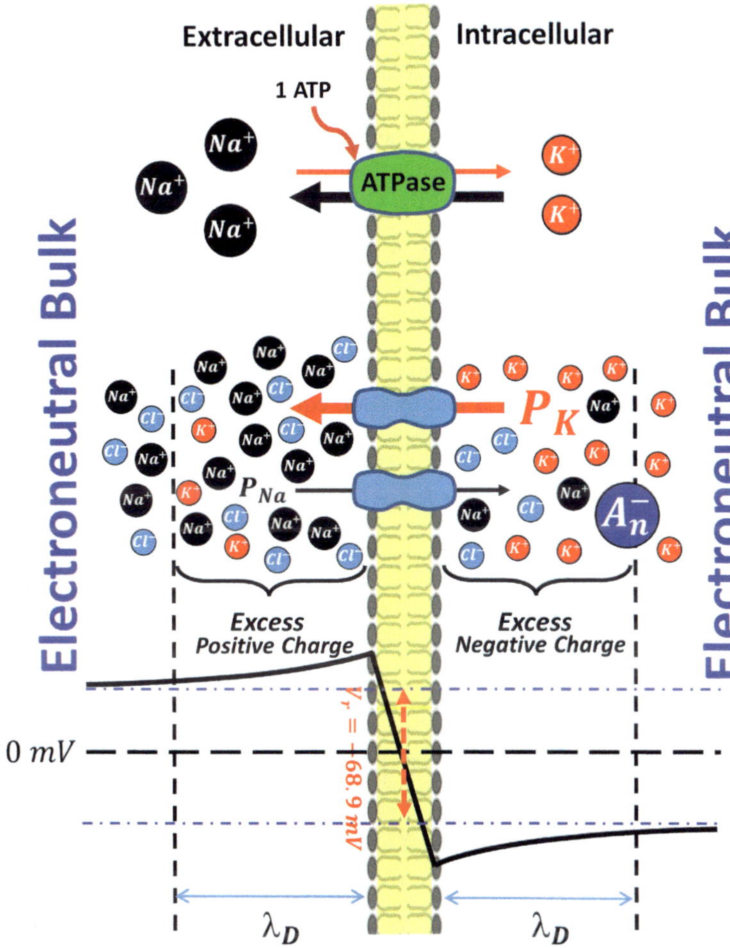

FIGURE 5.4 Definition ion gradients established by a Na-K ATPase pump and redistributed in the electrolytes according to the Debye–Hückel theory.

Note: The parameter $\lambda_D = \sqrt{\epsilon RT / F^2 \sum_k z_k^2 \bar{c}_k}$ (m) is known as the Debye length.

activity of a Na-K ATPase pump in combination with voltage-dependent differences in the membrane permeability to these ions.

i) If hypothetically we have an ECS and a cytosol with exactly the same molar concentrations of the **NaCl** and **KCl** solutes, the resting membrane potential (5.19) will be zero regardless the permeability of the membrane to these three ions. However, in reality the neuronal cytosol contains macromolecule A_n^- with n negative charges that cannot permeate the membrane. These impermeable charges will exert electrostatic forces on the ions, creating a redistribution of them across the membrane. This is a very important property in electrophysiology as it creates a resting membrane potential different from zero in the absence of active transport. The Gibbs-Donnan equilibrium (Frederick G. Donnan, 1870–1956) establishes the final distributions of ions and the related electric potential difference E_{GD} across the membrane, which should satisfy $E_{GD} = E_{Na} = E_K = E_{Cl}$. Hence, the following relationship between the molar concentrations must be held at the Gibbs-Donnan equilibrium:

$$R_{GD} = \frac{[Na^+]_e}{[Na^+]_i} = \frac{[K^+]_e}{[K^+]_i} = \frac{[Cl^-]_i}{[Cl^-]_e} \tag{5.21}$$

R_{GD} is known as the Gibbs-Donnan ratio. In addition, because of the bulk electroneutrality in both the ECS and cytosol, equations (5.22) and (5.23) must be also satisfied.

$$n[A_n^-]_i + [Cl^-]_i = [Na^+]_i + [K^+]_i \tag{5.22}$$

$$[Cl^-]_e = [Na^+]_e + [K^+]_e \tag{5.23}$$

After some mathematical steps, it can be demonstrated that:

$$R_{GD} = \frac{2([Na^+]_e + [K^+]_e)}{n[A_n^-]_i + \sqrt{(n[A_n^-]_i)^2 + 4([Na^+]_e + [K^+]_e)^2}} \tag{5.24}$$

$$V_{GD} = \frac{RT}{F} ln(R_{GD}) \tag{5.25}$$

Using typical molar concentrations in neurons (Table 5.1) and considering that cells contain approximately $[A_n^-]_i \approx 5\,mM$ with $n = 50$, we can demonstrate the $V_{GD} \approx -20.3\,mV$ at $T = 310\,K$. Therefore, $-20.3\,mV$ out of the resting membrane potential $V_r = -68.9\,mV$ are just due to the existence of large impermeable anions A_n^- inside the neurons.

ii) The Na-K ATPase pump, in combination with differences in the voltage-dependency of ion permeability $P_k(V_m)$ across the membrane (Chapter 6), is responsible for the remaining $-48.6\,mV$. Increases of $[Na^+]_i$ and $[K^+]_e$ above its value at rest causes activation of the pump, following the 9-step sequence in Figure 5.5. The pump consumes about 75–85 % of the total ATP hydrolysis. It has been suggested that the pump transport K^+ ions simultaneous rather than "*ping pong*" effect (Sachs, 1977).

Figure 5.5 shows the Albers-Post scheme for the Na-K ATPase pump. Characterization of this pump was possible thanks to the work by Jens Christian Skou, for which he shared the Nobel prize in Chemistry in 1997. The pump has nine sequential steps:

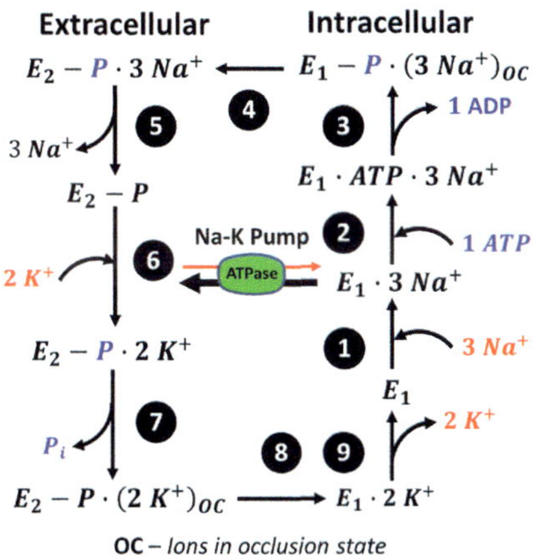

FIGURE 5.5 Na-K pump ATP hydrolysis.

(1) Binding to three Na^+ ions with high-affinity on the intracellular side of the pump in state E_1, a site that has low-affinity for K^+ ions; **(2)** Binding of ATP (high-affinity) for the phosphorylation of aspartate; **(3)** The occlusion of the $3\,Na^+$ ions; **(4)** The protein suffers a conformational change $E_1 \rightarrow E_2$ with extracellular binding sites with high-affinity for K^+ ions and low-affinity for Na^+ ions; **(5)** Dissociation of the Na^+ ions to the extracellular space; **(6)** Binding of K^+ ions; **(7)** Hydrolysis of aspartyl phosphate to occlude the $2\,K^+$ ions; **(8)** Conformational changes $E_2 \rightarrow E_1$ with the inward transport of K^+; and **(9)** Dissociation of the K^+ ions to the intracellular space. The α-unit of the pump is sensitive to cardiotonic steroids, e.g., ouabain and digoxin. The full stoichiometry reaction of the Na-K ATPase pump is described in equation (5.26).

$$3Na^+_{cyt} + 2K^+_{ECS} + 1\,ATP_{cyt} \leftrightarrow 3Na^+_{ECS} + 2K^+_{cyt} + 1\,ADP_{cyt} + 1\,P_{i\,cyt} \tag{5.26}$$

Calculating the fluxes of Na^+ and K^+ ions due to active pumping by the Na-K ATPase constitutes a very challenging task. However, a simplified expression (5.27) can be obtained if we assume rapid equilibrium partial reactions (Smith and Crampin, 2004).

$$i^p_{Na/K} = \upsilon^p_{Na/K} F \tag{5.27}$$

With $\upsilon^p_{Na/K} = \dfrac{\alpha_1^+\alpha_2^+\alpha_2^+\alpha_4^+ - \alpha_1^-\alpha_2^-\alpha_3^-\alpha_4^-}{\Sigma}$

$$\alpha_1^+ = \frac{k_1^+\left(\overline{[Na^+]}_i\right)^3}{\left(1+\overline{[Na^+]}_i\right)^3 + \left(1+\overline{[K^+]}_i\right)^2 - 1} \qquad \alpha_2^+ = k_2^+$$

$$\alpha_3^+ = \frac{k_3^+\left(\overline{[K^+]}_e\right)^2}{\left(1+\overline{[Na^+]}_e\right)^3 + \left(1+\overline{[K^+]}_e\right)^2 - 1} \qquad \alpha_4^+ = \frac{k_4^+\overline{[MgATP]}}{1+\overline{[MgATP]}}$$

TABLE 5.4
A summary of parameters Na-K ATPase pump model under equilibrium partial reactions

Parameter	k_1^+, s^{-1}	$k_1^-, s^{-1}mM^{-1}$	k_2^+, s^{-1}	k_2^-, s^{-1}	k_3^+, s^{-1}	$k_3^-, s^{-1}mM^{-2}$	k_4^+, s^{-1}	k_4^-, s^{-1}
Value	1050	172.1	481	40	2000	79.3×10^3	320	40
Parameter	$K^e_{d,Na}, mM$	$K^i_{d,Na}, mM$	$K^e_{d,K}, mM$	$K^i_{d,K}, mM$	K^e_{MgATP}, mM	Δ		
Value	15.5	2.49	0.213	0.5	2.51	-0.031		

$$\alpha_1^- = k_1^-\left[MgADP\right] \qquad \alpha_2^- = \frac{k_2^-\left(\overline{\left[Na^+\right]_e}\right)^3}{\left(1+\overline{\left[Na^+\right]_e}\right)^3+\left(1+\overline{\left[K^+\right]_e}\right)^2-1}$$

$$\alpha_3^- = \frac{k_3^-\left[P\right]\left[H^+\right]}{1+\overline{\left[MgATP\right]}} \qquad \alpha_4^- = \frac{k_4^-\left(\overline{\left[K^+\right]_i}\right)^2}{\left(1+\overline{\left[Na^+\right]_i}\right)^3+\left(1+\overline{\left[K^+\right]_i}\right)^2-1}$$

$$\overline{\left[Na^+\right]_i} = \left[Na^+\right]_i / K^{i,0}_{d,Na} \qquad \overline{\left[Na^+\right]_e} = \left[Na^+\right]_e / K^{e,0}_{d,Na}$$

$$\overline{\left[K^+\right]_i} = \left[K^+\right]_i / K^i_{d,K} \qquad \overline{\left[K^+\right]_e} = \left[K^+\right]_e / K^e_{d,K}$$

$$\overline{\left[MgATP\right]} = \frac{\left[MgATP\right]}{K_{MgATP}} \qquad K^e_{d,Na} = K^{e,0}_{d,Na}\, e^{(1+\Delta)FV_m/3RT}$$

$$K^i_{d,Na} = K^{i,0}_{d,Na}\, e^{\Delta FV_m/3RT}$$

Δ is the voltage dependence partition factor and Σ is a sum of 16 positive product terms. In Smith and Crampin (2004), all kinetic parameters in the Na-K pump were fitted to experimental data and adjusted to $T = 310K$. We provide a summary in Table 5.4. A modification to the kinetic model for K^+ proposed by Castillo et al. (2015) may be also considered.

5.5 NEURONAL MEMBRANE: BASIC ELECTRIC CIRCUIT

Using the GHK equation (5.18) to predict the transmembrane flux of an ion J_k requires the knowledge of the membrane electric potential V_m, the permeability of the membrane to the ion (P_k) and the molar concentrations of the ion at both sides (c_k^i, c_k^e) of the membrane. In the steady-state condition of a neuron at rest (Goldman equilibrium), these magnitudes are constant, with $V_m = V_r$, P_k representing mainly leak channels and the molar concentrations as those in Table 5.1. However, during neuronal activation, these magnitudes departure from equilibrium following the evolution of an electrodiffusion system with high-nonlinear dynamics. Due to the existence of voltage-gated ion channels, $P_k(V_m)$ will change with time in a manner directly linked to the membrane potential (Chapter 6). Except in pathological conditions (e.g., cortical spreading depression), the electric currents i_k induced by changes in the ion

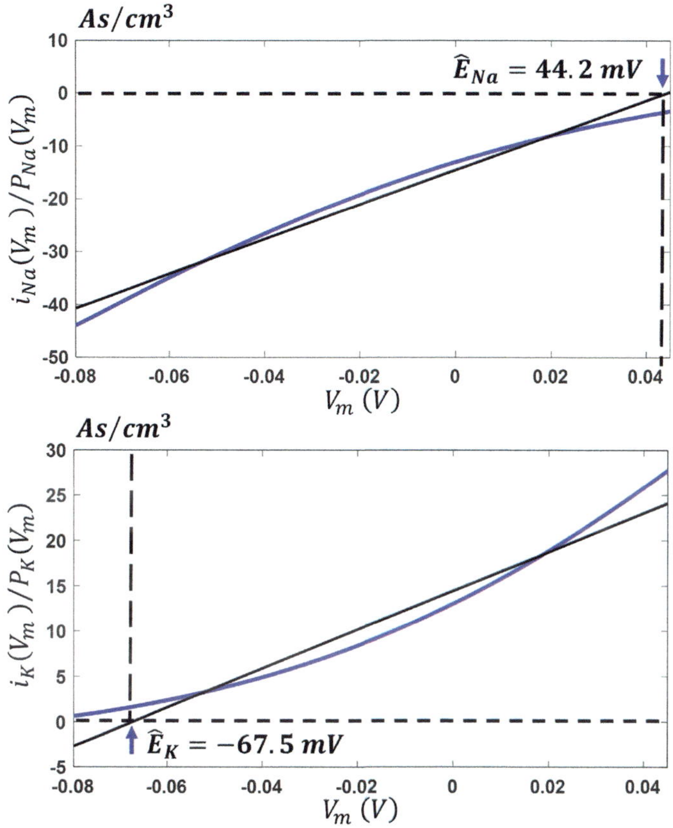

FIGURE 5.6 Linearity of the permeability-normalized GHK fluxes $(i_k(V_m)/P_k(V_m)$ for Na^+ and K^+ in the physiological range of the transmembrane electric potential $(-80 \leq V_m \leq 45mV)$.

permeability cause negligible variations in the molar concentrations of these ions at both sites of the membrane. Hence, the molar concentrations (c_k^i, c_k^e) in equation (5.18) can be considered constant. In the case of Na^+ ions, the dependency of the ratio $i_{Na}(V_m)/P_{Na}(V_m)$ with the membrane potential V_m (Figure 5.6, blue) is approximately linear (black, $R^2 = 0.98$) in the physiological range for APs $(-80 \leq V_m \leq 45mV)$.

$$i_{Na}(V_m) \approx \overbrace{P_{Na}(V_m) f_{Na}}^{g_{Na}(V_m)} (V_m - \hat{E}_{Na}) \tag{5.28}$$

with $f_{Na} = 0.33C/mVcm^3$ and $\hat{E}_{Na} = 44.2mV$

As for K^+, a similar linear relationship ($R^2 = 0.96$) is found (5.29).

$$i_K(V_m) \approx \overbrace{P_K(V_m) f_K}^{g_K(V_m)} (V_m - \hat{E}_K) \tag{5.29}$$

with $f_K = 0.24C/mVcm^3$ and $\hat{E}_K = -67.5mV$

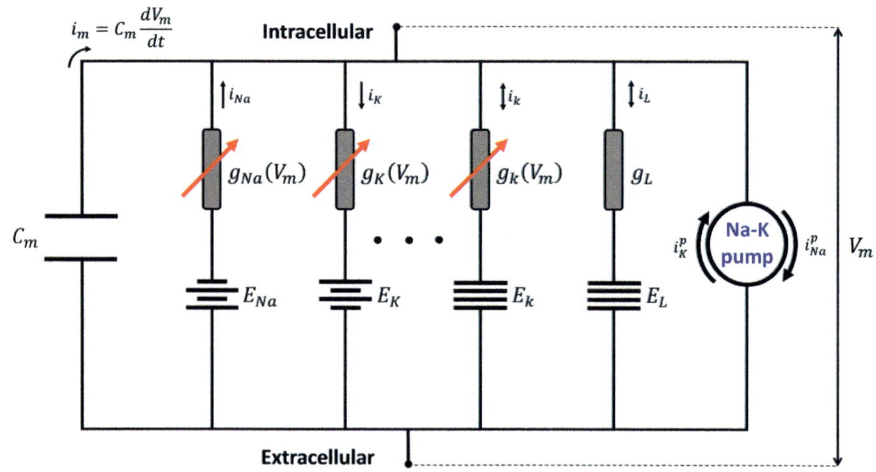

FIGURE 5.7 The parallel conductance model.

The voltage-dependent parameters $g_{Na}(V_m)$ and $g_K(V_m)$ are the conductances of the Na^+ and K^+ channel (mS/cm^2).

One of the most significant properties of the membranes is its capacity to temporarily store differential changes in electric charges dQ_m after a differential voltage difference dV_m is applied to it, referred as membrane capacitance $c_m = dQ_m / dV_m$. Charge is accumulated at both sides of the hydrocarbon chains of the phospholipids. c_m depends on the relative electric permittivity ($\epsilon_r \approx 2.2$) of the hydrocarbon chains and their average separation ($d \approx 2\,nm$); $c_m = \epsilon_r \epsilon_0 / d \approx 0.97 \mu F / cm^2$. The hydrocarbon chains are hydrophobic; hence, they also act as isolators ($10^9 \Omega cm^2$). However, electric currents leak across the membrane as the results of a variety of factors: thermal movements of proteins/lipids, ion leak channels, and water channels. Therefore, the membrane can also be characterized by a transmembrane resistance at rest $r_m \approx 40{,}000\ \Omega cm^2$.

The *parallel conductance model* (PCM) proposes that the transmembrane electric potential could be represented by an equivalent electric circuit with four current components in parallel (Figure 5.7): **1)** ionic currents i_k for each ion that result from their specific electromotive forces E_k and is regulated by a variable resistance $r_k = 1/g_k$ ($K\Omega cm^2$)(e.g., Na and K). **2)** a capacitive current $i_m = c_m dV_m / dt$, **3)** a leak current i_L, and **4)** the current generated by the Na-K ATPase pump (5.27). Applying the Kirchhoff laws for circuits, we can demonstrate that if the neuron is at rest (i.e., $dV_m / dt = 0$), only Na^+ and K^+ ion species are present, and we ignore the effect of the Na-K ATPase pump, the following equation can be used to calculate the resting membrane potential V_r.

$$V_r = \frac{g_{Na} E_{Na} + g_K E_K}{g_{Na} + g_K} \tag{5.30}$$

Defining the ratio $r_{Na/K} = g_{Na} / g_K = (P_{Na} f_{Na}) / (P_K f_K)$ and taking into account that $P_K : P_{Na} = 1 : 0.04$ at rest, we obtain:

$$V_r = \frac{r_{Na/K} E_{Na} + E_K}{r_{Na/K} + 1} = -61.94\,mV$$

The electric current generated by the Na-K ATPase pump ($i^p_{Na/K}$) is usually not included in the calculation of V_r; neither using the GHK model (5.18) nor the PCM model (5.30). To include the pump electric current, a correction to equation (5.19) was proposed by Mullins and Noda (1963) by assuming the system reaches the steady state quickly.

$$V_r = \frac{RT}{F} ln\left(\frac{P_{Na}\left[Na^+\right]_e + r_p P_K \left[K^+\right]_e}{P_{Na}\left[Na^+\right]_i + r_p P_K \left[K^+\right]_i} \right) \tag{5.31}$$

$r_p = 3/2$ represents the pump coupling ratio (i.e., number of Na^+ ions pumped out per K^+ ions pumped in). It can be alternatively written as (Armstrong, 2003):

$$V_r = \frac{RT}{F} ln\left(\frac{P_{Na}\left[Na^+\right]_e + P_K \left[K^+\right]_e + i^p_{Na/K} / F}{P_{Na}\left[Na^+\right]_i + P_K \left[K^+\right]_i} \right) \tag{5.32}$$

5.6 ION HOMEOSTASIS

The presence of impermeable macromolecules A_n^- inside the neurons will tend to drive water into the cell as a consequence of the osmotic pressure difference $\Delta\pi$ ($atm = 760\,mm(Hg)$), which can be linearly related to the transmembrane difference in the molar concentration of all solutes in the solutions Δc_s by the van't Hoff's law (Jacobus "Henry" van't Hoff Jr., 1852–1911): $\Delta\pi = \sigma RT \Delta c_s$. The factor σ describes the relation between the membrane permeability to the macromolecules with respect to its permeability to water $\sigma = 1 - P_s / P_{H_2O}$, and it is called the *reflection coefficient*. If the membrane is absolutely impermeable to all solutes in the solution $\sigma = 1$ and if solute's permeability is equal to that of the water, $\sigma = 0$. Whenever impermeable macromolecules A_n^- are present in the neuron cytosol, the final distributions of ions in two aqueous solutions separated by a semipermeable membrane are such that a difference in total solute concentration will be established (Gibbs-Donnan equilibrium):

$$\Delta c_s = n\left[A_n^-\right]_i \left(1 + \sqrt{1 - \frac{\left(\left[Na^+\right]_e + \left[K^+\right]_e\right)}{n\left[A_n^-\right]_i}} \right) \tag{5.33}$$

Hence, water will flow to the cytosol via water pores/channels and the cell will swell. According to the Starling equation (Ernest H Starling, 1866–1927), water volume flow J_v (cm/s) into the neuron will cease when the osmotic pressure $\Delta\pi$ equals the hydrostatic pressure difference ΔP.

$$J_v = L_p(\Delta\pi - \Delta P) \tag{5.34}$$

L_p is the hydraulic conductivity $(cm/atms)$. Based on Laplace's law, the hydrostatic pressure in the neuron is: $\Delta P = 4\pi k_m(1 - r_a/r)H(r - r_a)$ for changes in soma's radius r with respect to its resting value r_a. $H(\cdot)$ is the Heaviside function and k_m the spring constant of the membrane. To keep neurons away from the Gibbs-Donnan equilibrium, avoiding them to swell, the Na-K ATPase pumps continuously translocate leaking Na^+ and K^+ ions across the membrane at a cost of metabolic energy usage. Together with cation-chloride cotransporters (CCCs), local impermeant anions help establish the final chloride concentration in the cytosol.

There are many pores/channels in cell membranes that facilitate molecule diffusion. In the contest of ion homeostasis, the porins and aquaporins are the most important. A porin is β-barrel in the cell membrane that acts as a pore for molecules diffusion across the membrane. In contrast to most β-barrels, the hydrophobic residues of the β-barrels in porins are in contact with the surrounding lipids and their hydrophilic residues are oriented toward the aqueous interior of the pore. In general, porins are water-filled pores that participate in the passive transportation of hydrophilic molecules across the membrane. Porins are characterized by four parameters: **1)** the tilting angle β relative to the membrane, **2)** shear number S (i.e., the number of amino acid residues in each β strands), **3)** the number of β strands n, and **4)** the barrel opening radius R. These parameters are related by the following equations: $4\pi R cos(\alpha) = nb$ and $nb tan(\alpha) = Sa$, with peptide parameters $a - 3.3\dot{A}$ and $b - 4.4\dot{A}$.

On the other hand, aquaporins are water channel proteins. Therefore, aquaporins form pores in the membranes for the transport of water molecules. Aquaporins were discovered by Peter Agre (1949–present) and Roderick MacKinnon (1956-present), receiving the Nobel prize in Chemistry for it in 2003. Aquaporins allows the moving in and out of pure water molecules in cells. They are tetramers in the cell membrane formed from the clustering of four monomers. Each monomer acts as a water channel. The monomers are composed of membrane-spanning α-helices connected by five loops. They typically have a conserved asparagine–proline–alanine (NPA) motifs embedded in the membrane.

5.7 PROBLEMS AND EXERCISES

Problem 5.1 – Calculate the amount of $\mathbf{CaCl_2\text{-}2H_2O}$ to be added to a $1L$ of the internal solution referred in this chapter to reach the basal Ca^{2+} concentration in the cytosol $[Ca^{2+}] = 80\,nM$. What is the associate change in the concentration of Cl^-?

Problem 5.2 – Assuming that the Goldman equilibrium has been reached for a cellular membrane separating two $K_x^+ Na_{1-x}^+ Cl^-$ solutions, demonstrate equation (5.19) from (5.18).

Problem 5.3 – Assuming that the membrane is an infinite planar capacitor, demonstrate the relationship $c_m = \epsilon_r \epsilon_0 / d$.

Problem 5.4 – Demonstrate equation (5.30) using the Kirchhoff laws for circuits.

Problem 5.5 – Calculate the osmotic pressure across the neuron membrane separating the CSF and cytosol solutions (Table 5.1), if $5\,mM$ of impermeable macromolecules A_n^- are added to the cytosol. Note, each macromolecule has 50 negative charges.

REFERENCES

Armstrong CM. The Na/K pump, Cl ion, and osmotic stabilization of cells. *PNAS* 100(10), 6257–6262, 2003.

Castillo JP, Rui H, Basilio D, Das A, Roux B, Latorre R, Bezanilla F, Holmgren M. Mechanism of potassium ion uptake by the Na/K-ATPase. *Nat Commun* 6(7622), 1–8, 2015.

Gabriel S, Lau RW, Gabriel C. The dielectric properties of biological tissues: II. Measurements in the frequency range 10 Hz to 20 GHz. *Phys Med Biol* 41, 2251–2269, 1996.

Goldstein SA, Bayliss DA, Kim D, Lesage F, Plant LD, Rajan S. International Union of Pharmacology. LV. Nomenclature and molecular relationships of two-P potassium channels. *Pharmacol Rev* 57, 527–540, 2005.

Goto T, Hatanaka R, Ogawa T, Sumiyoshi A, Riera J, Kawashima R. An evaluation of the conductivity profile in the somatosensory barrel cortex of Wistar rats. *J Neurophysiol* 104, 3388–3412, 2010.

Hodgkin AL, Katz B. The effect of sodium ions on the electrical activity of the giant axon of the squid. *J Physiol* 108, 37–77, 1949.

Keh HJ, Hsu WT. Electric conductivity of a suspension of charged colloidal spheres with thin but polarized double layers. *Colloid Polym Sci* 280, 922–928, 2002.

Mori Y. A multidomain model for ionic electrodiffusion and osmosis with an application to cortical spreading depression. *Physica D: Nonlinear Phenom* 308, 94–108, 2015.

Mori Y, Peskin C. A numerical method for cellular electrophysiology based on the electrodiffusion equations with internal boundary conditions at membranes. *Commun Appl Math Comput Sci* 4(1), 85–134, 2009.

Mullins LJ, Noda K. The influence of sodium-free solutions on the membrane potential of frog muscle fibers. *J Gen Physiol* 47, 117–132, 1963.

Ren D. Sodium leak channels in neuronal excitability and rhythmic behaviors. *Neuron* 72(6), 899–911, 2011.

Sachs JR. Kinetic evaluation of the Na-K pump reaction mechanism. *J Physiol* 273, 489–514, 1977.

Schwann HP. Determination of biological impedances. *Phys Tech Biol Res* 6, 323–407, 1963.

Smith NP, Crampin EJ. Development of models of active ion transport for whole-cell modelling: Cardiac sodium–potassium pump as a case study. *Prog Biophys Mol Biol* 85, 387–405, 2004.

Software II.1 – PHREEQC (Version 3, www.usgs.gov/software/phreeqc-version-3) – A computer program for speciation, batch-reaction, one-dimensional transport, and inverse geochemical calculations. The diffusion coefficients are taken from the thermodynamic database "phreeqc.dat".

Solbrå A, Bergersen AW, van den Brink J, Malthe-Sørenssen A, Einevoll GT, Halnes G. A Kirchhoff-Nernst-Planck framework for modeling large scale extracellular electrodiffusion surrounding morphologically detailed neurons. *PLoS Comput Biol* 14(10), e1006510, 2018.

6 Neuronal Excitability

6.1 OVERVIEW

Chapter 5 introduced useful concepts to understand neurons at the equilibrium. However, the brain is in a permanent state of ongoing activation with neurons having merely transitory periods at the equilibrium. In fact, the brain of an individual under sensory deprivation will be never able to return to normal functioning if by chance all its neurons are at the equilibrium at a given moment of time. There are approximately 86 billion neurons in the human brain; hence, the probability for this situation to happen is practically zero. It is the continuous interaction among neurons through highly complex local and global connectivity webs, what keep them far from the equilibrium; and it is in that process that cognition and consciousness emerge. In this chapter, we will study a plethora of ionic channels that neurons count on to respond (i.e., depart from equilibrium) to stimulations by other neurons in the brain connectivity webs. These responses are manifested as changes in the transmembrane electric potential that depending on their spatiotemporal characteristics could be classified into: **i)** graded potentials and **ii)** action potentials (AP). Graded potentials are associated with postsynaptic currents in the branchy dendritic trees of neurons. APs take place basically in the funnel-type of ending segment of neurons soma. APs have a defined shape (i.e., 100 mV peak-to-peak amplitude and 2–3 msec duration), which is mostly conserved across neuron types, brain circuits, and species. We will present the mathematical foundations used to describe transient changes in the transmembrane electric potentials. We will also study mechanisms for ions active transport (e.g., co/counter-transport and pumps), which are critical to reestablish transmembrane ionic gradients after these postsynaptic and somatic currents take place.

6.2 HODGKIN–HUXLEY MODEL – THE ACTION POTENTIAL (AP)

The first indirect sign of changes in the transmembrane electric potential due the nerve impulse conduction was provided in 1848 by Emil du Bois-Reymond, who used an astatic galvanometer in an experiment with an "electric fish". One of his students, Julius Bernstein (1839–1917), demonstrated in 1902 that the AP was caused by sudden changes in the membrane permeability to ions (*"Membrane Theory of Electrical Potentials"*). Fifty years later, Alan L Hodgkin (1914–1998) and Andrew

DOI: 10.1201/9780367901509-6

F Huxley (1917–2012) combined the voltage-clamp technique and manipulations of ionic concentrations to measure these changes in membrane permeability to ions in the squid giant axon (Hodgkin and Huxley, 1952a), with particular interest in the intriguing overshoot (Huxley, 2002). Most importantly, they introduced a quantitative framework, known today as the Hodgkin–Huxley (H–H) model, to demonstrate that APs result from nonlinear voltage-dependent changes in the membrane conductance to Na^+ and K^+ ions in a particular temporal sequence. According to their data and model (Figure 6.1), APs in the nerve were initiated by a voltage-dependent Na^+ ion inward current that rapidly depolarizes the axons. This inward current was inactivated in about $1-2\,ms$. The axon repolarized to its resting membrane potential by the action of a voltage-dependent K^+ outward current, which also causes a subsequent temporary hyperpolarization. This pioneer work led them to receiving the

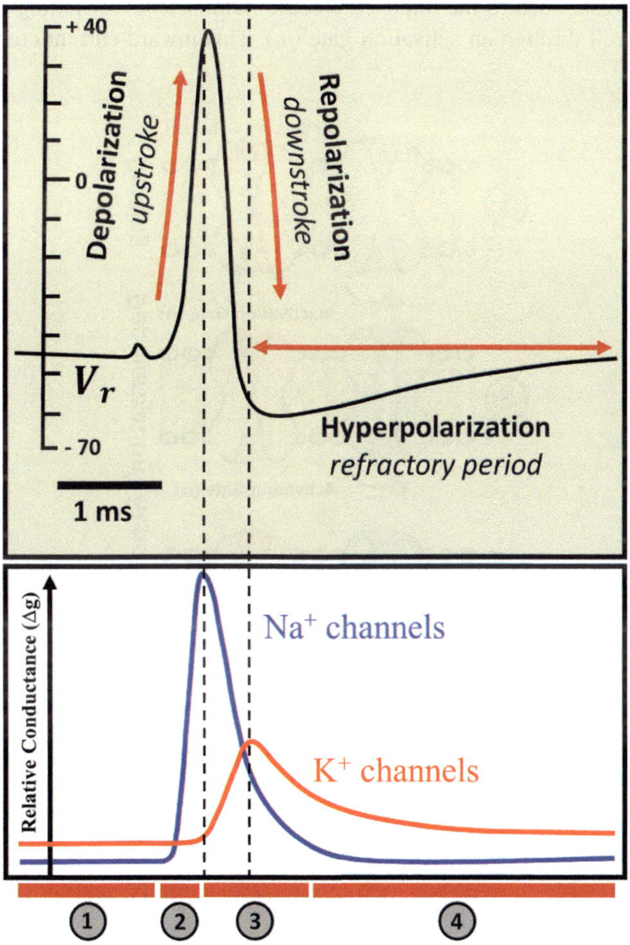

FIGURE 6.1 The AP phases with the associated Na^+ and K^+ ion currents.

Nobel Prize for physiology or medicine in 1963. Today, we know that the voltages-dependent channels referred by them were the fast Na^+ (Nav1.2) and delayed-rectifier K^+ (Kv2.1) voltage-gated channels. These two ion channels are members of more general families of ion channels later discussed in this chapter and have been well-known in electrophysiology for some time. Hence, we will use them to introduce what is known today as the more general H-H theoretical framework.

Based on voltage-clamp data and empirical reasoning, Hodgkin and Huxley proposed the existence of independent voltage-dependent gating in the membrane to Na^+ and K^+ ions (Figure 6.2). In their framework, the channel for Na^+ ions has a very fast activation gate and a relative slower inactivation gate. The channel for K^+ ions was considered slow and without an inactivation gate. At the equilibrium, although the inactivation gate was open the channel remains impermeable to Na^+ ions because the activation gate was closed. As soon as the neuron membrane was depolarized above certain threshold, a rapid conformational change in the channel occurs, as charged particles in it responded to the imposed electric field, selectively letting Na^+ ions to flow into the cell through an activation gate (m). This inward current created further

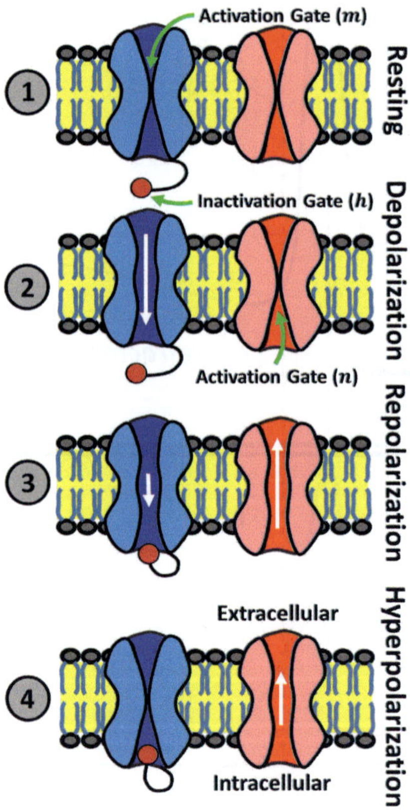

FIGURE 6.2 A schematic representation of the gate opening/closing sequence for Na^+ (blue) and K^+ (red) ions during the four different phases (Figure 6.1) of AP.

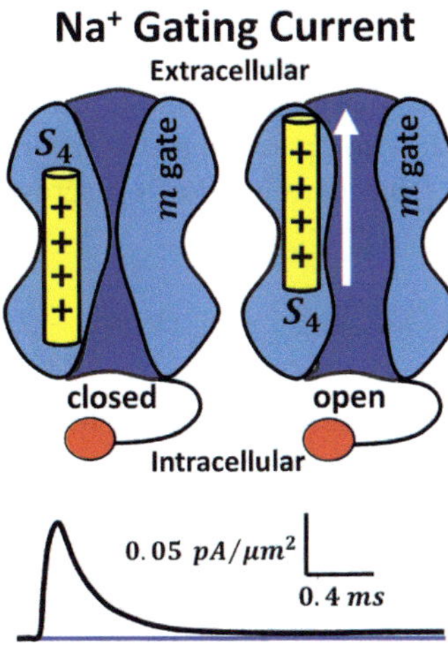

FIGURE 6.3 An illustration of the mechanism originating the Na^+ gating current (inset below, modified from Armstrong and Bezanilla, 1973) due to the displacement of the helices $S4$ in the Nav1.2 channel.

depolarization of the neuron membrane. As the membrane potential approaches the equilibrium potential for Na^+, this current gradually ceases, and it is finally terminated by the closing of the inactivation gate (h) for this channel. At the same time, the activation gate (n) for K^+ ions opens, letting this cation to flow out of the cell causing a repolarization of the membrane. The activation gate for K^+ remains open for certain period because of the absence of inactivation and a long activation time constant. At the same time, the inactivation gate for Na^+ has not yet returned to the open equilibrium state, resulting in the hyperpolarization of the neuron. During the hyperpolarization, the neurons cannot fire new APs, a phase known as the refractory period of the neuron.

As the result of charged particles movement inside the channel, Hodgkin and Huxley predicted the existence of gating currents. We know today that gating current are produced by displacement of charges in transmembrane helices (i.e., segment $S4$), which comprises the voltage sensor (Figure 6.3). It took twenty years for the first observation of the Na^+ channel gating current (Armstrong and Bezanilla, 1973).

In the parallel conductance model (Figure 5.7), the total transmembrane current I has four different components: **a)** the capacitive i_c, **b)** the ionic i_k, **c)** the leak i_L and **d)** the Na-K ATPase pump $i^p_{Na/K}$.

$$I = i_c + \sum_k i_k + i_L + i^p_{Na/K} \tag{6.1}$$

With $k = \{Na^+, K^+, Cl^-, Ca^{2+}, \cdots\}$

By considering the neuron a closed system, we should expect $I = 0$. From the Kirchhoff current law for circuits, the parallel conductance model implies:

$$C_m dV_m / dt = -\left(\sum_k i_k + i_L + i^p_{Na/K}\right) \tag{6.2}$$

Assuming that changes in molar concentrations of ionic species does not change considerably during neuronal activation, we can use the linear approximation for the ionic fluxes (Eq. 5.28 and 5.29) and neglect the effect of the Na-K ATPase pump $i^{p}_{Na/K} \approx 0$, which is also very slow compared to that of the ion channel dynamics. Also, considering only Na^{+} and K^{+} ion currents flowing across the membrane, equation (6.2) can be written:

$$C_m dV_m / dt = -g_{Na}(V_m)(V_m - \hat{E}_{Na}) - g_K(V_m)(V_m - \hat{E}_K) - g_L(V_m - \hat{E}_L) \quad (6.3)$$

The voltage-dependent conductance of a channel for an ion k is represented by $g_k(V_m) = \bar{g}_k p_k(V_m)$, with $\bar{g}_k$ the maximum conductance density of the channel (mS / cm^2) and $p_k(V_m)$ the probability of being open at a membrane potential V_m. It was assumed that an ion can only cross the channels when a certain number of particles approach the channels. Three m-type particles were needed close to the activation gate of the Na^{+} channel for it to open, while a single h-type particle nearby the inactivation gate will open it. Four n-type particles were required to be in close proximity to the activation gate for the K^{+} channel Kv2.1 to open as it does not have inactivation. Interaction between these particles was negligible; hence, the approaching events for these particles were considered independent (Eq. 6.4 and 6.5).

$$p_{Na}(V_m) = p_{Na}^{m_1,m_2,m_3,h}(V_m) = p_{Na}^{m_1}(V_m)p_{Na}^{m_2}(V_m)p_{Na}^{m_3}(V_m)p_{Na}^{h}(V_m) = (p_{Na}^{m}(V_m))^3\, p_{Na}^{h}(V_m) \quad (6.4)$$

$$p_K(V_m) = p_K^{n_1,n_2,n_3,n_4}(V_m) = p_K^{n_1}(V_m)p_K^{n_2}(V_m)p_K^{n_3}(V_m)p_K^{n_4}(V_m) = (p_K^{n}(V_m))^4 \quad (6.5)$$

Unit state variables m, h and n, each satisfying the law of mass action, were introduced to represent the voltage-dependent probabilities $p_{Na}^{m}(V_m)$, $p_{Na}^{h}(V_m)$ and $p_K^{n}(V_m)$, respectively. For example (Figure 6.4), under this mathematical construct, $m-1$ represents the proportion of the m-type particles far from the Nav1.2 activation gate and m the proportion in close proximity to it. The rate of transfer from a distant to a close position $\alpha_m(V_m)$ (ms^{-1}) was supposed to depend on the membrane potential V_m. The rate of transfer from a close to distant position $\beta_m(V_m)$ (ms^{-1}) must similarly depend on V_m.

$$\frac{dm}{dt} = \alpha_m(V_m)(1-m) - \beta_m(V_m)m = \frac{m_\infty(V_m) - m}{\tau_m(V_m)} \quad (6.6)$$

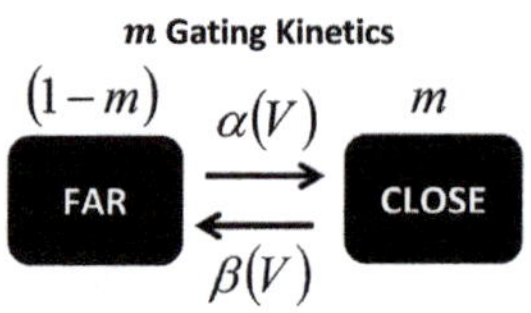

FIGURE 6.4 The law of mass action for the m particles underlying the kinetics of the activation gate in the Nav1.2 channel is shown.

With $m_\infty(V_m) = \alpha_m(V_m)/(\alpha_m(V_m)+\beta_m(V_m))$ and $\tau_m(V_m) = 1/(\alpha_m(V_m)+\beta_m(V_m))$.
Equivalent equations were established for state variables h and n.

$$\frac{dh}{dt} = \alpha_h(V_m)(1-h) - \beta_h(V_h)h = \frac{h_\infty(V_m)-h}{\tau_h(V_m)} \tag{6.7}$$

$$\frac{dn}{dt} = \alpha_n(V_m)(1-n) - \beta_n(V_m)n = \frac{n_\infty(V_m)-n}{\tau_n(V_m)} \tag{6.8}$$

With $h_\infty(V_m) = \alpha_h(V_m)/(\alpha_h(V_m)+\beta_h(V_m))$ and $\tau_h(V_m) = 1/(\alpha_h(V_m)+\beta_h(V_m))$
And $n_\infty(V_m) = \alpha_n(V_m)/(\alpha_n(V_m)+\beta_n(V_m))$ and $\tau_n(V_m) = 1/(\alpha_n(V_m)+\beta_n(V_m))$
Under all these considerations, equation (6.3) transforms into:

$$C_m dV_m / dt = -\bar{g}_{Na} m^3 h(V_m - \hat{E}_{Na}) - \bar{g}_K n^4 (V_m - \hat{E}_K) - g_L (V_m - \hat{E}_L) \tag{6.9}$$

6.3 INTRACELLULAR RECORDING: VOLTAGE AND CURRENT CLAMP MODES

For the H-H model to be completely specified, the rates of transfer α_s and β_s in equations (6.6-6.8) must be determined using intracellular recording s={m,h,n}, which allows measuring either voltage difference or electric current across the membrane of a neuron. For intracellular recording, both a sharp electrode inserted inside the cytoplasm of a neuron and a reference bath electrode outside are connected to an amplifier. The intracellular recording technique provides two types of measurements from neurons: **1)** membrane potential responses to electric current injection I_{Inj} (*current-clamp*) and **2)** transmembrane electric currents originating from holding the membrane potential at a fixed value $V_m(t) = V_c$ (*voltage-clamp*). As the intracellular recording technique requires high precision in placing sharp electrodes inside the cytoplasm of neurons, first observations were performed using the giant axon of the *Loligo forbesi* squid, which was discovered in 1930 by John Zachary Young (1907–1997). The squid giant axon provided sufficient area ($500\,\mu m$ in diameter, $60\,cm$ in length) to place the electrodes.

Using intracellular recording under a current-clamp approach, Hodgkin and Huxley (1939) provided the first known observation of an AP from this squid axon. The current clamp approach is represented mathematically by equation (6.10).

$$C_m dV_m / dt = -\bar{g}_{Na} m^3 h(V_m - \hat{E}_{Na}) - \bar{g}_K n^4 (V_m - \hat{E}_K) - g_L (V_m - \hat{E}_L) + I_{Inj} \tag{6.10}$$

It has been suggested that they used a current-clamp technology developed by Cole and Curtis (1939) to measure dynamic changes in membrane impedance in the squid giant axon. During the summer of 1938, Hodgkin learned with Kenneth S Cole (1900–1984) to dissect squid giant axons at Woods Hole (USA). However, membrane potential changes quickly during a course of an AP; hence, it was very

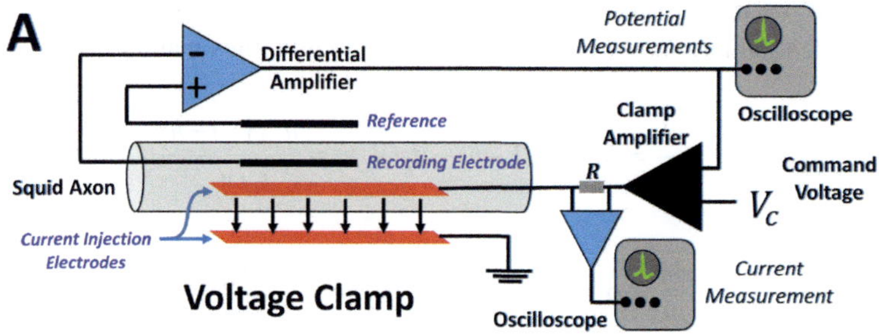

FIGURE 6.5 A schematic representation of the voltage-clamp experimental design.

difficult to study current–voltage properties of the membrane with a current-clamp approach.

War World II prompted Hodgkin and Huxley to put their research on hold as they were sent to work on aircraft monitoring with radar and predictors in targeting for gunnery development. These tasks endowed them with knowledge about feedback control systems and underlying mathematics crucial for the post-war development of the voltage-clamp approach (Figure 6.5). The voltage-clamp technique consisted of a feedback system to control the potential difference across the membrane, i.e., holding it a given value V_c, by injecting a current into the neuron. Departures of the membrane potential from the clamped value are amplified and dynamically injected into the neuron using a different pair of electrodes. The injected current I_m needed to keep the membrane potential constant over time is measured with an oscilloscope. As the voltage is constant, i.e., $dV_m / dt = 0$, the mathematical equation describing the voltage-clamp approach is:

$$I_m = -\bar{g}_{Na} m^3 h(V_m - \hat{E}_{Na}) - \bar{g}_K n^4 (V_m - \hat{E}_K) - g_L (V_m - \hat{E}_L) \tag{6.11}$$

As the rates of transfer α_s and β_s does not depend on time, equations (6.6-6.8) can be solved analytically.

$$m(t) = m_\infty(V_c) - \left(m_\infty(V_c) - m_0\right) e^{-t/\tau_m(V_c)} \tag{6.12}$$

$$h(t) = h_\infty(V_c) - \left(h_\infty(V_c) - h_0\right) e^{-t/\tau_h(V_c)} \tag{6.13}$$

$$n(t) = n_\infty(V_c) - \left(n_\infty(V_c) - n_0\right) e^{-t/\tau_n(V_c)} \tag{6.14}$$

To separate Na^+ and K^+ currents, they used seawater and sodium-deficient solutions ($[Na^+]_e \approx 0$, Figure 6.6). We can obtain similar results by using tetrodotoxin (TTX) discovered in 1964. Using this methodology (Figure 6.7), a family of Na^+ and K^+ currents were recorded by Hodgkin and Huxley (1952b) by setting different values of V_c.

The rates of transfer α_s and β_s were fitted to voltage-clamp experimental data using one of the most popular mechanical calculators, the Brunsviga 20 (Brunsviga

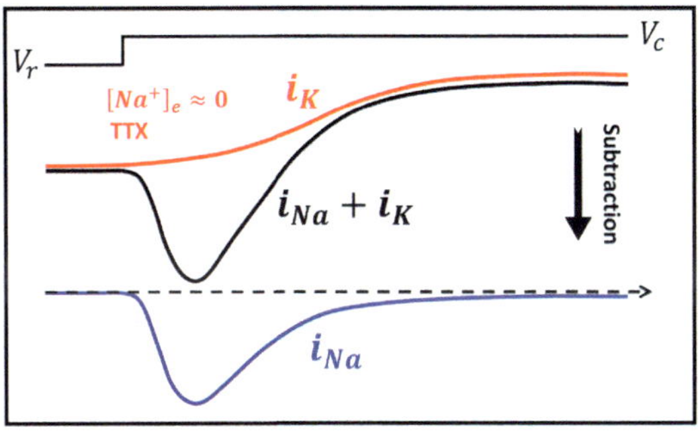

FIGURE 6.6 Transmembrane electric currents observed from the squid giant axon after clamping the membrane potential at $51 mV$ above V_r (black – seawater and red – $[Na^+]_e \approx 0$).

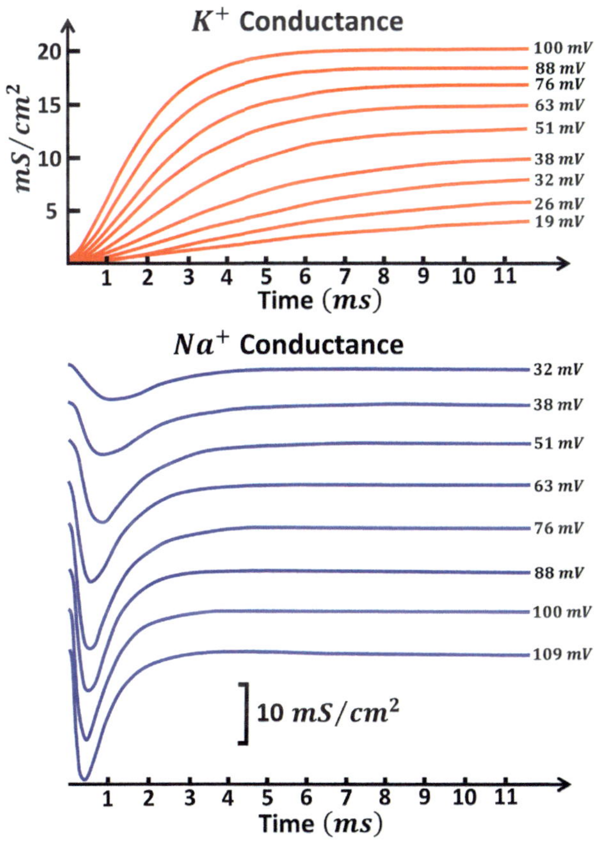

FIGURE 6.7 The fitted traces to the voltage-clamp experimental data for a set of V_c values (adapted from Hodgkin and Huxley, 1952a).

Note: Data were obtained at temperature $6-7$ °C. The membrane potential is shown from its resting value.

Maschinenwerke, Grimme, Natalis & Co.). These rates satisfy the following parametric equations:

$$\alpha_m(V_m) = \frac{0.1(V_m + 25 - V_r)}{e^{(V_m+25-V_r)/10} - 1} \quad \beta_m(V_m) = 4(V_m - V_r)/18$$

$$\alpha_h(V_m) = 0.05(V_m - V_r)/20 \quad \beta_h(V_m) = \frac{1}{e^{(V_m+30-V_r)/10} + 1}$$

$$\alpha_n(V_m) = \frac{0.01(V_m + 10 - V_r)}{e^{(V_m+10-V_r)/10} - 1} \quad \beta_n(V_m) = 0.125(V_m - V_r)/80$$

6.4 VOLTAGE-GATED ION CHANNELS

The pioneer work by Hodgkin and Huxley led to an escalation for the interest in electrophysiology. With the development of more sophisticated experimental techniques (e.g., whole-cell, patch-clamp and sharp modalities, *see next section*), electrophysiologists were able to characterize many ionic channels in different types of neurons. Nowadays we know that ion channels, like Nav1.2 and Kv2.1, are integral proteins that under certain conditions let specific ions flow across the membrane in the direction of their electrochemical potential gradient. These ionic fluxes occur through water-filled pores exposed by conformational changes in these proteins whenever a force is applied to them. These forces could be linked to: **a)** voltage changes, **b)** ligand binding, **c)** mechanical deformations, and **d)** temperature/chemical variations. Gating is the process by mean of which these channels open and close. Once the pore is open, selectivity of an ion to flow through the channel is determined by either its solvation/desolvation characteristics or its formation/breaking dynamics during the ion-protein interactions. Malfunction of the channels could lead to a series of channelopathies. In this chapter, we will focus our attention only on ion channels that open/close in response to changes in the transmembrane voltage (voltage-gated), both with and without the need for an intracellular signaling. This type of channel is the most important in the genesis and transmission of neuronal activity. Ligand-gated channels, like ionotropic receptors, will be studied in Chapter 7. Detailed discussions about the role of ion channels in sensory transduction can be found in Sperelakis (2012).

Typically, *voltage-gated channels* (VGC) are formed by units containing arrays of helical transmembrane segments (TM) linked by P-loops. These channels comprise a selectivity filter and a pore-domain gate control. Some voltage-gated channels might need the presence of an additional intracellular signaling (e.g., Ca^{2+}, PIP_2). In some case, they might also be regulated by either extracellular K^+ or intracellular Mg^{2+}. The mechanisms underlying permeation and ion selectivity in most of these channels are still elusive. Let us start by examine the structure and function of the two most important channels in the AP genesis and propagation, i.e., Na^+ and K^+ channels.

***Na⁺* channels** (Nav) are composed of an α subunit and two or more transmembrane glycoproteins (β_1, β_2 and/or β_3 subunits) with an extracellular N-terminus and a cytoplasmic C-terminus. The large α subunit, formed by four repeated motifs (I, II, III, and IV), carries out most important functions of the channel. Each motif has six linked TMs ($S_1,\cdots,S_6$). The voltage sensor is in the hydrophobic S4. In addition, the channels have a selectivity filter and an intracellular inactivation gate that "swings" shut, turning off the channel. Of the nine isoforms in the Nav family (Nav1.1 – Nav1.9), six have a binding site for TTX, a selective blocker of the channel, in the short P-loop between S5 and S6. Na^+ ion selectivity, as described in the four-barrier model (Hille, 1975), involves two sequential steps: **1)** a partial stepwise dehydration of the ion as it interacts with an ionized carboxylie acid at an outer site with high field-strength, and **2)** the conduction/rehydration of the ion as it moves within two inner sites of the channel. A detailed structural characterization of the bacterial Na^+ channels can be found in Catterall (2014). Later in this chapter (Table 6.1), we provide the gating kinetics for the transient (Nav1.2, TTX-sensitive) and persistent (Nav1.6, nonselective TTX-sensitive) Na^+ channels, which are very important for neuronal synaptic integration and AP firing in the brain.

***K⁺* channels** (Kv) are also formed by a α subunit responsible for ion conduction that is regulated by β and other auxiliary subunits. As in the Na^+ channel, the α subunit has four domains (I, II, III and IV). Different subfamilies of K^+ channels are characterized by the way these two generic channel components are arranged inside each domain: **a)** 4TM – four transmembrane helices/segments (S1–S4) and **b)** 2TM/P – two inner helices/segments and a P-loop. For example, each domain in the voltage-gated K^+ channels (Kv superfamily) are composed of six transmembrane helices/segments (TM1–TM6): a 4TM component and a 2TM/P component arrange in sequence to form a 6TM/P channel (Figure 6.8).

The voltage-sensing domain, located in the most external region, is formed by the TM1-TM4 helices. An outer TM5 and an inner pore lining TM6 form the channel pore. TM5 and TM6 joined by a re-entrant P-loop (2TM/P). The selectivity filter is in the P-loop. Although the unhydrated radius of Na^+ (50 pm) is smaller than that of K^+ (80 pm), pores of Kv channels have 100- fold more preference for K^+ over Na^+. It is still unclear whether the ions pass the selective filter hydrated or dehydrated. Three hypotheses have been postulated for the high K^+ selectivity by Kv channels: **1)** the snug-fit, **2)** the field-strength, and **3)** over-coordination. The snug-fit hypothesis is based on the favorability of specific size of an ion to conform the K^+ channel (induced-fitting). The field-strength and the over-coordination hypotheses are related to the core-ligand configuration of these cations. Coordinating ligands tends to jam-pack around smaller ions like Na^+, causing an electrostatic repulsion among them and with the ion-binding sites (backbone carbonyl oxygen in $S1-S4$). Therefore, the dehydration energetic cost and the energy needed to enter the pore for Na^+ are greater than for K^+. This balance between the repulsive (dehydration) and attractive (ion-protein) interaction might explain the pore selectivity in several voltage-gated ion channels. Earlier, Hodgkin and Keynes (1955) postulated that K^+ channels conduct unidirectional and fast as they have a long and narrow pore, which was able to hold 2-3 ions (knock-on mechanism). There are 12 super-families of Kv channels with

FIGURE 6.8 The Birth of an Idea (2007) by Julian Voss-Andreae (www.JulianVossAndreae.com) is a sculpture that illustrates the Kv protein structure.

α-subunit voltage gating. The gating kinetics for the fast, non-inactivating K^+ current (Kv3.1), the transient rapidly inactivating I_A "or K_t" (Kv4.2-3), the slowly inactivating K^+ current K_{slow} "or K_p" (*not identified*), and the M-current (Kv7.2–Kv7.5) channels are provided later on (Table 6.1). These channels were selected as they are crucial in AP signaling and synaptic integration by dendrites.

***Ca^{2+}* channels** (Cav) play key roles in a variety of cellular functions in the brain (e.g., Ca^{2+} homeostasis, gene expression, protein phosphorylation, muscle contractibility, and astrocytic signaling). There are three major super-families of voltage-gated Ca^{2+} channels (VGCCs): **1)** high-voltage activated (HVA, Cav1.1-4, Cav2.1-3), **2)** low-voltage activated (LVA, Cav3.1-3), and **3)** intermediate voltage activated. These channels can be found in smooth muscle cells, glial cells, endothelial cells, and neurons. We focus this chapter only on those channels involved in neuronal excitability, neurotransmitter release and synaptic integration. Cav channels are hetero-multimeric protein complexes, which comprises an $\alpha 1$ subunit, as well as some secondary units that modulate the channel kinetics (β, $\alpha 2-\delta$, and perhaps γ). The pore, a selective filter for Ca^{2+} and a voltage sensor are in the $\alpha 1$ subunit. Divalent cation selectivity (DCS) in VGCCs happens because of the existence of a set of four-conserved glutamate/aspartate residues coordinating Ca^{2+} ions in the pore (EEEE locus). T-type Ca^{2+} channels (Cav3.3) constitutes the main LVA subtype in neurons.

This channel becomes de-inactivated when the membrane is hyperpolarized, and from this stage can open to depolarization. The long-lasting activation HVA channels (L-Type, Cav1.1-4) is being involved in dendritic integration. It plays a crucial role in the $Na^+ - Ca^{2+}$ resonance phenomenon linking the soma and the distal trunk in neocortical pyramidal cells (PCs) (Larkum et al., 2009). L-Type channels is sensitive to 1,4-dihydropyridines (DHPs). The role of two HVA channels, P/Q-type (Cav2.1) and N-type (Cav2.2), in neurotransmitter release from axon terminal and dendritic integration in many neurons have been well-studied. The intermediate voltage activated (R-type, Cav2.3) Ca^{2+} channel has been linked to synaptic plasticity and nociception. Table 6.1 provides the gating kinetics for the L-type and T-type Ca^{2+} channels. Gating kinetics for the Cav2.1-3 channels are provided in Chapter 7.

There are several other K^+ channels that, in addition to a change in membrane potential, require intra/extra-cellular signaling to open. Here, we introduce the two more important families: **1)** the inwardly-rectifier K^+ channels (K_{IR}) and **2)** the Ca^{2+} -activated K^+ channels. However, the specifics about their functions are discussed in further chapters.

Inwardly-rectifier K^+ channels (K_{IR}) are also important for multicellular signaling in the brain as it lets K^+ current flow preferably in the inward direction (into the cells). In this way, it helps stabilize the resting membrane potential of the cell. These channels are formed of two TM and a single P-loop between them (2TM/P) channel. They are present in astrocytes, endothelial cells, and smooth muscles cells. This type of channels plays important roles in the neurovascular coupling and K^+ ion homeostasis. K_{IR} channels underlie the inhibitory character of the $GABA_B$ receptors (Chapter 7). The K_{IR} channel is activated by phosphatidylinositol 4,5-bisphosphate (PIP_2). We will discuss them in more details, as well as will provide gating kinetics, in Chapters 9 and 10.

Ca^{2+}-activated K^+ channels have an α-subunit that forms the pore, the voltage sensor, and the calcium sensor. It could have 6-7 TM with the voltage sensor in the S4 segment and the pore between the S5 and S6 segments. The channels have also regulatory β subunits. The Ca^{2+}-binding domain is in the cytoplasmic carboxyl domain of the α-subunit. Additional Ca^{2+}-binding proteins (e.g., calmodulin) might also be part of the α-subunit. These channels are classified into three main groups: **1)** small conductance KCa channels (SK1, SK2, SK3), **2)** large conductance KCa channels (BK, or Slo1), and **3)** intermediate conductance IK. SK channels are blocked by apamin. They are selectively activated by $[Ca^{2+}]_i$ increases. Therefore, they are responsible for a long-lasting hyperpolarization sometime observed after an AP, known as slow after-hyperpolarization (sAHP), which reduce the current-frequency response of certain neurons. They have been also associated with neurotransmission and vascular relaxation. BK channels are expressed in a variety of cells, e.g., neurons, astrocytes, smooth muscle cells and are involved in the hyperemic response and neuronal excitability. These channels are also essential for the regulation of vascular tone. Due to the channel large conductance ($\sim 200 - 300\,pS$), small $\left[Ca^{2+}\right]_i$ changes, as those related to spikes in cellular somas and sparks in microdomain clusters, can cause a significant efflux of K^+ from the cells to the extracellular space with impact in volume transmission. IK channels are activated by very low concentrations of Ca^{2+} by four resident calmodulin molecules, but they are voltage insensitive. IK channels are not discussed in this book. Gating kinetics and functions for SK and BK are discussed in Chapters 9 and 10.

More details about all channels discussed in this section can be found in the *Channelpedia* of the "Blue Brain Project" (https://channelpedia.epfl.ch/).

6.5 WHOLE-CELL, PATCH-CLAMP, AND SHARP MODALITIES

Intracellular recording is a key electrophysiological technique to characterize a plethora of voltage-gated ion channels in particular cells. The large size of the electrodes used to study the squid giant axons limited the application of the voltage/current-clamp approach to study smaller mammalian excitable cells, which was possible thank to the introduction of the pulled glass micropipette electrodes (or sharp electrodes, $tip \approx 0.01-0.1\mu m$) by Ling and Gerald (1949). However, this clamping technique suffer of two major problems: **a)** it was not possible to record the electric currents from a single ion channel and **b)** it required the use of two electrodes inside the cells (current injecting and voltage reading electrodes). Neher and Sakmann (1976) developed the patch-clamp technique that allows first recordings of single ion channels from small mammalian cells. These currents were very tiny to be observed with a valve-based amplifiers. Therefore, differential amplifiers (voltage-clamp) and voltage-followers (current-clamp) based on modern transistors are required. The patch-clamp technique also used pulled glass microelectrodes, but of larger diameters ($tip \approx 1-2\mu m$, $< 20\,M\Omega$). Under patch-clamp, the micropipette electrodes are not inserted inside the cell, but sealed onto the surface of the cell membrane. The tip is usually heated in a microforge to make it smooth, which helps form the seal with the cells. For this groundbreaking electrophysiological method, Neher and Sakmann received the Nobel Prize in Physiology or Medicine in 1991.

Once the tip is adhered to the membrane by a low resistance seal ($< 50\,M\Omega$), there are several particular settings that could be created (Figure 6.9):

- **Perforated patch** – A small amount of an antifungal/antibiotic agent (e.g., nystatin, gramicidin, amphothericin-B) is added to the electrode solution to generate small pores in the membrane, establishing an electrical connection with the cytoplasm.
- **Cell attached** – A suction is applied to the micropipette to create a *gigaseal* ($10-100\,G\Omega$). A portion of the cell membrane is suctioned into the micropipette forming an omega-like shaped. This provides not only mechanical stability for the recording, but perfect isolation of a single membrane path as any current leakage through the tip-cell gap is minimized. Under this condition, disruptions of the interior of the cell are avoided; hence, extracellular acting agents can be applied to the membrane through the micropipette to study ligand-gated ion channels (e.g., neurotransmitter receptors).
- **Excised patch** *"inside-out"* – By gently pulling the micropipette once the gygaseal is formed, the membrane patch breaks off and separates from the rest of the cell. Taking the micropipette to the solution/air interface, or placing in a low-Ca^{2+} solution, will help breaking the vesicle of membrane formed in

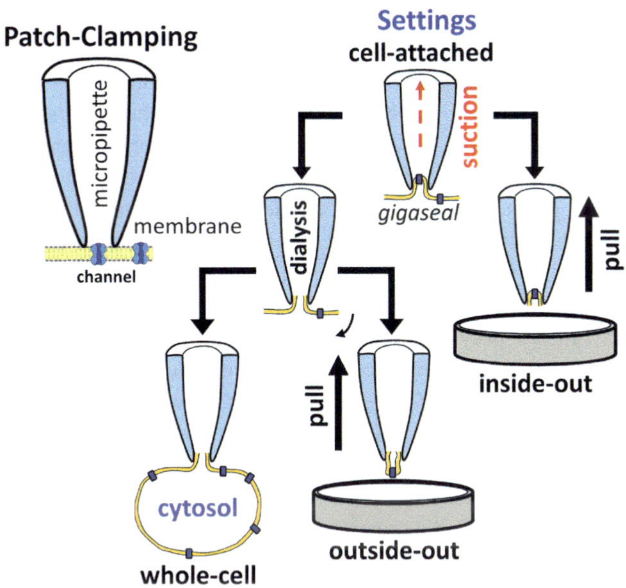

FIGURE 6.9 A variety of patch-clamp settings: cell-attached, whole-cell and excised (outside-out and inside-out).

Note: The procedures to create each setting are illustrated.

the tip after pulling the micropipette, exposing out the intracellular wall of the membrane. In this configuration, a cytosolic internal solution must be placed in the bath.

- **Whole-cell patch** – Under this approach, a dialyzing of the cell membrane is produced after the gigaseal is formed. Membrane rupture can be created by applying either a stronger suction or large current pulses through the micropipette. The content in the micropipette must be equivalent to the cytosolic internal solution. As for the perforated patch setting, the whole-cell patch technique allows researcher to record transmembrane currents flowing through extended portions of the membrane, or even the entire membrane surface for small cell ($< 20\,\mu m$).
- **Excised patch** *"outside-out"* – If the micropipette is gently pulled after the membrane has been dialyzed using same methods as for whole-cell path), the cytoplasmic bridge will collapse forming this way the outside out excised path. Under this technique, the solution in the micropipette must be equivalent to cytosolic internal solution. The content of the extracellular CSF can be modified.

All these path-clamp settings are very useful to characterize cellular signaling in isolated living cells, acute tissue sections, and cultured cells. More recently, in vivo patch clamping have been developed using automatic pressure and positioning control systems The technique is based on three pressure types: **1)** high pressure $(800-1000\,mbar)$ used to prevent clogs in the micropipette, **2)** low pressure $(20-30\,mbar)$ used to search for cells, and 3) small vacuum $(15-150\,mbar)$ used to establish the gigaseal. These three pressures, and atmospheric pressure, are switched and controlled by electronic valves and pressure regulators. A controlled piezoelectric micromanipulator moves automatically the pipette $(2-3\,\mu m$ steps) until the cell is found and the gigaseal established. Another technological advancement has been possible by the introduction of the patch chip used in cultured cells, or even using suspension methods. Cultured cells migrate to existing orifices in these chips where the gigaseal is spontaneously created as the cells grow. Chips are made from silicon, glass, PDMS, and/or polymide. In the electrolyte-metal interface of micropipette electrodes Cl^- ions in the electrolytic solution are exchanged for electrons in the silver wire following this reaction: $Cl^- + Ag(s) \leftrightarrow AgCl + e^-$. For patch-clamp recording in voltage-clamp mode (Figure 6.10), researchers could use a field effect transistor (FET)-input operational amplifier (op-amp), which uses feedback to hold the pipette voltage at the clamped potential (V_C).

In Figures 6.11 and 6.12, the traces on the top correspond to single channels observation with an excised inside-out patch. The equivalent whole-cell recordings are shown on the bottom panels, which are equivalent to the average of many single-channels data.

Electric currents through single Na^+ (Figure 6.11) and K^+ (Figure 6.12) channels while the membrane potential is clamped, as those symbolically illustrated, can be recorded by using the patch-clamp technique, for example, the inside-out setting. To obtain reliable transmembrane current I_m data, it is crucial to compensate for the micropipette capacitance C_p. Note that a whole-cell patch-clamp recording would be equivalent to the average of all single channel electric currents.

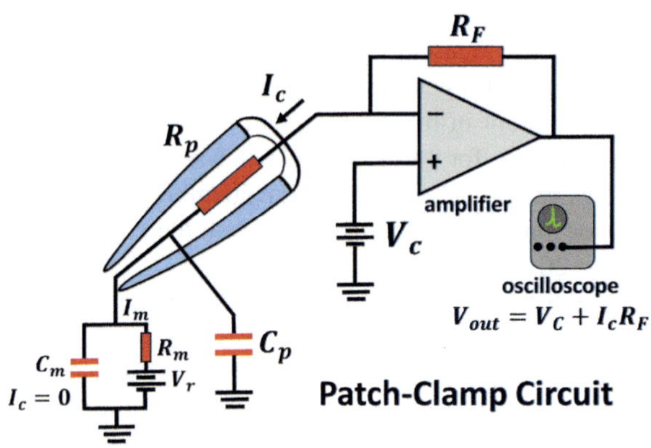

FIGURE 6.10 A schematic representation of the circuit underlying the path-clamp technique with a single electrode.

Note: Parameters C_m, R_m and V_r represent the capacitance, resistance and resting membrane potential for the section of the membrane in the respective patch-clamp setting.

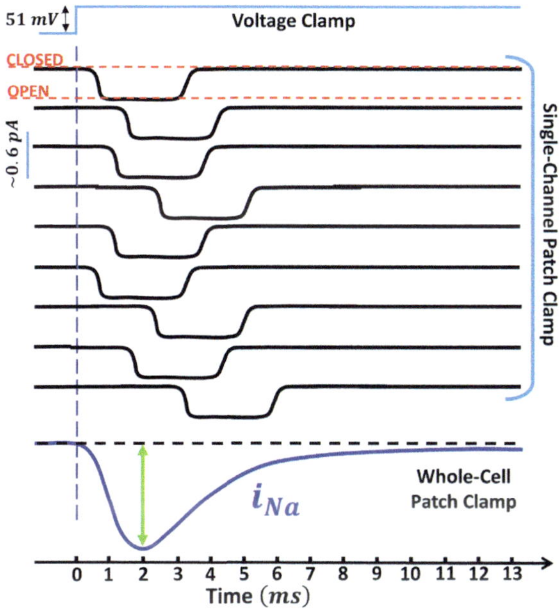

FIGURE 6.11 Symbolic patch-clamp recording under a voltage-clamp paradigm for Na^+.

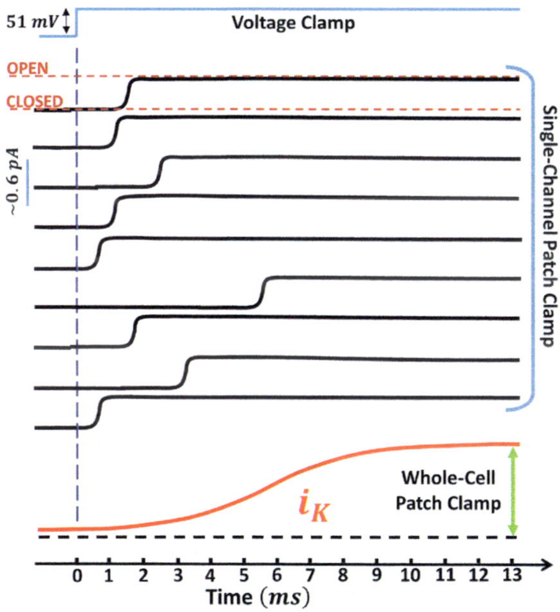

FIGURE 6.12 Symbolic patch-clamp recording under a voltage-clamp paradigm for K^+.

TABLE 6.1
Gating kinetics of a variety of ion channels

Code	Name	Exponent	α	β
Nav1.2	Na_t	$p=3$	$\frac{0.182\,T_A(V_m+38)}{1-e^{-(V_m+38)/6}}$	$\frac{-0.124\,T_A(V_m+38)}{1-e^{(V_m+38)/6}}$
		$q=1$	$\frac{-0.015\,T_A(V_m+66)}{1-e^{(V_m+66)/6}}$	$\frac{0.015\,T_A(V_m+66)}{1-e^{-(V_m+66)/6}}$
HCN1	I_h	$q=1$	$\frac{0.00643\,(V_m+154.9)}{e^{(V_m+154.9)/11.9}-1}$	$0.00193\,e^{V_m/33.1}$
Kv7.2-5	M-current	$p=1$	$0.0033\,T_A\,e^{0.1(V_m+35)}$	$0.0033\,T_A\,e^{-0.1(V_m+35)}$
Cav1.1-4	L-type	$p=2$	$\frac{1.6}{e^{-0.072\,(V_m-5)}+1}$	$0.02\,(V_m+8.69)(e^{(V_m+8.69)/5.36}-1)$
			m_∞ and h_∞	**τ_m and τ_h**
Kv4.2-3	I_A	$p=4$	$\frac{1}{1+e^{-(V_m+10)/19}}$	$\frac{0.34+0.92\,e^{-((V_m+83)/23)^2}}{T_A}$
		$q=1$	$\frac{1}{1+e^{(V_m+76)/10}}$	$\frac{8+49\,e^{-((V_m+81)/59)^2}}{T_A}$
unclear	K_{slow}	$p=2$	$\frac{1}{1+e^{-(V_m+11)/12}}$	$\frac{1.25+175.03\,e^{0.026(V_m+10)}}{T_A}$, $V_m<60$; $\frac{1.25+13\,e^{-0.026(V_m+10)}}{T_A}$, otherwise
		$q=1$	$\frac{1}{1+e^{-(V_m+64)/11}}$	$\frac{360+(1010+24(V_m+65))e^{-((V_m+85)/48)^2}}{T_A}$
Cav3.3	T-type	$p=2$	$\frac{1}{1+e^{-(V_m+40)/6}}$	$5+\frac{20}{T_A\left(1+e^{(V_m+35)/5}\right)}$
		$q=1$	$\frac{1}{1+e^{(V_m+90)/6.4}}$	$20+\frac{50}{T_A\left(1+e^{(V_m+50)/7}\right)}$
Nav1.6	Nap	$p=3$	$\frac{1}{1+e^{-(V_m+52.6)/4.6}}$	$6/\left(T_A\left(\frac{0.182(V_m+38)}{1-e^{-(V_m+38)/6}}-\frac{0.124(V_m+38)}{1-e^{(V_m+38)/6}}\right)\right)$
		$q=1$	$\frac{1}{1+e^{(V_m+48.8)/10}}$	$10^6/\left(T_A\left(\frac{6.94(V_m+64.4)}{1-e^{-(V_m+64.4)/2.63}}-\frac{2.88(V_m+17)}{1-e^{(V_m+17)/4.63}}\right)\right)$
Kv3.1	Fast, NI	$p=1$	$\frac{1}{1+e^{-(V_m-18.7)/9.7}}$	$\frac{4}{T_A\left(1+e^{-(V_m+56.56)/44.14}\right)}$

Notes: The factor T_A accounts for changes in the relative rates as a consequence of a variation in the temperature $T_A = Q_{10}^{(T_2-T_1)/10} = 2.3^{(34-21)/10}$. An ionic current is represented as: $I_k = \bar{g}_k m^p h^q (V_m - \bar{E}_k)$. References and $\bar{g}_k$ values are in Hay et al. (2011) and Herrera et al. (2020). NI – non-inactivated.

Figure 6.13 shows the voltage-current curves for Na^+ (blue) and K^+ (red) ions. The curves are created from data in Figures 6.11 and 6.12 (*green arrows*): the Na^+ current peak (Figure 6.11) and the K^+ current steady-state value (Figure 6.12) for each V_c.

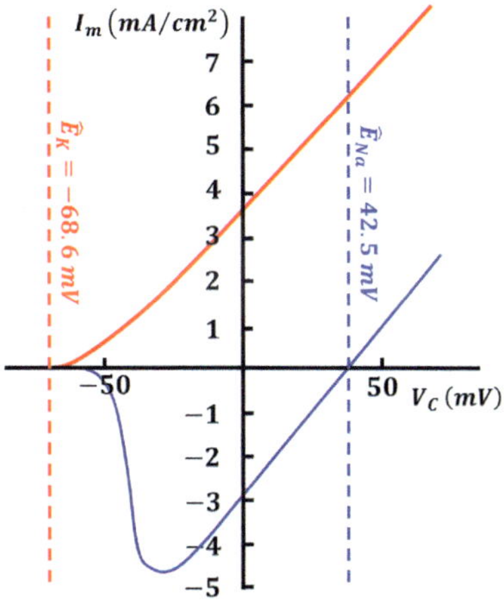

FIGURE 6.13 The Na^+ and K^+ voltage-current curves.

6.6 BORG-GRAHAM'S GENERALIZATIONS

We used the law of mass action to model the transition of charged particles underlying the channel gating kinetics. However, transitions between states of the channel due to conformational changes can be modeled by using the theory of reaction rates in the context of thermodynamics (Eyring, 1935). Under this framework, the rate for the transition between any channel conformational state depends exponentially on the Gibbs free energy barrier existing for that transition to happen. Let us assume two conformational states in a channel gate: close (c) and open (o). The forward and backward transitional rates can be defined as:

$$\alpha\left(V_m\right) = \alpha_0 e^{-\Delta G^{c\to o}\left(V_m\right)/RT} \tag{6.15}$$

$$\beta\left(V_m\right) = \beta_0 e^{-\Delta G^{o\to c}\left(V_m\right)/RT} \tag{6.16}$$

Approximating the Gibbs free energy barrier by means of the linear terms in Taylor's expansion and assuming that conformational changes in a channel happen as a consequence of the movement of a gating particle with charge q, Borg-Graham (1991) demonstrated that these rate constants could be expressed as a function of: **a)** the relative position of the energy barrier ($0 \le \gamma \le 1$) and **b)** a half-activation voltage V_H. To provide more flexibility to the linear model, we assumed that the rate strength coefficients were different for the forward α_0 and backward β_0 directions.

$$\alpha\left(V_m\right) = \alpha_0 e^{z\gamma F\left(V_m - V_H\right)/RT} \tag{6.17}$$

$$\beta\left(V_m\right) = \beta_0 e^{-z(1-\gamma)F\left(V_m - V_H\right)/RT} \tag{6.18}$$

TABLE 6.2
Results from fitting the Borg-Graham model to the gating kinetics of ion channels

Name	Exponent	z	γ	α_0	β_0	V_H
Na_t	$p=3$	3.14	0.52	9.04	0.81	−20.12
	$q=1$	−2.46	0.56	0.54	0.18	−80.94
I_h	$q=1$	−2.68	0.69	0.0001	0.0005	−41.26
M-current	$p=3$	5.34	0.50	0.02	0.0045	−27.29
L-Type	$p=2$	1.83	0.54	0.24	0.27	−20.00
I_A	$p=4$	1.02	1.00	1.09	2.32	−19.99
	$q=1$	−2.97	0.38	0.004	0.60	−21.44
K_{slow}	$p=2$	1.59	1.00	0.06	0.069	−19.96
	$q=1$	−1.78	0.65	0.0002	0.0037	−20.04
Nap	$p=3$	2.80	0.35	0.26	1.68	−68.56
	$q=1$	−1.82	0.29	1.47×10^{-4}	1.02×10^{-3}	−20.20
Coulter et al. (1989)						
T-Type	$p=3$	3.43	0.5	0.06	0.06	−63.0
	$q=1$	−4.24	0.75	0.008	0.008	−83.5

Note: It was not possible to fit the model for Kv3.1. Borg-Graham model parameters for the T-type Ca^{2+} channels were obtained from rat thalamocortical relay neurons (Coulter et al., 1989).

To estimate the Borg-Graham model parameters for the ion currents in Table 6.2, we applied the following procedure. First, artificial data was generated using the activation and/or inactivation gating kinetics model for these channels (Table 6.1). Variations of magnitudes m_∞ (or h_∞) and τ_m (or τ_h) were used as data for model fitting, which were sampled every 5 mV. A normally distributed noise was added to each simulated data set to make it looks similar to those experimentally observed. The Borg-Graham's model was then fitted (fmincon function, method: *interior point with factorization*, $Tol = 1.0\times10^{-6}$, MATLAB) to the simultaneous m_∞ (or h_∞) and τ_m (or τ_h) data sets using equations (6.17 and 6.18). Results are presented in Table 6.2. By linearly approximated the Gibbs free energy barrier, it is assumed that contributions to this barrier from interactions between the electrical field and charges/rigid-dipoles in the protein are well-modeled by a scaled membrane potential.

6.7 ION ACTIVE TRANSPORT – CO/COUNTER-TRANSPORT VS. PUMPS

Active transport helps maintain the electrochemical potential difference for both inorganic and organic ions across the membrane of cells. Active transport could be performed by pumps (*primary active transport*) and co/counter transporters (*secondary active transport*). Because of its importance for the establishment of V_r in neurons, one of the most important pumps in living organisms, the Na-K ATPase, was described in Chapter 5. In this section, we will introduce the general equation for the equilibrium in secondary active transport and provide a formalism to estimate the fluxes of ions across the membranes in primary/secondary active transport using chemical kinetics. At the end, we focus on the application of these theoretical frameworks to model two important active transport mechanisms for Ca^{2+} ions in neurons: **1)** the Ca^{2+} pumps and **2)** the Na-Ca counter-transporter (or exchanger). In Chapters 7–10, other important transport mechanisms for brain imaging will be discussed. In addition, transport mechanisms across the cell membrane of organic molecules (e.g., glucose) via facilitated diffusion (*uniporter*) are studied in other chapters. Pumps and (co/counter)-transporters are large integral proteins. These types of active transport depend on association/dissociation mechanisms with the transported ions, which creates little conformational changes in these integral proteins. Conformational changes might either directly require energy in the form of ATP hydrolysis to be completed (pumps) or not being directly linked to it (co/counter-transporters). The rate of these conformational changes depends on the affinity that the integral protein has to the ion at both side of the membrane.

6.7.1 Electrochemical Potential-Driven Transporters: The Symporters and Antiporters

A co-transport (or symport) is a reversible unidirectional flow of ions across the membrane in which some ions are transported against their electrochemical gradient (*secondary active*) using energy from the strong gradients of other ions being co-transported with the formers. In contrast to symporters, a counter-transport (antiport) moves some ions in the opposite direction. Antiporting is also a reversible secondary active transport, which uses the energy in the electrochemical gradients of certain ions to move others against their gradients. At the thermodynamic equilibrium, symporters and antiporters transport substance steady-stately across membranes separating two regions (1) and (2) satisfying a zero net electrochemical potential (or zero Gibbs free energy variation).

$$\Delta G = \overline{\Delta\mu}_{net} = \sum_{k=1}^{N} \Delta\bar{\mu}_k = \sum_{k=1}^{N}\left(RTln\frac{c_k^1}{c_k^2} + z_k F\Delta\varphi \right) = 0 \tag{6.19}$$

Where $\Delta\varphi = \varphi_1 - \varphi_2$ is the potential difference across the membrane separating the two regions. Note that these regions could be either cytosol, ECS or any intra-organelle domain (e.g., mitochondrion, endoplasmic reticulum, or a synaptic vesicle). Let us assume that a number n_k of the k-th organic/inorganic ion are transported in each cycle. In such case, we will refer to a transporter having $n_1 : n_2 \cdots : n_N$ coupling stoichiometry. Then, the following relationships will be held at the equilibrium:

$$\prod_{k=1}^{N}\left(\frac{c_k^1}{c_k^2}\right)^{s_k n_k} = e^{-\frac{F\Delta\varphi}{RT}\sum_{k=1}^{N} s_k n_k z_k} \tag{6.20}$$

For some transporters, the order these ions bind the protein is relevant, but for others binding could happen randomly. The secondary active transport can be electrogenic ($\sum_{k=1}^{N} s_k n_k z_k \neq 0$) or non-electrogenic ($\sum_{k=1}^{N} s_k n_k z_k = 0$). The factor s_k define the

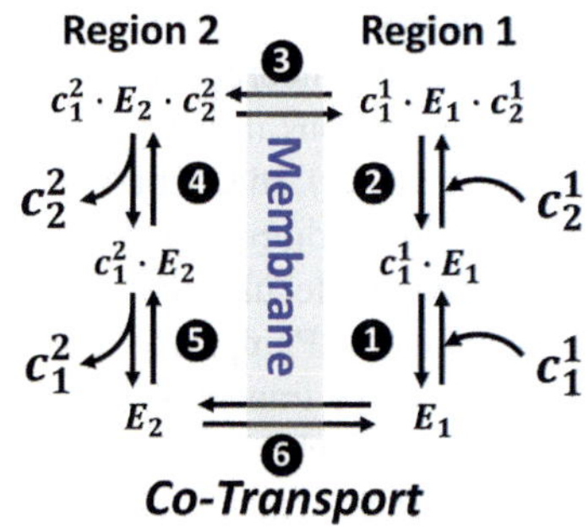

FIGURE 6.14 Diagrams representing the protein cycles for the simplest forms of co-transports.

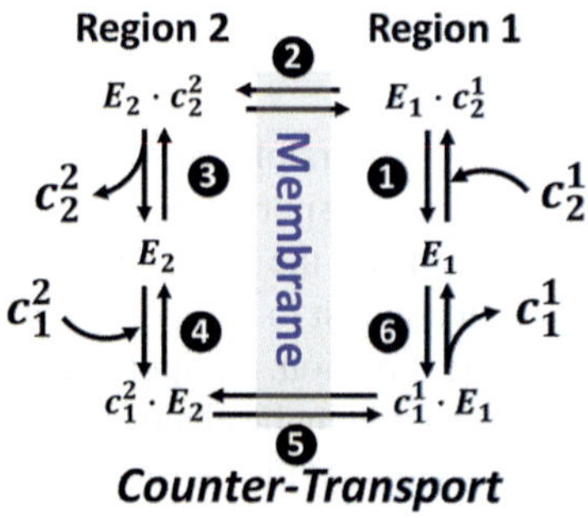

FIGURE 6.15 Diagrams representing the protein cycles for the simplest forms of counter-transport.

direction of the flux, i.e., co-transporting ($s_k = 1$) and counter-transporting ($s_k = -1$). Depending on the type of ions (substrates) loaded from regions 1 and 2 at a single cycle, antiporters could be either homoexchange (*same ion*) or hetereoexchange (*different ions*). We illustrate the protein binding cycle for the simplest forms of co/counter-transports, i.e., only two organic/inorganic ions are involved (Figures 6.14 and 6.15).

In what follow, we provide details for the particular case of the Na-Ca exchanger (NCX). Experimental evidence suggests that $3\,Na^+$ ions extrude a single Ca^{2+} ion in a modified exchange diffusion model. Taking into account that $n_{Na} = 3$, $n_{Ca} = 1$, $s_{Na} = 1$ and $s_{Ca} = -1$, equation (6.20) transforms to:

$$c_{Ca}^{e} = c_{Ca}^{i} \left(\frac{c_{Na}^{e}}{c_{Na}^{i}} \right)^{3} e^{-\frac{FV_r}{RT}}$$

Using values in Table 5.1 for the resting membrane potential $V_r = -71.47\,mV$, and the Na^+ intra/extracellular concentrations ($c_{Na}^{i} = 10\,mM$, and $c_{Na}^{e} = 145\,mM$), we demonstrate that the NCX generates a ~210-fold gradient for Ca^{2+} concentration across cell membranes at $310\,K$. Calculating ion fluxes for each transporter is challenging. However, we provide later in this chapter some general rules useful to estimate them in particular cases. For the case of the Na-Ca exchanger, a previous work by Wong and Bassingthwaighte (1981) provided theoretical equations for the unidirectional fluxes of these ions at steady state. For the calculation, it was assumed that: **1)** it involved three phases (intracellular, membrane, extracellular), **2)** carrier was confined to the membrane and ions were allowed in the membrane only combined with the carrier, **3)** the carrier were trivalent anions, with a monovalent for single Na^+ ion and divalent sites for which one Ca^{2+} and two Na^+ can compete, **4)** equilibrium is reached quickly, **5)** an alkali metal ion (or dextrose) can substitute Na^+ in the divalent site, and **6)** fluxes depend linearly on the carrier-cation concentration. Under these assumptions, the fluxes were described by the following equations:

$$j_{Ca} = \frac{1}{\lambda} C_T \gamma_e c^i_{Ca} \left(\frac{P_{Ca} \psi_i}{K_{i,Ca}} + \frac{P'_{Ca} c^i_{Na}}{K''_{i,Ca}} \right) \tag{6.21}$$

$$j_{Na} = \frac{1}{\lambda} C_T \gamma_i \left(c^e_{Na}\right)^2 \left(\frac{P_{Na} \psi_e}{K_{e,Na}} + \frac{P'_{Na} c^e_{Na}}{K''_{e,Na}} \right) \tag{6.22}$$

With, $\lambda = \gamma_e \delta_i + \gamma_i \delta_e$

$$\delta_j = 1 + \frac{c^j_{Ca}}{K_{j,Ca}} + \frac{c^j_{Ca} c^j_{Na}}{K^{"}_{j,Ca}} + \frac{\left(c^j_{Na}\right)^2}{K_{j,Na}} + \frac{\left(c^j_{Na}\right)^3}{K^{"}_{j,Na}} \quad \text{concentration-distributed factors}$$

$$\gamma_j = P_C \psi^3_j + \left(c^j_{Na}\right)^2 \left(\frac{P_{Na} \psi_j}{K_{j,Na}} + \frac{P'_{Na} c^j_{Na}}{K^{"}_{j,Na}} \right) + c^j_{Ca} \left(\frac{P_{Ca} \psi_j}{K_{j,Ca}} + \frac{P'_{Ca} c^j_{Na}}{K''_{j,Ca}} \right) \quad \text{concentration-dependent rates}$$

$$\psi_i = \frac{\xi e^{-\xi}}{1 - e^{-\xi}} \qquad \psi_e = \frac{\xi}{1 - e^{-\xi}} \qquad \xi = zFV_m / RT$$

z is the valence. C_T and P_C represent the total concentration of the carrier and its transport rate constant (s^{-1}). P_{Na}, P_{Ca}, P'_{Na}, and P'_{Ca} are the transport rate constants (s^{-1}) for the carrier-ion complexes $Na_2 - C$, $Ca - C$, $Na_3 - C$, $Ca - Na - C$, respectively. $K_{j,Na}$ and $K_{j,Ca}$ are the K's for divalent binding. $K^{"}_{j,Na}$ and $K^{"}_{j,Ca}$ are the binding rate to the carrier – divalent-ion complex. Different sets of model parameters ("*lumped*") were fitted to Na^+ and Ca^{2+} efflux/influx data obtained from the squid axon under different experimental manipulations (Wong and Bassingthwaighte, 1981).

6.7.2 ATP-Driven Ion Pumpss

There are mainly two types of pump families with relevance to brain imaging: the F-ATPases and the P-ATPases. The ATP synthase, a particular F-ATPase, is a

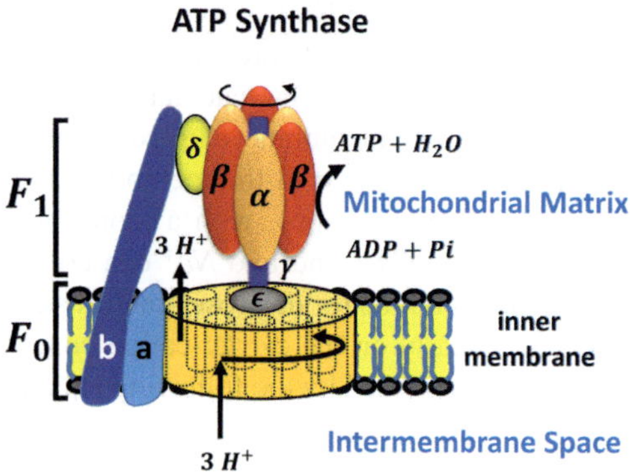

FIGURE 6.16 Schematic representation of the ATP synthase structure.

crucial component in the electron-transport/oxidative-phosphorylation pathway. This enzyme is involved in an oxidative phosphorylation as ADP is phosphorylated to ATP, an endergonic chemical reaction. The energy needed for this reaction comes from the H^+ proton electrochemical gradient created by the redox reactions of the electron transport chain (ETC), which is known as the *chemiosmotic coupling hypothesis*. Peter D Mitchell proposed the existence of two terms contributing to the proton electrochemical gradient: a negative-inside membrane potential $\Delta\varphi$ and an acid-outside gradient of H^+ (ΔpH). For this contribution, Mitchell received the Nobel Prize in Chemistry in 1978. In each cycle, $3H^+$ protons propel a rotational mechanism that brings tightly together ADP and Pi. The ATP synthase is an integral protein located in the mitochondrial inner membrane (Figure 6.16). It comprises two main sections F_0 and F_1. The ATP synthase is a molecular machine because of the presence of a rotational subunit allowing for ATP production. F_0 contains c-rings and subunits **a** and **b**. F_1 is comprised of five subunits (α, β, δ, γ and ϵ).

The reaction occurring in the ATP synthase is represented by the following stoichiometry:

$$ADP + Pi + (3H^+)_{interspace} \leftrightarrow ATP_{cyt} + H_2O + (3H^+)_{matrix}$$

Paul D Boyer (1918–2018) received 1/2 Nobel Prize in Chemistry in 1997 (shared with John E Walker) for proposing a rotational mechanism in the synthesis of ATP by means of an asymmetric central γ subunit, which involves three active β sites of F_1. The β subunits could be in three different conformational states: $\beta - ADP$, $\beta - ATP$, or $\beta - empty$. In the $\beta - ADP$ state available Pi is forced to bind ADP, causing the subunit to change conformation to a stable $\beta - ATP$ as a result from the γ subunit rotation. A following rotation makes the subunit to change to $\beta - empty$ conformation, which has a very low affinity for ATP; hence, easily releasing it to the mitochondrial matrix.

The Na-K ATPase pump, one of the most important P-ATPases, was studied in Chapter 5. In this section, we introduce another important P-ATPase, i.e., the Ca-ATPase pumps and illustrate the use of chemical kinetics while estimating their Ca^{2+} ion flow rates. Intracellular Ca^{2+} signaling plays a crucial role in many cellular functions in brain cells. For example, it has been linked to synaptic neurotransmitter release, the eicosanoid pathway in the neurovascular coupling, smooth muscle cell contractions/dilations, metabolic modulation, gene expression and cellular differentiation. It is the cytosolic ion with the higher molar concentration gradients (~ 2.5×10^4-fold with the extracellular space and ~ 2.5×10^3 with the endoplasmic reticulum, ER). Hence, intracellular Ca^{2+} concentration can increase in several order of magnitude during specific cellular events in the brain. Depending on the specific underlying mechanism (e.g., channels/receptors in the plasma membrane, intercellular gap-junctions, and IP_3R/RyR in the ER membrane), signal magnitude/waveform and cell involved, intracellular Ca^{2+} events can be classified into: **a)** sparklets, **b)** flashes, **c)** slow elevations with plateaus, **d)** sparks, **e)** puffs and **f)** global waves. After these events take place, Ca^{2+} ions must be quickly removed from the cytoplasm. As discussed before, the NCX contribute to reestablish the basal intracellular Ca^{2+} concentration, a mechanism with low affinity but high capacity. Ca^{2+} ions may also be temporally buffered by different mechanisms (e.g., calmodulin and the mitochondrion pump). However, Ca-ATPase pumps constitute the mechanism with the highest affinity for Ca^{2+} ion clearance from the cytosol. Ca-ATPase pumps can be either extrude Ca^{2+} to the extracellular space or pumped it back to the endoplasmic reticulum. The α subunit of the Ca-ATPase pumps is similar to that of the Na-K ATPase pump, and they both have a similar folded-catalytic polypeptide composition. According to location and amino-acid sequence, Ca-ATPase pumps in the brain are classified into two families: a) plasma membrane Ca^{2+} pumps (PMCA) and sarcoplasmic/endoplasmic reticulum Ca^{2+} pumps (SERCA). Interaction with calmodulin is present in PMCA pumps, but not in SERCA pumps. During large increases in intracellular Ca^{2+}, the calmodulin-Ca^{2+} complex binds to the PMCA, releasing the auto-inhibitory mechanism. As for the Na-K ATPase pump, Ca-ATPase pumps uses the Albers-Post mechanism (Figure 6.17). The general stoichiometry for Ca-ATPase pumps is:

$$nCa^{+}_{cyt} + mH^{+}_{ECS/L} + 1\,ATP_{cyt} \leftrightarrow nCa^{+}_{ECS/L} + mH^{+}_{cyt} + 1\,ADP_{cyt} + 1\,Pi_{cyt} \quad (6.23)$$

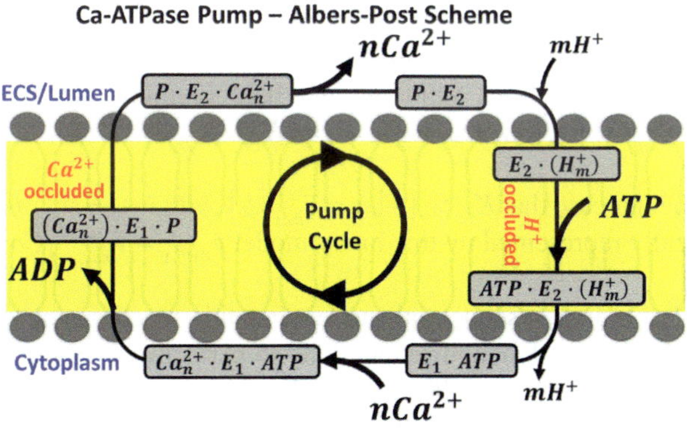

FIGURE 6.17 The Albers-Post scheme for a generic Ca-ATPase pump.

However, the affinity and stoichiometry ratio for PMCA and SERCA are different. It has been suggested that the PMCA pump operates with a 1:1 Ca^{2+}/ATP stoichiometry, with a $K_D \approx 10-20\mu M$ at rest. As the result from the calmodulin-Ca^{2+} disinhibition, affinity could significantly increase to $K_D \approx 1\mu M$. The stoichiometry for the SERCA pump is 2:1 Ca^{2+}/ATP and the affinity is higher ($K_D \approx 0.1-1.0\mu M$).

Let us introduce some general formalism to represent the chemical kinetics underlying pump/transporter function. We will limit our analysis to two types of chemical reactions:

i) Single binding – Michaelis-Menten kinetic model: In this case, a substrate S binds an enzyme E to form an ES complex in a reversible chemical reaction with forward K_{Df} and reverse K_{Dr} rates (**Step 1**). Then, the substrate irreversibly dissociates from the enzyme with a rate K_{cat} to form product P (**Step 2**). For the PMCA pump, the substrate would be the cytosolic Ca^{2+} ion and the product would be the Ca^{2+} ion pumped to the ECS/Lumen.

$$E+S \overset{K_{D(f,r)}}{\leftrightarrow} ES \overset{K_{cat}}{\rightarrow} E+P \tag{6.24}$$

$$\frac{d[S]}{dt} = K_{Dr}[ES] - K_{Df}[E][S] \quad \frac{d[E]}{dt} = (K_{Dr} + K_{cat})[ES] - K_{Df}[E][S]$$

$$\frac{d[ES]}{dt} = K_{Df}[E][S] - (K_{Dr} + K_{cat})[ES] \quad \frac{d[P]}{dt} = K_{cat}[ES]$$

It is assumed that $[ES]$ quickly reaches a steady-state value $d[ES]/dt \approx 0$. This steady-state approximation implies: $K_{Df}[E][S] = (K_{Dr} + K_{cat})[ES]$. The enzyme is conserved over time; hence, $[E_0] = [E] + [ES]$. After some mathematical manipulations, it can be proved that:

$$v = K_{cat}[ES] = \frac{v_{max}[S]}{K_M + [S]} \quad v_{max} = K_{cat}[E_0] \text{ and } K_M = \frac{(K_{Dr} + K_{cat})}{K_{Df}}$$

This is known as the Briggs-Haldane kinetic model. In the Michaelis-Menten model, it is assumed that S binding and dissociation occurred faster than P formation ($K_{cat} \ll K_{Dr}$); hence, $K_M \approx K_D = K_{Dr} / K_{Df}$.

In the particular case that $K_D \gg [S]$, $v = \bar{v}_{max}[S]$, with $\bar{v}_{max} = v_{max} / K_D$. The PMCA pump is typically represented by this linear model $v_{PMCA} = \bar{v}_{max}^{PMCA}[Ca^{2+}]_{cyt}$.

ii) Double binding – sequential model: For this reaction, the enzyme needs a second substrate S bound sequentially to perform the conformational change from the ES_2 complex. Therefore, the first reversible reaction (**Step 1**) is followed by another reversible reaction in which the ES complex binds a second S to form ES_2, right before the catalytic change happens (**Step 2**). K_{D1} and K_{D2} represent the equilibrium dissociation constants for the first and second reaction, respectively.

$$E + S \overset{K_{D1}}{\leftrightarrow} ES + S \overset{K_{D2}}{\leftrightarrow} ES_2 \overset{K_{cat}}{\rightarrow} E + P \tag{6.25}$$

To obtain the reaction rate v in this case, we use the definition of the equilibrium dissociation constants: $[ES] = K_{D1}[E][S]$ and $[ES_2] = K_{D2}[ES][S]$. Similarly, $[E_0] = [E] + [ES] + [ES_2]$. After some mathematic manipulations, it can be demonstrated that:

$$v = K_{cat}[ES_2] = \frac{v_{max}[S]^2}{\left(K_{D1}K_{D2} + K_{D2}[S] + [S]^2\right)}$$

If the enzyme is catalyzed quickly after the second S binds the ES complex $K_{D2f} \gg K_{D2r}$, then K_{D2} is relative small: $v = v_{max}[S]^2 / \left(\bar{K}_M + [S]^2\right)$, with $\bar{K}_M = K_{D1}K_{D2}$. The SERCA pump is typically represented by this 2nd-order Hill's model:

$$v_{SERCA} = v_{max}^{SERCA}\left(\left[Ca^{2+}\right]_{cyt}\right)^2 / \left(\bar{K}_{SERCA} + \left(\left[Ca^{2+}\right]_{cyt}\right)^2\right)$$

6.8 DENDRITIC MULTI-COMPARTMENTAL MODELS – MORPHOLOGICAL CHARACTERIZATION

The first knowledge of the neuronal morphology originated from "La Reazione Nera", a method created by Camilo Golki (1843–1926) to stain neurons by fixing silver chromate particles inside them from a reaction of silver nitrate and potassium dichromate. Based on first observations, Golki postulated the "*reticular theory*" by means of which the brain was formed by a network of connected wires. A Spanish pathologist, Santiago Ramon y Cajal (1852–1934) used Golki's method to demonstrate for the first time in 1887 the existence of independent units in the brain, i.e., the neurons. The *"neuron doctrine"* was initially criticized by Golki. For their crucial work on the structure of the nervous system, they both shared the Nobel Prize in Physiology or Medicine in 1906. The neuron doctrine establishes: **a)** neurons are the structural/functional units of the nervous system, **b)** neurons are individual cells comprising three parts (dendrites, soma and axon), and **c)** conduction takes place in the direction: dendrites → soma →axon. For the purpose of this book, neurons in the brain are considered to share three common morphological features: **1)** they have a large either spheroidal- or pyramidal-like soma (or body), **2)** their very branchy cylindrical-like dendritic trees collect synaptic inputs from other neurons and spatiotemporally convey them into the soma, a process referred as dendritic integration, and **3)** soma-generated all-or-none responses (i.e., APs) are propagated to other neurons though a cable-like axon. Many theoretical models used in brain imaging represent somas and dendrites by a single compartment with an equal membrane potential, which is referred to as the *space-clamp* approach. With the development of sophisticated electrophysiological and optical techniques, the role of dendritic compartmentation in neuronal computation and brain imaging has become clearer and should not be neglected in many practical situations. Therefore, we dedicate this section to describe the most important theories and biophysical models for dendritic integration and AP genesis/propagation.

Together with microfilaments and intermediate filaments, microtubules (MTs) running in parallel constitute a mechanical support for dendritic cytoskeleton. MTs are polymers of tubulin organized with plus-ends (β-subunits) and minus-ends (α-subunits). In dendrites, MTs contain mixed polarity orientations. MT lattices are used by dynein and kinesin motor proteins (*cargo transport*) to move vesicles, macromolecules and organelles along the dendritic trees. Dendritic cross-sectional areas are linearly correlated to the number of microtubules. Using serial focused ion beam scanning electron microscopy (FIB-SEM) images of dendrites, single MTs have been tracked within a tissue section and co-registered across sections. Thanks to this technique, MT reorganization at a dendritic bifurcation has been well-characterized. It can be assumed that a microtubule from a mother dendrite (diameter a_m) at a bifurcation always run into one of the daughter dendrites (diameters a_{d_k}, $k = 1, 2, \cdots$). Therefore, a variety of scaling laws has been proposed to represent changes in dendritic caliber at these bifurcations. Dendritic tapering has been mainly associated with changes in calibers due to bifurcating dendrites. Based on the relationship between the input conductances of dendritic branches at a bifurcation, Rall (1959) proposed the *3/2 power law* (6.26)

$$a_m^{3/2} = \sum_k a_{d_k}^{3/2} \tag{6.26}$$

Based on the 3/2 power law and several additional assumptions (i.e., similar boundary condition and electrotonic distance, equal transmembrane/intracellular resistances, and homogeneous synaptic inputs), the *dynamic lumping model* was proposed to represent a complex dendritic arborization by a single equivalent cylinder with a variable diameter calculated using equation (6.26) (Figure 6.18).

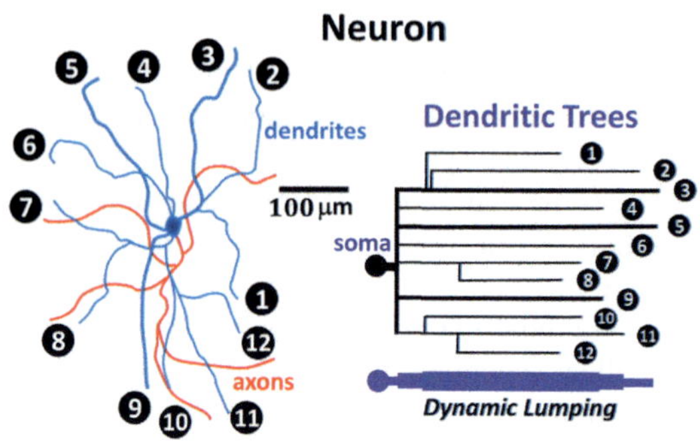

FIGURE 6.18 The schematic representation of a neuron with complex dendritic trees (blue) and axonal branches (red).

Note: An alignment of the dendritic branches according to the radial distance from the soma is used to create a dynamic lumping model with the help of the 3/2 power law.

However, recent studies have proven that this scaling law is not always satisfied by the dendrites of many neurons. To model more general dendritic tapering, the use of a series of electrotonically connected-cylinders following dendritic branches with decreasing diameters has been proposed. The combination of these cylinder-based representations of dendrites with a spheroidal/pyramidal characterization of the soma is referred as the *multi-compartment models* of neurons (Figure 6.19). Each compartment in the model is represented by the elementary membrane circuit, which comprises a capacitor C_m, a leaky resistor R_m, and a transmembrane current source I_m in parallel. To illustrate inter-compartment connections, the electrical diagram for a bifurcation (red) connecting a mother $\{k\}$ with two daughters $\{b_1, b_2\}$ compartments is shown in Figure 6.19. Two serially connected compartments $\{k_1, p\}$ is also a case of interest for a theoretical analysis. Capacitances and resistances for these compartments are functions of the total surface membrane area A of the compartment: $C_m = c_m A$ and $R_m = r_m / A$, with $c_m \approx 0.97 \mu F / cm^2$ and $r_m \approx 40{,}000\,\Omega cm^2$ described in Chapter 5. The intracellular and extracellular sides of consecutive elementary membrane circuits are connected via ½ resistors as illustrated (Figure 6.19). The intracellular resistor for each cylindrical compartment is calculated as $R_i = r_{cyt} l / \Delta S$, where ΔS and l are the transversal area and length of the cylinder, respectively. The relative resistance of the cytosol is $r_{cyt} \approx 166\,\Omega cm$ The resistance of the ECS is approximately equal to that of the cytosol (i.e., $r_{ECS} \sim= r_{cyt}$). The intracellular resistance for the soma is typically neglected because of its shape and large size. Assuming an extracellular/intracellular volume fraction $V_e / V_i \approx 0.3$, it is demonstrated that the resistance of the extracellular space is $R_e \approx 1.3 R_i$.

$$\frac{R_e}{R_i} = \frac{r_e l / A_e}{r_i l / A_i} = \frac{r_e}{r_i}\frac{A_i l}{A_e l} = \frac{r_e}{r_i}\frac{V_i}{V_e} = \frac{r_e}{r_i}\frac{1}{0.3} \approx 1.3$$

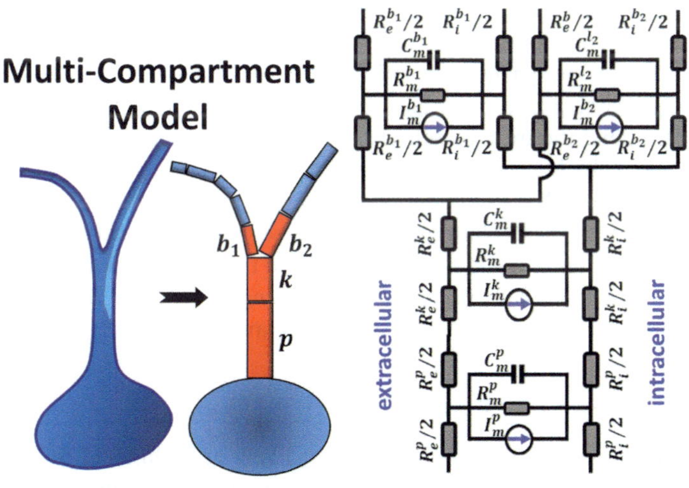

FIGURE 6.19 A symbolic neuron with a single bifurcating dendrite is used to illustrate multi-compartment modeling.

For the general case of non-zero extracellular resistance ($R_e \neq 0$), the mathematics underlying the multi-compartment model of neurons is not trivial (Wang et al., 2013). However, if the extracellular space is assumed to be space clamped ($R_e \approx 0$), a simple theoretical equation can be used to calculate the membrane potential in each compartment. We present it for the most general case in the multi-compartment model: one dendritic compartment (k) serially connected to a preceding compartment (p) and bifurcating into two parallel branches (b_1 and b_2) (6.27).

$$C_m^k \frac{dV_m^k}{dt} = -\frac{V_m^k}{R_m^k} - I_m^k - \frac{2\left(V_m^k - V_m^p\right)}{\left(R_i^k + R_i^p\right)} + 2\sum_l \frac{\left(V_m^{b_l} - V_m^k\right)}{\left(R_i^{b_l} + R_i^k\right)} \tag{6.27}$$

The transmembrane current I_m^k could comprise different components: **a)** currents through voltage-gated ion channels (Table 6.2), and **b)** current clamp injections I_{Inj}. Gating variables for each ionic current may depend nonlinearly on the membrane potentials and/or ligand molar concentrations.

$$I_m^k = \sum_{l=1}^{M} \bar{g}_l m^{p_l} h^{q_l} \left(V_m - \hat{E}_l\right) + I_{Inj} \tag{6.28}$$

A system of differential equations $d\boldsymbol{v}_m / dt = A\boldsymbol{v}_m + \boldsymbol{b}(\boldsymbol{v}_m, \boldsymbol{m}, \boldsymbol{h})$, $d\boldsymbol{m} / dt = \boldsymbol{f}_{BG}^m(\boldsymbol{v}_m, \boldsymbol{m})$, and $d\boldsymbol{h} / dt = \boldsymbol{f}_{BG}^h(\boldsymbol{v}_m, \boldsymbol{h})$, with matrix $\boldsymbol{A}$ and vector function $\boldsymbol{b}(\boldsymbol{v}_m, \boldsymbol{m}, \boldsymbol{h})$ constructed using the equations for all compartments (Eq. 6.27), must be solved simultaneously to obtain the membrane potential in the entire neuron. The nonlinear vector functions $\boldsymbol{f}_{BG}^m$ and $\boldsymbol{f}_{BG}^h$ are constructed using the Borg-Graham model discussed earlier in this chapter. Matrix $\boldsymbol{A}$ comprises resistors/capacitors for all compartments and is expected to have a semi-diagonal structure. Vector function $\boldsymbol{b}(\boldsymbol{v}_m, \boldsymbol{m}, \boldsymbol{h})$ depends nonlinearly on $\boldsymbol{v}_m$ and the gating variables ($\boldsymbol{m}$,$\boldsymbol{h}$). In this case, the membrane potential is represented by a vector $\boldsymbol{v}_m = (V_m^1, V_m^2, \cdots)'$. Gating variables for each compartment k and ion current l are comprised in two other vectors ($\boldsymbol{m} = \{m_{k,l}\}$, $\boldsymbol{h} = \{h_{k,l}\}$). Current I_m^k could also include synaptic inputs I_{syn} via neurotransmission receptors (Chapter 6).

Image processing platforms have been very useful to represent parametrically the complex dendritic trees of neurons. In this chapter, readers are referred to three of them (**Neurolucida**, https://mbfbioscience.com/neurolucida, **Imaris**, https://imaris.oxinst.com, and **NEMO** (http://centropiaggio.unipi.it/nemo-neuron-morphological-analysis-tool). These platforms allow the segmentation and rendering of complex neuronal dendritic trees. More sophisticated theoretical frameworks to represent dendritic computation have been formulated using hierarchical decompositions and graph theory (Vormberg et al., 2017; Aliaga Maraver et al., 2018).

6.9 AP PROPAGATION – CABLE THEORY

Once an AP is generated at the soma, most likely at its hillock, following spatio-temporal integration of dendritic activity, it propagates along the axons. Most axons of neurons in the central nervous system have a myelin sheath surrounding and isolating them electrically from the ECS. In the peripheral and central nervous systems, myelin sheaths are formed by Schwann cells and oligodendrocytes, respectively. An oligodendrocyte produces extension to many different axons. Each peripheral axon is wrapped with many Schwann cells. Myelin sheaths are distributed along the axons leaving small gaps ($\sim 2\mu m$) called nodes of Ranvier (Louis-Antoine Ranvier, 1835–1922). Ion flow across the membrane occurs only at the nodes of Ranvier. Therefore, conduction along myelinated axons occurs in a saltatory fashion, where AP are regenerated in each node but dissipate as heat during a fast propagation between nodes. There are few non-myelinated axons in the brain, for which conduction happens continuously along the axonal axis. The vertebrate C-fiber ($1-3\mu m$), very important in slow and strong sensory transmission, are also unmyelinated. Mossy fibers in the hippocampus, involved in encoding short-term memory, are also unmyelinated. The distant between consecutive nodes of Ranvier varies with neurons, but it is in the order of $300-2000\mu m$. The nodes of Ranvier contain Na-K ATPase pumps, NCXs and high-density of voltage-gated Na^+ channels (Nav1.6) and some K^+ channels (Kv1.2 and IK). The diameter of myelinated axons is not typically tapered, and depending on the type of neuron it could range from $1-20\mu m$. Therefore, intracellular R_i and extracellular R_e longitudinal resistances are constant along the axon. The cable theory for AP propagation along axons was motivated by the theoretical framework underlying telegraph communication via transatlantic undersea cable by Lord Kelvin in 1855. A few years earlier (1849), Hermann von Helmholtz was the first to measure the velocity of the electric signal ($\sim 24-39\, m/s$) propagating along the dissected sciatic nerve of the frog. However, it was not until mid-20th century that Hodgkin and Huxley (1952a) adapted Kelvin's theory to explain the conduction of the AP along the giant axon of the squid (Figure 6.20). In a remarkable finding, they reported a theoretical velocity for AP propagation in the unmyelinated giant axon of the squid of $18.8\, m/s$, while the experimentally observed one was $21.2\, m/s$.

Applying Ohm's law for current flow in the intracellular and extracellular areas results in:

$$V_i(x)-V_i(x+\Delta)=I_i(x)R_i\Delta \quad \text{and} \quad V_e(x)-V_e(x+\Delta)=-I_e(x)R_e\Delta$$

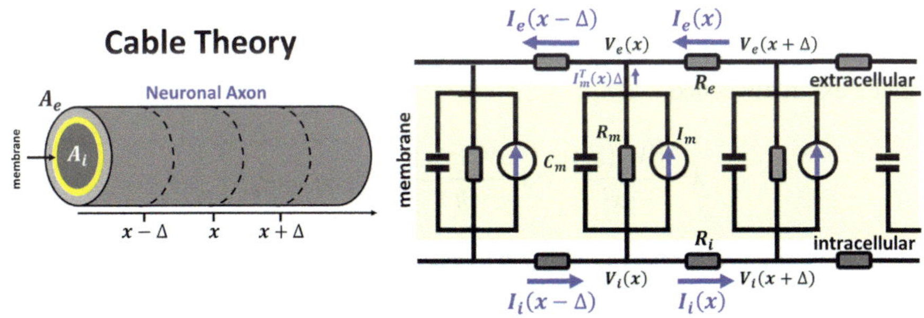

FIGURE 6.20 An illustration of the neuron axon (left) with effective intracellular A_i and extracellular A_e areas.Three positions along the longitudinal axis x of the axon are highlighted for which the equivalent electric circuit is shown (right).

$$\lim_{\Delta\to 0}\frac{\left(V_i(x+\Delta)-V_i(x)\right)}{\Delta}=\frac{\partial V_i(x)}{\partial x}=-I_i(x)R_i \qquad \text{and} \qquad \frac{\partial V_e(x)}{\partial x}=I_e(x)R_e$$

Applying the Kirchhoff law for charge conservation at the nodes of the circuit results in:

$$I_i(x-\Delta)-I_i(x)=I_m^T(x)\Delta \qquad \text{and} \qquad I_e(x-\Delta)-I_e(x)=I_m^T(x)\Delta$$

$$\lim_{\Delta\to 0}\frac{\left(I_i(x)-I_i(x-\Delta)\right)}{\Delta}=\frac{\partial I_i(x)}{\partial x}=-I_m^T(x) \quad \text{and} \qquad \frac{\partial I_e(x)}{\partial x}=-I_m^T(x)$$

Therefore, taking the second derivative of the intracellular and extracellular potentials respectively, we can obtain the following relationships:

$$\frac{\partial^2 V_i(x)}{\partial x^2}=-\frac{\partial I_i(x)}{\partial x}R_i=I_m^T(x)R_i \quad \text{and} \qquad \frac{\partial^2 V_e(x)}{\partial x^2}=\frac{\partial I_e(x)}{\partial x}R_e=-I_m^T(x)R_e$$

The membrane potential is defined as $V_m = V_i - V_e$; hence, the following equation holds:

$$\frac{\partial^2 V_m(x)}{\partial x^2}=I_m^T(x)\left(R_i+R_e\right) \tag{6.29}$$

The total transmembrane current I_m^T has three components: **i)** the capacitive, **ii)** the leaky and iii) the transmembrane current source I_m. Depending on the particular situation, the latter could include currents via ion channels, currents by active transport, synaptic currents, and experimentally clamped currents. Although the dependency with time was implicit in the previous deduction, we will include it explicitly henceforth.

$$I_m^{T(x,t)}=C_m\frac{\partial V_m(x,t)}{\partial t}+\frac{V_m(x,t)}{R_m}+I_m^{(x,t)} \tag{6.30}$$

The final cable equation (6.31) is obtained by substituting (6.30) into (6.29), with time t explicitly added.

$$\lambda^2\frac{\partial^2 V_m(x,t)}{\partial x^2}=\tau\frac{\partial V_m(x,t)}{\partial t}+V_m(x,t)+I_m^{(x,t)}R_m \tag{6.31}$$

With the time and length constants for the axon defined as $\tau = R_m C_m$ and $\lambda^2 = R_m/\left(R_i + R_e\right)$, respectively. As transmembrane currents are not permissible for myelinated axons, the cable equation take a simplified linear form:

$$\lambda^2\frac{\partial^2 V_m(x,t)}{\partial x^2}=\tau\frac{\partial V_m(x,t)}{\partial t}+V_m(x,t) \tag{6.32}$$

Equation (6.31) and (6.32) are partial differential equations of the parabolic class. Solution of these equations require knowing initial and boundary conditions. As initial condition, we can assume the entire axons is at its resting membrane potential $V_m(x,0) = V_r$. During AP propagation along myelinated axons, equation (6.32) can be solved assuming the following boundary condition at the initial node of Ranvier $V_m(0,t) = \varphi(t)$, where $\varphi(t)$ represent the temporal course of an AP generated at the node. A generalization of the cable equation for multiple passive dendrites with exponential tapering has been proposed in the past (Evans and Kember, 1998). Conduction velocities are different in myelinated and unmyelinated fibers (Rushton, 1951). In myelinated axons, it depends linearly on the myelin thickness. In unmyelinated axons, conduction velocity is proportional to the square root of the axon diameter.

There are several theoretical platforms useful to implement multi-compartment and cable equation models for neurons, some linked to extensive databases for ion channel kinetics, active transport models, and neuronal morphology. We refer here to some of the most relevant: a) Neural Ensemble (http://neuralensemble.org), b) Neuron (https://neuron.yale.edu/neuron), c) Genesis (http://genesis-sim.org/GENESIS), and NEST (https://nest-simulator.org). A more integrated platform has been recently proposed (BMTK, Welcome to the Brain Modeling Toolkit — Brain Modeling Toolkit 1.0.7 documentation (alleninstitute.github.io)).

6.10 PROBLEMS AND EXERCISES

Problem 6.1 – Assuming that the soma of a neuron is approximately spherical (radius, $r = 15\,\mu m$), provide an approximate estimation of the variation in the intracellular concentration of Na^+ and K^+ $3\,ms$ after clamping its membrane potential at $V_c = V_K + 25\,mV$. Needed additional data can be found in this and Chapter 5: V_r, $\hat{E}_{Na}$, $\hat{E}_K$, and gating kinetics (Nav1.2 and Kv2.1). The maximum conductivity values of the membrane to these ions are: $\bar{g}_{Na} = 2\,mS/cm^2$ and $\bar{g}_K = 0.5\,mS/cm^2$.

Problem 6.2 – Calculate the errors made in the estimation of $[Na^+]_i$ and $[K^+]_i$ using the parallel conductance model in **Problem 6.1**. Base your analysis on the molar concentrations for these two ions provided in Table 5.1. Assume the Na-K ATPase has been inhibited and the ECS is infinite; hence, molar concentrations for these ions are constant in it.

Problem 6.3 – Estimate the parameters in the Borg-Graham model for Nav1.2 and Kv2.1.

Problem 6.4 – Given the following layer VI pyramidal cell (Figure 6.21): **i)** dendrite branch 1 is a passive cylinder of diameter $d_1 = 2\,\mu m$ and length $l_1 = 10\,\mu m$, **ii)** dendrite branch 2 is a passive cylinder of length $l_2 = 6\,\mu m$ and with MTs following a 3/2 power scaling law, **iii)** apical dendrite is an active cylinder (CaL type with $\bar{g}_{CaL} = 0.2\,mS/cm^2$ and K_{slow} with $\bar{g}_{K_{slow}} = 0.01\,mS/cm^2$) of diameter $d = 5\,\mu m$ and length $l_A = 10\,\mu m$, **iv)** trunk is a passive cylinder of diameter $d = 2\,\mu m$ and length $l_T = 30\,\mu m$, and **v)** soma is an active sphere (H-H type Na^+ and K^+ ion channels with $\bar{g}_{Na} = 2.5\,mS/cm^2$ and $\bar{g}_K = 0.5\,mS/cm^2$) of radius $a = 10\,\mu m$.

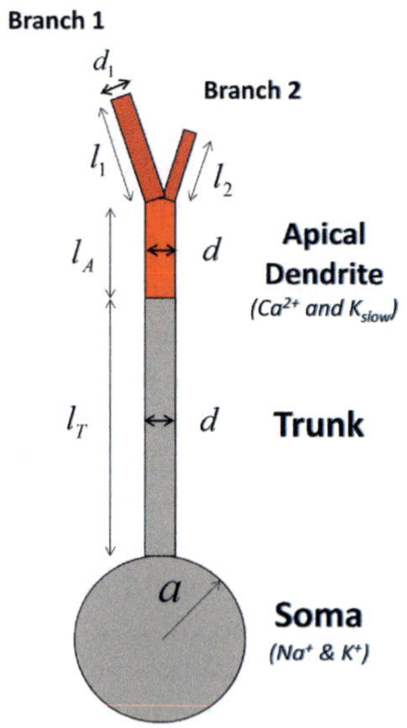

FIGURE 6.21 An illustration of the compartmentalized layer VI pyramidal cell.

a) Calculate absolute values for total resistances, capacitances and maximum transmembrane conductance for all compartments.
b) Create the circuit assuming the conductivity of the extracellular space is infinite and obtain the set of differential equations for this neuron (*feel free to use the dynamic lumping approach*).
c) Obtain the back-propagating response of this neuron in all compartments to a somatic pulse of current $t=0-200\,ms$, $200\,pA$ performed with a glass micropipette.

Problem 6.5 – Given the following concentrations for K^+ in the intra (i) and extra (e) cellular spaces of a pyramidal cell, its resting membrane potential, and the electric conductance for sodium ($[K^+]_i = 140\,mM$, $[K^+]_e = 7\,mM$, $V_r = -65\,mV$, and $\bar{g}_{Na} = 3.5\,mS/cm^2$), calculate the intracellular concentration of Na^+ ions $[Na^+]_i$ if the electric conductance for K^+ $\bar{g}_K$ is three times large than that for Na^+ ions $\bar{g}_{Na}$ and the extracellular concentration for Na^+ is $[Na^+]_e = 145\,mM$. Use the Nernst and Goldman equilibrium equations.

REFERENCES

Aliaga Maraver JJ, Mata S, Benavides-Piccione R, DeFelipe J, Pastor L. A method for the symbolic representation of neurons. *Front Neuroanat* 12(106), 1–19, 2018.

Armstrong CM, Bezanilla F. Currents related to movement of the gating particles of the sodium channels. *Nature* 242, 459–461, 1973.

Borg-Graham LJ. Modeling the nonlinear conductances of excitable membranes. In: Wheal H, Chad J (Eds.), *Cellular and Molecular Neurobiology: A Practical Approach.* Oxford University Press, New York, pp. 247–275, 1991.

Catterall WA. Structure and function of voltage-gated sodium channels at atomic resolution. *Exp Physiol* 99(1), 35–51, 2014.

Cole KS, Curtis HJ. Electric impedance of the squid giant axon during activity. *J Gen Physiol* 22 (5), 649–670, 1939.

Coulter DA, Huguenard JR, Prince DA. Calcium currents in rat thalamocortical relay neurones: Kinetic properties of the transient low-threshold current. *J Physiol Land* 4(14), 587–604, 1989.

Evans JD, Kember JC. Analytical solutions to a tapering multicylinder somatic shunt cable model for passive neurons. *Math Biosci* 149, 137–165, 1998.

Eyring H. The activated complex in chemical reactions. *J Chem Phys* 3, 107–115, 1935.

Hille B. Ionic selectivity, saturation, and block in sodium channels. A four-barrier model. *J Gen Physiol* 66, 535–560, 1975.

Hodgkin A, Huxley A. Action potentials recorded from inside a nerve fibre. *Nature* 144(3651), 710–711, 1939.

Hodgkin AL, Huxley AE. A quantitative description of membrane current and its application to conduction and excitation in nerve. *J Physiol (London)* 117, 500–540, 1952a.

Hodgkin AL, Huxley AF. Currents carried by sodium and potassium ions through the membrane of the giant axon of Loligo. *J Physiol* 116, 449–472, 1952b.

Huxley A. From overshoot to voltage clamp. *Trends Neurosci* 25(11), 553–558, 2002.

Larkum ME, Nevian T, Sandier M, Polsky A, Schiller J. Synaptic integration in tuft dendrites of layer 5 pyramidal neurons: A new unifying principle. *Science* 80(325), 756–760, 2009.

Neher E, Sakmann B. Single-channel currents recorded from membrane of denervated frog muscle fibres. *Nature* 260(5554), 799–802, 1976.

Rall W. (1959) Branching dendritic trees and motoneuron membrane resistivity. *Exp Neurol* 1, 491–527.

Rushton WAH. A theory of the effect of fibre size in medullated nerve. *J Physiol* 115, 101–122, 1951.

Sperelakis N. *Cell Physiology Sourcebook – Essentials of Membrane Biophysics* (4th Edition). Elsevier, Amsterdam, 2012.

Vormberg A, Effenberger F, Muellerleile J, Cuntz H. Universal features of dendrites through centripetal branch ordering. *PLOS Comput Biol* 13(7), e1005615, 1–25, 2017.

Wang K, Riera J, Enjieu-Kadji H, Kawashima R. The role of extracellular conductivity profiles in compartmental models for neurons: Particulars for layer 5 pyramidal cells. *Neural Comput* 25, 1807–1852, 2013.

Wong AYK, Bassingthwaighte JB. The kinetics of Ca-Na exchange in excitable tissue. *Math Biosci* 53(3–4), 275–310, 1981.

7 Inter-Neuronal Communication

7.1 OVERVIEW

Chapter 6 presented important principles underlying the genesis and propagation of APs in neurons. According to the neuron doctrine, some mechanisms for inter-neuronal communication are necessary to transmit the APs traveling along axons to other neurons. This communication occurs at small intercellular junctions called the synapses. A synapse involves a presynaptic neuron from where a signaling is transmitted, a postsynaptic receiving neuron, and an astrocyte acting as an external intermediator. Although it quite varies considerably, the typical number of synapses per neuron can be several thousands. Therefore, there are more than 86 trillion synapses in the human brain. Exposition to daily experiences causes dynamic reconfigurations of the connection strengths between neurons (*synaptic plasticity*), a complex process underlying adaptive development, personality, and learning/memory. Dysfunctions of synaptic communication might cause different types of brain and cognitive disorders. This chapter revisits main biophysical principles underlying neuronal communication.

7.2 SYNAPTIC TRANSMISSION TYPES: ELECTRICAL AND CHEMICAL

Initial studies of the motor synapses of the crayfish suggested that the propagation of an AP between two neurons might occur in a form of an electrical spark. However, the paralytic action of curare suggested the possibility of a chemical transmission. Today, we know that most of the synapses in the central nervous system are chemical, but some synapses are electrical. In electrical synapses (Figure 7.1), electrotonic currents caused by a presynaptic AP are transmitted to postsynaptic neurons through gap-junction channels. These gap-junctions comprise two hemi-channels connecting the presynaptic and postsynaptic membranes, which must be in very close proximity $\sim 3.5\,nm$. Each hemi-channel, named connexon, consists of an annular array of six-peptide subunit (i.e., the connexin). Any molecule/ion with dimension $\leq 1000\,daltons$ can travel along the channel. In general, electrical synapses are bidirectional; hence, they typically play a crucial role in neuronal synchronization. The conductance of the gap-junction is regulated by two gating principles: **i)** the fast intercellular

DOI: 10.1201/9780367901509-7

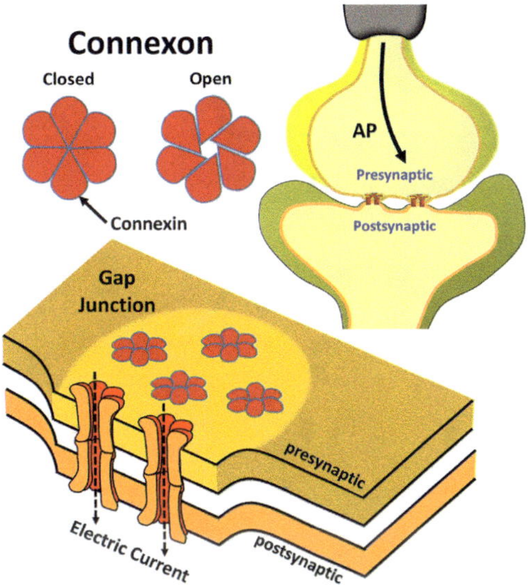

FIGURE 7.1 An illustration of the electrical synapses.

voltage-dependent V_j-gating and **ii)** the slow chemical dependent (H^+ and Ca^{2+}) transitions. The latter has been suggested to play an important role in neuronal uncoupling under extreme physiological conditions (abnormal pH or reactivity).

In contrast, chemical synapses (Figure 7.2) are slower and unidirectional. The discovery by Bernard Katz (1911–2003) of the fusion of synaptic vesicles (SV) with the plasma membrane in the presynaptic terminal (also referred as a bouton) for a quantal neurotransmitter release constitutes a fundamental building-block for our understanding of chemical synapses. In 1970, Katz received the Nobel Prize in Physiology or Medicine for this pioneering work. In a chemical synapse, membrane depolarization at the presynaptic terminal due to the arrival of an AP causes the quantal release of neurotransmitters. Katz's quantal theory is described with two parameters: **a)** the probability for the release of a neurotransmitter quantum p and **b)** the number of sites n that can release a quantum. Assuming independence between the quantum releasing sites, the probability a number x of quantal are releases follows a binomial distribution:

$$p(x:n) = \frac{n!}{x!(n-x)!} p^x (1-p)^{n-x} \tag{7.1}$$

Data fitting from different types of neurons indicates that the *quantal content* $1 < pn < 300$, with the lowest values reported for neurons in the central nervous system.

The process by mean of which neurotransmitters are released from the presynaptic terminal comprises five major steps: **1)** SV pool establishment at the terminal either

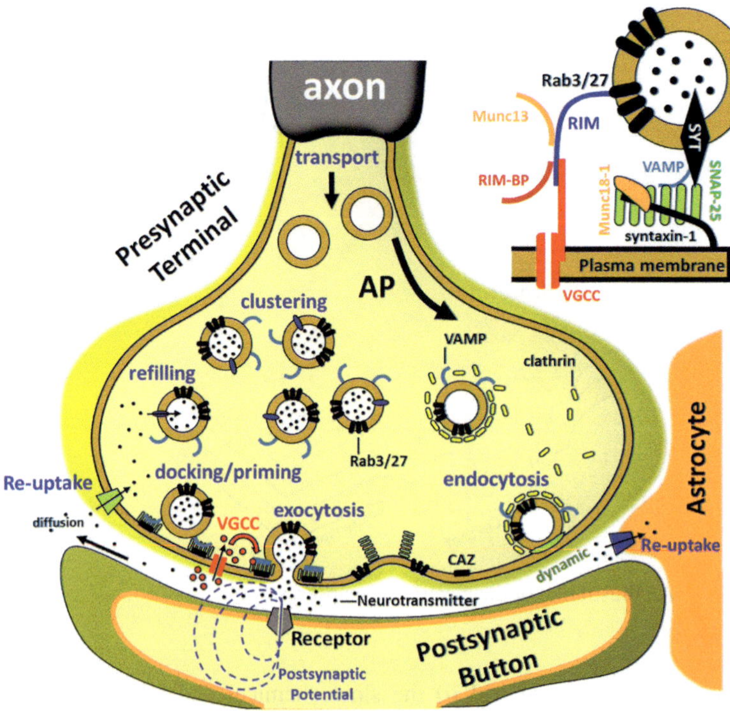

FIGURE 7.2 An illustration of the chemical synapses.

by endocytosis or by cargo-transport from neuron's somas, **2)** SV clustering, **3)** SV refilling, **4)** docking/priming SVs, and **5)** neurotransmitter release (exocytosis). We focus this section on the four most relevant steps.

<u>*SV Refilling*</u> – Neurotransmitters released into the synaptic cleft are taken either back to the presynaptic terminal or recycled by surrounding astrocytes. Reuptake/recycling depends on secondary active transport in both neurons and astrocytes. Together with spontaneous diffusion of neurotransmitters out of the cleft and their degradation by enzymes, reuptake /recycling helps terminate the synaptic signaling. Details about co/counter-transporters in presynaptic terminals involved in the reuptake of each specific neurotransmitter are provided in Chapter 8. The astrocytic-dependent recycling of two amino acid neurotransmitters is also discussed in Chapter 8. Long-term efficiency in synaptic transmission depends on the speed of refilling SVs with neurotransmitters at high concentrations. Therefore, once neurotransmitters are back in the terminal, they must be quickly transported back into the secretory SVs. Classical neurotransmitters are injected into the SVs using the H^+ electrochemical driving force across the vesicle membrane, which is created by the vacuolar (V-type) H^+ -ATPase (Blakely and Edwards, 2012). The refilling process depends also on the cytosolic concentration of the particular neurotransmitter and the ionic coupling

with the counter-transporter. These are the counter-transporters for the classical neurotransmitters: **a)** VMAT ($K_M < \mu M$) – the vesicular monoamine transporter, **b)** VAChT ($K_M \sim mM$) – the vesicular ACh transporter, **c)** VGAT ($K_M \sim mM$) – the vesicular γ-aminobutyric-acid(GABA)/glycine transporter, and **d)** VGLUT1-3 ($K_M \sim mM$) – the vesicular glutamate transporter. The role of chlorine in SV refilling has been also suggested.

Docking/Priming SVs – A very complex molecular machinery underlies the docking and priming of SVs. We will describe only some of the most important aspects in here. Release occurs at specialized areas of the membrane comprising a dense collection of proteins named cytomatrix at the active zone (CAZ). The organization of these release sites is mediated by three multidomain proteins tightly coupled to form a large protein complex (RIM, RIM-BP, and Munc13). Facilitated by an N-terminal domain, RIM binds to Rab3/27 GTP-binding proteins on the SV and to MUNC-13. In addition, together with RIM-BP, RIM recruits VGCCs into the active zones forming microdomains for a fast $[Ca^{2+}]_i$ increase. The SNARE/SM protein complex, the molecular machinery involved in the core fusion, is composed of the SNARE proteins and the SM protein Munc18-1. The SNARE proteins are of two types: the vesicular SNARE protein (synaptobrevin/VAMP) and the plasma membrane SNARE proteins (syntaxin-1 and SNAP-25). MUNC-13 participates in syntaxin-1 conformational change from closed to open, which promotes SNARE complex assembly. In this state, Munc18-1 remains attached to syntaxin-1 in a form of interaction with the SNARE complex. Under these conditions, the SVs are primed into an energized and metastable state, ready to be released. The force generated by SNARE complex assembly is essential for the fusion of the SV and plasma membrane. It has been speculated that the SM proteins facilitate lipid mixing during fusion (Südhof, 2013).

Exocytosis – The participation of basal Ca^{2+} signaling to promote the formation of transitory t-SNARE complexes, composed of "open" syntaxin-1 and SNAP-25, remains unclear. At the arrival of an AP, opening of VGCCs causes a rapid increase in $[Ca^{2+}]_i$. Most synapses in the brain involve two specific HVA (P/Q-type "Cav2.1" and N-type "Cav2.2") subtypes of VGCCs. SVs have a Ca^{2+}-sensor protein (synaptotagmin, SYT), which can be anchored (primed SVs) to the membrane via an N-terminal domain. The Ca^{2+}-sensor is formed by two C-terminals (C_2-domains) to which Ca^{2+} ions can bind. Thanks to the RIM/RIM-BP binding, VGCCs are densely distributed in clusters around primed SVs. The distribution and density of VGCCs in each terminal to form these microdomains ($< 100\,nm$) is dynamically controled by the neuron's transcription apparatus, an important aspect of synaptic plasticity. However, free Ca^{2+} ions inside these microdomains can be rapidly buffered by several endogenous mechanisms. Upon binding Ca^{2+}, the SYT C_2-domains become activated, initiating the fusion of the membrane of the SV and that of the neuron due to interactions of the SYT with the SNARE complex assembly and phospholipids in the plasma membrane. A protein, the complexin adaptor (Priming I), assists SYT by activating/clamping the core fusion machinery. PIP_2 binding to RIM may be also crucial for efficient exocytosis (Priming II). During the fusing process, neurotransmitters are quickly ejected ($<< 100\,\mu s$) to the synaptic cleft.

SV Endocytosis – SVs can be transported from the soma using microtubule cargo highways. However, presynaptic terminals have also developed endocytic mechanisms to recapture/reuse membranes that have been fused. Three mechanisms have been proposed in the past: **a)** clathrin-coated pits (CCP), **b)** "kiss and run", and **c)** bulk endocytosis. However, the prevalent pathway of SV protein internalization has been suggested to be that mediated by clathrin (Saheki and De Camilli, 2012). CCP nucleation is initiated with interactions of adaptors and/or accessory factors. Nucleation leads to the initial quick development of the coat with important SV components incorporated, creating a deeply invaginated bud with a narrow neck connected to the plasma membrane. Final fission of this neck is performed by the GTPase dynamin. Once the SVs are formed, the clathrin-based coat is rapidly removed.

7.3 THE TRIPARTITE SYNAPSIS – HEBB AND STDP RULES FOR PLASTICITY

The synapse was initially thought to have two main elements, a presynaptic terminal and a postsynaptic site. They are surrounded by astroglia cells, which was assumed to provide physical support. Nowadays, we know that the perisynaptic processes of protoplasmic astrocytes, together with pre/postsynaptic neuronal components, form a tripartite unit involved in synapse functioning. Although there have been also calls for *tetrapartite* (extracellular matrix) and *quad-partite* (microglia) synapses, this book focuses only on the role of the tripartite synapse. Details about the multiple roles of these astrocytes in synapses, such as homeostasis regulation, metabolic support, aminoacid neurotransmitter recycling, are discussed in Chapters 8 and 9. Their role in synaptic activity integration, process and modulation is one of the most important for modern cognitive neuroscientists. Astrocytes respond to synaptic activation in the form of spontaneous intrinsic Ca^{2+} oscillations, whose strength holds nonlinear relationships with synaptic activity (Perea and Araque, 2005). In addition, astrocytes release a series of modulatory molecules that influence synaptic activity, a process called gliotransmission (Volterra and Bezzi, 2002). Gliotransmission occurs in a Ca^{2+} -signaling dependent manner via vesicle and lysosome exocytosis. These are among the most important gliotransmitters identified to date: glutamate, GABA, D-serine, ATP, TNFα, and prostaglandins. As neurons and astrocytes dynamically interact, short and long –term adjustments of synapse strengths occur. In this context, this section first introduces the concept of synaptic strength from a theoretical viewpoint, which is followed by a short discussion of the underlying molecular bases.

In his book, "The Organization of Behavior (1949), Donald Hebb (1904–1985) referred to an increase in the efficacy of an axon to activate a nearby neuron if these two neurons fire APs with a high degree of correlation. He even went far by postulating that some growth process or metabolic change might take place to facilitate what is known today as the *Hebbian learning* theory. In neuroscience, it, referred to as the *Hebb synapse*, has been linked to long-term potentiation (Bliss and Lomo, 1973) and kindling (Goddard et al., 1969). It is known as the *Hebb rule* in neural/artificial network models. A series of experiments in the 80's motivated the generalization of the Hebbian learning theory to be temporally asymmetric, a theoretical framework known as the spike-timing-dependent plasticity (STDP). Under the STDP, changes in the synaptic connection strength $\Delta\omega_{j\rightarrow i}$ from neuron j to neuron i depends on their relative firing timings (7.2)

$$\Delta\omega_{j\rightarrow i} = \sum_{l}\sum_{m} W\left(t_i^m - t_j^l\right) \tag{7.2}$$

With $W(x) = \begin{cases} A_+ e^{-x/\tau_+} & x > 0 \\ -A_- e^{x/\tau_-} & x < 0 \end{cases}$

According to this theory, there was a critical window ($\tau_\pm \approx 10ms$) for the induction of either a synaptic potentiation or depression. The non-negative scaling factors $A_\pm$ could, in principle, depend on the value for the synaptic strength $\omega_{j\rightarrow i}$ during that critical window. Superscripts m and l refers to the spiking events in the postsynaptic and presynaptic neurons, respectively. Note that under this theory, the synaptic connection strengthens if presynaptic spikes occur on average within the critical window but before the postsynaptic spikes. In many synapse types, this might lead to the long-term potentiation (LTP). However, the synaptic connection weakens, a type of long-term depression (LTD), if presynaptic spikes occur typically after the postsynaptic ones. The STDP depends on the synapse types and brain regions. Some studies have suggested nonlinearities on the postsynaptic spike rate and a multiple postsynaptic spiking requirement for the STDP to take place. Both external and internal factors affect synaptic plasticity in the form of STDP, which underlie learning, development, and information storage in the brain.

The concept of dendritic spines is very useful to discuss potential molecular bases of LTP and LTD. Spines are small protrusions in the dendrites of some neurons, called spiny neurons, which receive glutamatergic synapses. Although their shape might dynamically change as defined by F-actin accumulation and its conversion into G-actin, spines are in general classified according to morphology into three types: thin (filopodia), mushroom and stubby. Mushroom spines, the most typically observe in mature brains, have a thin (~ $0.2\,\mu m$), long neck (~ $0.7\,\mu m$) and a large head ($0.07\,\mu m^3$). The heads comprise: **a)** a dense endoplasmic reticulum (ER) apparatus for protein synthesis and Ca^{2+} signaling, and **b)** a postsynaptic density (PSD) area densely filled with α-amino-3-hydroxy-5-methyl-4-isoxazolepropionic acid/kainate (AMPA/k) and dependent on N-methyl-D-aspartate (NMDA) receptors, as well as PSD95 molecules. The narrow neck of spines limits diffusion in and out of any intracellular signaling molecule. Because of its small volume, molar Ca^{2+} concentration, one of the most important intracellular signaling in spines due to its involvement in protein regulation by phosphorylation, can change dramatically in a very short time window. Upon the arrival of a large and sustained amount of glutamate, Ca^{2+} concentration can increase quickly inside the spine through a variety of mechanisms: **1)** NMDA receptor opening, **2)** the activation of the ryanodine (RyR) and inositol trisphosphate (IP_3R) receptors on the ER membrane, and **3)** the opening of several VGCCs (Cav2.1-3). The specific conditions required for each of these mechanisms to be involved depend on the glutamatergic signaling profiles in the synaptic gap and the postsynaptic membrane potential magnitude. The immediate impact of released glutamate into the synaptic cleft is the activation of AMPA/k receptors, which due to the opening of mainly Na^{2+} channels lead to the depolarization of the spine membrane. Local changes in membrane potentials cause the opening of VGCCs, consequently increases in $[Ca^{2+}]_i$. Spine membrane depolarization could also trigger the release of the Mg^{2+}-block of the ion channel from the NMDA receptor. Mg^{2+}-unblocking has been proposed as a key mechanism for STDP as it helps detect coincident pre- and postsynaptic AP events. Extracellular glutamate could also bind metabotropic glutamate receptors (mGluR, Group I), which are G-protein-coupled receptors (GPCR) that induce the production of inositol trisphosphate (IP_3) by the γ-phospholipase C (PLC_γ). Cytosolic IP_3 stimulate the release of Ca^{2+} from ER via IP_3R. Finally, cytosolic Ca^{2+} can cause RyR activation, leading to the Ca^{2+}-induce Ca^{2+} release (CICR) response. Different types of LTP and LTD have been linked to the differential sensitivity to cytosolic Ca^{2+} of kinases and phosphatases involved in phosphorylation. The following kinases have been linked to LTP: Ca^{2+}/calmodulin-dependent protein kinase II (CaMKII, via activation/inhibition of its autophosphorylation – "the CaMKII hypothesis", Lisman, 1994), cAMP-dependent protein kinase (PKA), protein kinase C (PKC), cGMP-dependent protein kinase (PKG), casein kinase II (CKII), tyrosine kinases (Src and Fyn), and mitogen-activated protein kinase (MAPK). Certain protein phosphatases have been also identified with LTP, e.g., protein tyrosine phosphatase (PTP). As for LTD, these are phosphatases associated to it in the past: protein phosphatase 1 (PP1), protein phosphatase 2A/2B (PP2A/B). LTD has also been related with some protein kinase regulation, e.g., PKA and PKC. Phosphorylation of NMDA and AMPA/k receptors could occur at different sites. For example, phosphorylation of certain serine (S) and tyrosine (Y) residues on the NR1, NR2A, and NR2B subunits is a key factor for the regulation of NMDA receptors. Phosphorylation of GluR1-4 subunits is also crucial for AMPA/k receptors regulation and functioning. LTP and LTD types and mechanisms for induction are not the same for all neurons, brain regions and developmental stages. Finally, alterations in gene expression in the pre- or postsynaptic neurons have been also linked to synaptic plasticity.

7.4 LONG-TERM POTENTIATION/DEPRESSION

The magnitude and temporal profiles of $[Ca^{2+}]_i$ increases in the spine determine how kinase and phosphatases are recruited, which have an impact not only on whether LTP or LTD is induced, but also on its type (e.g., early/late LTP, homo/hetero-synaptic LTD, depotentiation/de-novo LTD). In an oversimplified view, brief and strong postsynaptic Ca^{2+} events are assumed to trigger LTP, while LTD has been associated with small, prolonged Ca^{2+} elevations (Sjöström et al., 2008 – the calcium hypothesis in synaptic plasticity). Two major aspects might affect Ca^{2+} dynamics inside the spine: the experimental paradigms used for LTP/LTD induction and Ca^{2+} diffusivity due to spine geometry/ER-location. Several experimental paradigms have been used to induce LTP, e.g., high-frequency stimulation ($1s$ train at $100\,Hz$) and θ-burst stimulation (TBS, bursts repeated at $5\,Hz$, ~ 10 bursts in one train separated typically by $10-60\,s$). LTD can be induced by low-frequency stimulation "LFS", by pairing baseline synaptic stimulation using depolarization, and by manipulating back-propagating AP timing. Chemical (receptor agonists) experimental paradigms can be also used to induce LTP and LTD. Accurate biophysical models can be used to evaluate the impact of spine geometry, ER locations and cytosolic Ca^{2+} -sources on Ca^{2+} dynamics inside the spine (Johenning et al., 2015; Breit et al., 2018). However, a simple calculation is useful to determine the mean time it takes Ca^{2+} ions to distribute inside the spine head. Based on the probability density function, Albert Einstein (1879–1955) introduced the n-D mean squared dispersion $\bar{x}$ of ions due to their Brownian motions: $\bar{x}^2 \cong 2nDt$. Given that the average diameter of spine head is ~ $0.25\,\mu m$ (approximating it by an sphere) and that the effective diffusion coefficient for Ca^{2+} ions inside the spine head is $D_{Ca} \approx 0.220\times10^{-9}\,m^2/s$, it will take approximately $47\,\mu s$ for a steady-state Ca^{2+} gradient to be a established inside the spine if its sources are homogeneously distributed inside the spine. Hence, the impact of Ca^{2+} diffusivity in LTP/LTD is negligible.

However, RyR and IP_3R kinetics are very important aspects for $[Ca^{2+}]_i$ dynamics inside the spine. Kinetic models for IP_3R are presented in Chapter 8. To conclude this section, a very useful four-state kinetic model of RyR, adjusted to cardiac cell data, is presented (Keizer and Levine, 1996).

$$I^{RyR}_{Ca^{2+}} = \rho_{RyR}(o_1 + o_2)I_{ref}\left([Ca^{2+}]_{ER} - [Ca^{2+}]_i\right) \tag{7.3}$$

With, RyR closed states c_1 and c_2 satisfying the following differential equations:

$$\frac{dc_1}{dt} = k_a^- o_1 - k_a^+\left([Ca^{2+}]_i\right)^4 c_1 \quad \text{and} \quad \frac{dc_2}{dt} = k_c^+ o_1 - k_c^- c_2$$

The RyR open states o_1 and o_2 satisfy the equations:

$$\frac{do_2}{dt} = k_b^+\left([Ca^{2+}]_i\right)^3 o_1 - k_b^- o_2 \quad \text{and} \quad o_1 = 1 - c_1 - o_2 - c_2$$

The fitted model parameters are: $k_a^- = 28.8\,s^{-1}$, $k_a^+ = 1500\,\mu M^{-4}\,s^{-1}$, $k_b^- = 385.9\,s^{-1}$, $k_b^+ = 1500\,\mu M^{-3}\,s^{-1}$, $k_c^- = 0.1\,s^{-1}$, $k_c^+ = 1.0\,s^{-1}$, and the single channel current $I_{ref} = 3.5\times10^{-18}\,mol/s$. The density of RyR on the ER membrane ρ_{RyR} must be determined experimentally for each particular dendritic spine. The Ca^{2+} concentration inside the ER could be considered constant $[Ca^{2+}]_{ER} = 250\,\mu M$. Ca^{2+} active transport and buffering might also have an important impact on its dynamics inside dendritic spines. Kinetics models for the PMCA and SERCA pumps were provided in Chapter 6. Models for Ca^{2+} buffering are discussed in Chapter 8.

Assisted by a plethora of proteins in the ER membrane (SERCA, RyR, and IP_3R) and plasma membrane (PMCA, VGCC, NCX, mGluR, and NMDAR) for $[Ca^{2+}]_i$ regulation, the spine can provide the optimal Ca^{2+} ion support for protein phosphorylation in the PSD region (Figure 7.3). The high density of proteins for active transport in the head ensures a very low steady state Ca^{2+} concentration in the proximity of the PSD, even after clamping it at $1\,\mu M$ in the dendrite (Figure 7.4). This property,

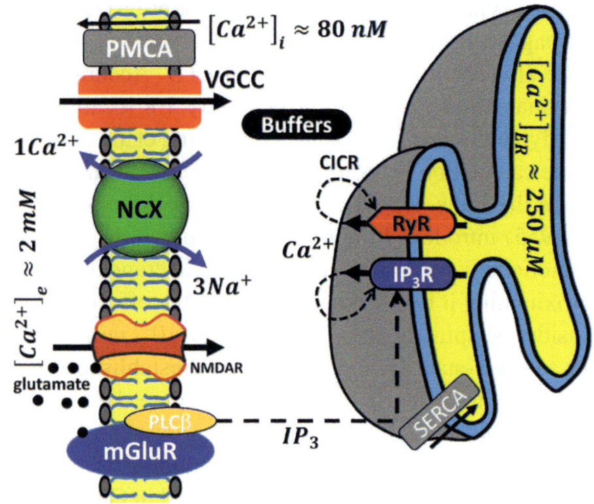

FIGURE 7.3 Different mechanisms for $[Ca^{2+}]_i$ control in spines.

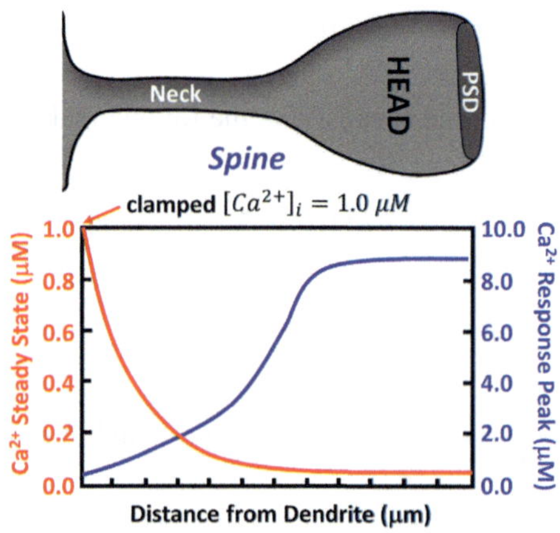

FIGURE 7.4 $[Ca^{2+}]_i$ spatial profiles in spines after clamping Ca^{2+} concentration at the dendrite ($1\mu M$, red) and during Ca^{2+} responses to a synaptic signaling (peak, blue).(Computer simulations adapted from Koch et al. (1992).)

together with the high Ca^{2+} diffusivity along spine neck, allows a maximum Ca^{2+} signal response (peak) at the PSD for optimal phosphorylation of neurotransmission receptors.

VGCCs (Cav2.1-3) are important for synaptic transmission, both in presynaptic terminals and spines. To identify the gating kinetics of these channels in terminal and spines, patch-clamp recording in very challenging experimental preparations are required. Table 7.1 summarizes some of the kinetic models proposed for these N-, R- and P/Q- Types of VGCCs.

TABLE 7.1
Kinetic models for N-, R- and P/Q-types of VGCCs

Name	**Exponent**	z	γ	α_0	β_0	V_H
N-Type	$p = 3$	3.0	0.5	0.1	0.1	–30.0
	$q = 1$	–4.0	0.7	0.025	0.025	–50.0

Name	**Exponent**	α	β
P/Q, N-, L-Type	$p = 2$	$\dfrac{-0.055\,(V_m + 27)}{e^{-(V_m+27)/3.8} - 1}$	$0.94\, e^{(V_m+75)/17}$
	$q = 1$	$0.000457\, e^{-(V_m+13)/50}$	$\dfrac{0.0065}{e^{-(V_m+15)/28} + 1}$

Name

$$C_0 \underset{\beta_1(V_m)}{\overset{\alpha_1(V_m)}{\rightleftarrows}} C_1 \underset{\beta_2(V_m)}{\overset{\alpha_2(V_m)}{\rightleftarrows}} C_2 \underset{\beta_3(V_m)}{\overset{\alpha_3(V_m)}{\rightleftarrows}} C_3 \underset{\beta_4(V_m)}{\overset{\alpha_4(V_m)}{\rightleftarrows}} C_4 \underset{\beta}{\overset{\alpha}{\rightleftarrows}} \mathbf{0} \quad \alpha_i(V_m) = \alpha_{i,0} e^{V_m/k_i} \quad \beta_i(V_m) = \beta_{i,0} e^{-V_m/k_i}$$

Li et al., (2007)

N-Type	$\alpha_{1,0} = 4.29$	$\alpha_{2,0} = 5.24$	$\alpha_{3,0} = 4.98$	$\alpha_{4,0} = 772.63$	$\alpha = 615.01$
	$\beta_{1,0} = 5.23$	$\beta_{2,0} = 6.63$	$\beta_{3,0} = 73.89$	$\beta_{4,0} = 692.18$	$\beta = 7.68$
	$k_1 = 68.75$	$k_2 = 39.53$	$k_3 = 281.62$	$k_4 = 18.46$	
R-Type	$\alpha_{1,0} = 9911.36$	$\alpha_{2,0} = 4.88$	$\alpha_{3,0} = 4.00$	$\alpha_{4,0} = 256.41$	$\alpha = 228.83$
	$\beta_{1,0} = 0.62$	$\beta_{2,0} = 21.91$	$\beta_{3,0} = 51.30$	$\beta_{4,0} = 116.97$	$\beta = 1.78$
	$k_1 = 67.75$	$k_2 = 50.94$	$k_3 = 173.29$	$k_4 = 16.92$	
P/Q-Type	$\alpha_{1,0} = 5.89$	$\alpha_{2,0} = 9.21$	$\alpha_{3,0} = 5.20$	$\alpha_{4,0} = 1823.18$	$\alpha = 247.71$
	$\beta_{1,0} = 14.99$	$\beta_{2,0} = 6.63$	$\beta_{3,0} = 132.80$	$\beta_{4,0} = 248.58$	$\beta = 8.28$
	$k_1 = 62.61$	$k_2 = 33.92$	$k_3 = 135.08$	$k_4 = 20.86$	

Notes: The Borg-Graham model for N-Type channels was proposed by Lytton and Sejnowski (1991). Reuveni et al. (1993) obtained the kinetics of HVA VGCCs in dendrites, but separation into their specific isomorphs were not possible. The most complete characterization of these channels in terminal is in Li et al. (2007), who modeled the activation gate of these channels using a macroscopic multistate gating model with five closed states and a single open state. Inactivated states were not included.

7.5 NEUROTRANSMITTERS: ROLES AND DISTRIBUTIONS

Chemical synapses are the most abundant in mammals. Depending on changes they produce in V_m on the postsynaptic site, chemical synapses can be either excitatory (depolarize, $V_m > V_r$) or inhibitory (hyperpolarize, $V_m < V_r$). Chemical synapses require neurotransmitters to diffuse from the releasing sites in the axon terminals to postsynaptic dendritic sites, where they bind different types of receptors. This type of neuronal communication takes place in axodendritic synapses. However, there are also axosomatic, dendrodendritic and axoaxonic synapses in the central nervous system. Therefore, axon terminals and dendritic spines are not always required for chemical synapses to occur. Typically, axosomatic synapses are projections of axons in the form of terminals directly onto the somas of neurons. In general, inhibitory synapses involve an axon terminal of a GABAergic neuron targeting a dendrite/soma of another neuron. Communication from axons can also reach neurons via varicosities, which are protuberances along the axons as they pass postsynaptic neurons. This type of synapses is called "*passant*". The release of neurotransmitters at a passant synapse may affect several postsynaptic neurons, which is referred to sometimes as volume transmission. As conventional chemical synapses involve a presynaptic site in close proximity to a postsynaptic site (cleft ~ 30 nm gap), they are also referred to as wiring transmission. Neurons can also communicate with other cells via unconventional neurotransmission, in which gases (nitric oxide "NO", carbon monoxide "CO") and peptides (e.g., prostaglandins) are released by the neurons without having them stored in SVs. Conventional chemical synapses release quantal of neurotransmitters from SVs by a Ca^{2+}-dependent exocytosis. Classified according to the size of the SVs, neurons release three major types of neurotransmitters: **a)** L-amino acids/acetylcholine (electron-lucent SVs of ~ $40\,nm$ in diameter), **b)** monoamines (electron-opaque SVs of ~ $40-70\,nm$ in diameter), and **c)** peptides (large dense core SVs of ~ $100-150\,nm$ in diameter). Glutamate (Glu) and GABA are the most important L-amino acid neurotransmitters. Glu is the most widespread and abundant neurotransmitter in the brain, with highest density in the cerebral cortex and brainstem. It causes postsynaptic membrane depolarization (excitatory) due to its action on AMPA/k (fast) and NMDA (slow, V_m-dependent) receptors. Glutamatergic synapses in spines are linked to memory and learning. About 35% of all synapses in the brain are GABAergic, with larger concentrations in the substantia nigra, globus pallidus, hypothalamus, periaqueductal gray, hippocampus, and cerebellum. GABA causes hyperpolarization (inhibitory) of the postsynaptic membrane by its action of $GABA_A$ (fast) and $GABA_B$ (slow, 2nd messengers) receptors. In the mature cerebral cortex, ~ 90% of all synapses are either glutamatergic or GABAergic, with the former outnumbering the latter ~ 5 : 1. Although its low density, GABA is very important in keeping the right excitatory/inhibitory balance in the cerebral cortex. Aspartic acid (excitatory) and glycine (inhibitory) are also L-amino acid neurotransmitters, but they are not discussed in this book as their density in the brain is very low. The first neurotransmitter, acetylcholine (ACh), was discovered in 1926 as a chemical signal underling the connection from vagus nerve to the heart. For this finding, Henry H Dale (1875–1968) and Otto Loewi (1873–1961) received the Nobel Prize in Physiology or Medicine 1936. ACh is synthetized by neurons from choline and acetyl-CoA by the choline-acetyltransferase. The concentration in the brain is low and spotty. ACh causes membrane potential depolarization by its action on nicotinic and muscarinic receptors. Muscarinic receptors are the most abundant in the brain. Free ACh in the synaptic cleft is metabolized back to choline by the acetylcholinesterase. ACh has been linked to several important cognitive functions. For example, ACh neurons in the pontomesencephalic tegmentum play a role in cortical activation during waking and in the genesis of rapid eye movement (REM) during sleep. There are also important ACh projections from the basal nucleus of Meynert to the cerebral cortex, the reticular thalamic nucleus, the amygdala, and the basal ganglia. Interneurons in the nucleus accumbens and striatum are cholinergic.

On the other hand, monoamines, with production linked to different midbrain nuclei, are involved in cognitive functions, mood cycles and emotion. They are subdivided into: **a)** the imidazoleamines (histamine), **b)** the catecholamines (epinephrine "E", norepinephrine "NE", dopamine "DA"), and **c)** the indolamines (serotonin "5-HT", melatonin

"MT"). In the brain, histamine promotes wakefulness. It is also involved in motivation, goal-directed behaviors, and homeostatic functions. It is found in the hypothalamus, brainstem, and cerebral cortex. NE can be found in the cerebral cortex, hypothalamus, brainstem, and cerebellum. It is involved in attention and consciousness, as well as in the regulation of the pituitary gland secretion. E causes an excitatory effect on the autonomic pathway. This neurotransmitter is found in the thalamus, hypothalamus, and midbrain. DA is involved in many different brain functions. Among the most notable are: movement control, pleasurable reward, neuroendocrine signal modulator, mood, sleep, attention, and memory/learning. DA is widely found in the brain, e.g., midbrain, amygdala, hippocampus, basal ganglia, cerebellum, thalamus, frontal cortex, and the hypothalamus. 5-HT is abundant in the hypothalamus, cerebellum, and limbic system. It is a neurotransmitter involved in mood/emotional states (impulsivity/sexual-behavior), cognition (e.g., memory/learning and imagination), and sleep. MT is produced by the pineal gland with a circadian rhythmicity. It is regulated by the suprachiasmatic nucleus (SCN) of the hypothalamus. It has been linked to sleep promotion and associated with neuroprotective and anti-excitatory role in the brain. Postsynaptic receptors for most important monoamines are presented in Table 7.2. However, monoamine receptors are also found in vessels (endothelial and smooth muscle cells) and glial cells, which have been linked to neurovascular coupling, metabolism homeostasis, hormones secretions and neuroprotection. The roles played by these non-neuronal monoaminergic systems will be sporadically discussed in this book.

Unconventional neuro- and glio- transmitters are also important for brain functions. Purines (adenosine, and adenosine triphosphate "ATP") are among the most important, e.g., ATP in astrocytic network communication (Chapter 8). Some peptides released by neurons (e.g., somatostatin, substance P, neuropeptide Y "NPY", vasoactive intestinal polypeptide "VIP") will be also discussed in the section for vascular control (Chapter 10).

7.6 LIGAND-GATED PRINCIPLES – THE IONOTROPIC AND METABOTROPIC RECEPTORS

Acting as ligands on specific receptors, neurotransmitters cause a variety of effects at the postsynaptic neuronal sites, from changes in membrane potentials by the opening of ion channels to modulation of synaptic transmission by the activation of intracellular second messengers. Postsynaptic receptors have been classified as either ionotropic or metabotropic.

Ionotropic receptors: The binding of a neurotransmitter to an ionotropic receptor (step ①) causes a protein-conformational change that leads to the direct formation of a channel (step ②) with selectivity to specific ions (Figure 7.5). The resulting ionic current by the channel causes small, graded changes in membrane potentials at localized sites in the postsynaptic neuron. In this chapter, five ionotropic receptors are discussed.

AMPA/k receptors: They are tetramers composed of subunits GluR1-4. The receptor has an extracellular N-terminus and an intracellular C-terminus. The receptor is phosphorylated at four sites on the GluR1

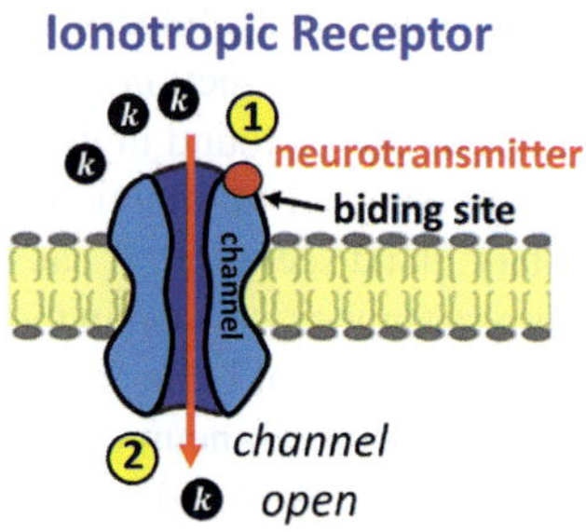

FIGURE 7.5 Mechanisms of action for the ionotropic receptors.

subunits. Channel localization and conductance is regulated by phosphorylation at three serine sites (PKC on S818, CaMKII/PKC on S831, and PKA on S845). Interneurons express more Ca^{2+}-permeable AMPA/k receptors. Permeability to ions is controlled by the GluR2 subunit.

NMDA receptors: They are formed by tetrameric assembly of two NR1 subunits and two NR2 subunits. There are four NR2 subunit types (NR2A, NR2B, NR2C, and NR2D). Signaling is mediated by a large NMDA receptor complex (NRC) of multiple interacting proteins. NMDA receptors are regulated by phosphorylation of its subunits, many of them on the intracellular carboxy-tail, which is crucial for synaptic plasticity. The Mg^{2+}-sensitivity and kinetics of the channel depend on the NR2 subunit. Allosteric binding sites for zinc (Zn^{2+}) ions and polyamines are also modulator of the channel. A D-aminoacid (D-serine), which is relevant in astrocyte function, act as another important modulator of NMDA receptors by binding to the glycine site on the NR1 subunit.

$GABA_A$ receptors: They are formed by five subunits (α, β and γ), each comprising four TMs with extracellular N and C terminus, organized around a central pore. There are 6 α, 3 β and 3 γ types of subunits. The most abundant $GABA_A$ receptor is formed by 2 α, 2 β and 1 γ subunits (pentamer α2β2γ). The activity of the receptor is modulated by a different allosteric binding sites in the α (benzodiazepines) and β (agonists/antagonists) subunits, which have been target for anesthetics.

Nicotinic $\alpha_{3,4,7}/\beta_{2,4}$ and 5-HT_3 receptors: These receptors are pentameric ligand-gated ion channel super-family, to which $GABA_A$ also belongs. Hence, the nicotine and 5-HT_3 receptors have a similar structure to the $GABA_A$ receptor, but are formed by different combinations of α, β, and γ subunits.

Metabotropic receptors: A metabotropic receptor causes changes in membrane potentials by a variety of complex mechanisms involving chemical reactions, intracellular signaling (cAMP, IP_3), and protein-conformational changes. Metabotropic receptors are GPCRs, which constitute the largest family of complex membrane proteins involved in intercellular signal transmission. In this section, the discussion is focused on those metabotropic receptors important for several brain functions. The common mechanisms of action for these particular

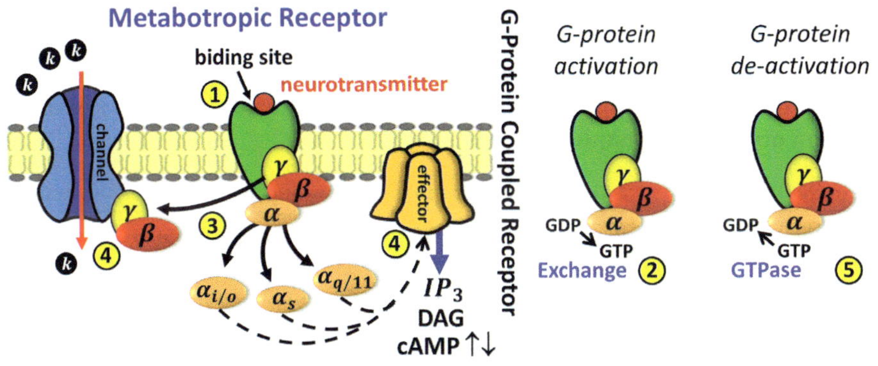

FIGURE 7.6 Mechanisms of action for the metabotropic receptors.

metabotropic receptors are illustrated in Figure 7.6. A GPCR is a cell-surface receptor which could have multiple components: a G-protein where the binding site for the neurotransmitter is located, combined with a membrane-bound effector protein, and/or an ion channel. G proteins comprise α, β and γ subunits. The G-protein is activated after a neurotransmitter binds the receptor (step ①), a state that occur by the exchange of the guanosine diphosphate (GDP) on the α-subunit to a guanosine triphosphate (GTP) (step ②). GTP-binding produces conformational changes (on-off switch), leading to the dissociation of the α-subunit and β_γ–subunit from the G-protein (step ③). An intracellular signaling is then emitted once different subtypes of α-subunits ($G_{i/o}$, G_s, and $G_{q/11}$) binds effector protein of different types in the GPCR (step ④). $G_{i/o}$ and G_s subtypes bind the adenylyl cyclase (AC) (effector) to regulate intracellular concentration of cyclic adenosine monophosphate (cAMP). $G_{i/o}$ inhibits AC (decreasing the cAMP levels), while G_s activates AC (increasing the cAMP levels). $G_{q/11}$ activate the phospholipase-β (PLC_β), which catalyzes the hydrolysis of PIP_2 to create diacyl glycerol (DAG) and IP_3. On the other hand, ion channels are regulated by the β_γ–subunit (step ④). GTP is hydrolyzed to GDP by monomeric GTPases activation (step ⑤). Table 7.2 shows a summary of the mechanisms of action for common ionotropic/metabotropic receptors in the brain.

7.7 EPSP AND IPSP DENDRITIC "GRADED" POTENTIALS – THE UP/DOWN STATES

Changes in membrane potential at the postsynaptic sites are a kind of graded potentials. An excitatory postsynaptic potential (EPSPs) is a graded depolarization of the membrane caused by the binding of neurotransmitters to receptor at the synaptic site. In contrast, if the membrane hyperpolarizes, the graded postsynaptic potential is called inhibitory (or IPSP). The long duration of the EPSPs and IPSPs facilitate their temporal summations over extended dendritic trees. In addition, they can propagate along the neuron dendrites either passively "electrotonically" or amplified by the presence of ion channels. Spatiotemporal integration of postsynaptic potentials in neurons is the most remarkable hallmark in brain computation.

TABLE 7.2
Neurotransmitter receptors in the brain: family, neurotransmitter (NT), receptor type (ionotropic, metabotropic), receptor name, transducers, and action on the postsynaptic neurons

Family	NT	X-tropic	Receptor	Transducers	Action
L-aminoacid	Glu	iono	AMPA/k	Na^{+}+ K^{+} ions, less (Ca^{2+})	excitatory
		iono	NMDA	Ca^{2+} ions, less (Na^{+})	excitatory
		metabo	$mGluR_{1,5}$	$G_{q/11}$→ PLC_{β} (IP_3,DAG)↑	$[Ca^{2+}]_i$↑
		metabo	$mGluR_{2\text{-}4,6\text{-}8}$	*Presynaptic* $G_{i/o}$ → cAMP↓ *Presynaptic* G_{β_γ} → channels	modulator
	GABA	iono	$GABA_A$	Cl^{-} ion	inhibitory
		metabo	$GABA_B$	G_{β_γ}→ open Kir (K^{+} efflux) $G_{i/o}$→ AC (cAMP ↓)	inhibitory
Cholinergic	ACh	Iono	$\alpha_{3,4,7}$ and $\beta_{2,4}$	Na^{+}+ K^{+} ions, less (Ca^{2+})	excitatory modulator
		metabo	M_1	$G_{q/11}$→ PLC_{β} (IP_3,DAG)↑ • PIP_2↓ → I_M↓ • DAG → PKC → I_M↓ less ($G_{i/o}$,G_s) → cAMP↑↓	excitatory modulator
			$M_{2,4}$	$G_{i/o}$→ AC (cAMP ↓) G_{β_γ} channel: K^{+}↑ +Ca^{2+}↓	inhibitory
			$M_{3,5}$	$G_{q/11}$→ PLC_{β} (IP_3,DAG)↑	excitatory
Monoamines					
imidazoleamine	histamine	metabo	H_3	$G_{i/o}$→ AC (cAMP ↓) *Presynaptic*-regulate N-type	inhibitory modulator
catecholamine	Adrenergic	metabo	α_{2A} type	$G_{i/o}$→ AC (cAMP ↓)	Inhibitory
	(E,NE)	metabo	β_2 type	G_s→ AC (cAMP ↑)	excitatory
	DA	metabo	$D_{1,5}$	G_s→ AC (cAMP ↑)	excitatory
		metabo	$D_{2\text{-}4}$	$G_{i/o}$→ (cAMP ↓)	inhibitory
indolamine	5-HT	metabo	$5\text{-}HT_{1\&5A}$	$G_{i/o}$→ AC (cAMP ↓)	inhibitory
		metabo	$5\text{-}HT_2$	$G_{q/11}$→ PLC_{β} (IP_3,DAG)↑	excitatory
		iono	$5\text{-}HT_3$	Na^{+}+ K^{+} ions	excitatory
		metabo	$5\text{-}HT_{4,6\&7}$	G_s→ cAMP↑	excitatory
	MT	metabo	$MT_{1,2}$	$G_{i/o}$→ cAMP↓	inhibitory

Given the extracellular concentration $[T]_e$ of a neurotransmitter T, the postsynaptic currents through a receptor R can, in general, be represented as: $I_R = \bar{g}_R p_R(V_m^{post},[T]_e)(V_m - E_R)$. Parameters $\bar{g}_R$ and E_R represent the maximum conductance of the receptor (mS/cm^2) and the receptor reversal potential, respectively. p_R is the probability of the channel associated to the receptor to be open at a given extracellular concentration $[T]_e$ of the neurotransmitter, which could depend on the membrane potential at the postsynaptic site (V_m^{post}). The size and sign of the postsynaptic graded potential depend dynamically on the effective difference between the membrane and reversal potentials for the receptor. As the reversal potentials for Glu receptors is $0mV$ and for GABA receptors is $\sim E_K$, a hyperpolarization ($V_m^{post} < -80mV$) of the postsynaptic membrane will cause larger inward current through both glutamatergic and GABAergic receptors; hence, GABA will act as an excitatory neurotransmitter. Similarly, depolarizing the postsynaptic membrane above the reversal potential for glutamatergic receptors ($V_m^{post} > 0mV$) will cause outward through both glutamatergic and GABAergic receptors; hence, Glu will act as an inhibitory neurotransmitter (repolarize the membrane) up to the neuron reaches the reversal potential for its receptor. In the brain, neurons interact with each other through a dense connectivity network, keeping them transiting from different global states of equilibrium. A large portion of the connectivity network involves glutamatergic and GABAergic neurons from the cerebral cortex, which are modulated/regulated by subcortical networks (e.g., thalamus, brainstem). Due to this dynamics and complex network, neurons in vivo exhibit high spontaneous firing rates. Based on a method called all-points histogram, it has been demonstrated that neurons have two preferred subthreshold membrane potentials. One of the preferred states is very hyperpolarized (*down-state*), while the other is more depolarized (*up-state*). Neurons in the up-state show spontaneous fluctuations in the membrane potential of higher amplitude. In this state, neurons are only a few millivolts away from the threshold to fire an AP. Neurons in the down-state are more silent and their spontaneous fluctuations in membrane potential are relatively low (free of noise).

7.8 MODELING NEUROTRANSMISSION AND SYNAPTIC EFFICACY

The current flow along gap-junctions in electrical synapses can be modeled by the Ohm's law: $I_{gao}^{i,j} = \overline{g}_{gap}(V_m^i - V_m^j)$, where $\overline{g}_{gap}$ stands for the maximum conductance of the gap-junction. The direction of the current depends on the membrane potentials of the *i*-th V_m^i and *j*-th V_m^j neuron, as defined for bidirectional synaptic transmission. However, modeling postsynaptic currents in chemical synapses is more complicated, requiring models for the neurotransmitter release from the presynaptic terminal and the receptor kinetics at the postsynaptic site. These two processes involve complex chemical reactions; hence, their kinetics can be described using Markov chain models. Simplified models for chemical synaptic transmission have proven to be useful as well. Yamada and Zucker (1992) proposed a simple model of Ca^{2+}-induced release of neurotransmitter from presynaptic terminal, which was based on: **a**) a Ca^{2+} ion influx at the arrival of an AP via HVA VGCC located at the terminal, **b**) an inexhaustible supply of docked SVs, and **c**) a Ca^{2+}-dependent SV release via protein activation, characterized by a kinetic reaction (7.4, Destexhe et al.,1994).

$$4Ca^{2+} + X \underset{k_u}{\overset{k_b}{\leftrightarrow}} \dot{X} \tag{7.4}$$

X represents a Ca^{2+}-binding protein and $\dot{X}$ its activated state. The associated forward and backward rate constants are $k_b = 10^5\, s^{-1}mM^{-4}$ and $k_u = 100\, s^{-1}$, respectively. $\dot{X}$ then reversibly binds to transmitter-containing vesicles, V_e, to release n neurotransmitters from an active stage V_e.

$$\dot{X} + V_e \underset{k_2}{\overset{k_1}{\leftrightarrow}} \dot{V}_e \xrightarrow{k_3} nT \tag{7.5}$$

with $k_1 = 10^6\, s^{-1}mM^{-1}$, $k_2 = 100\, s^{-1}$, and $k_3 = 4000\, s^{-1}$.

Finally, the neurotransmitter cleaning from the synaptic cleft by a variety of mechanisms (e.g., diffusion, enzymatic degradation, reuptake/recycling) is modeled as $T \xrightarrow{k_c} \cdots$, with $k_c = 10^4\, s^{-1}$. If all intervening reactions in the release process are relatively fast, the system can be considered in steady state. Under this assumption, the neurotransmitter release is represented by a static nonlinear transfer function (7.6) of the presynaptic membrane potential (Destexhe et al., 1994)

$$[T]\left(V_m^{pre}\right) = \frac{T_{max}}{1 + e^{-\left(V_m^{pre} - V_p\right)/K_p}} \tag{7.6}$$

With T_{max} – the maximal concentration of transmitter in the synaptic cleft, $V_p = 5mV$, and $K_p = 2mV$.

Using fast-perfusion methods and outside-out excised membrane patches, progresses have been made in the characterization of the molecular mechanisms underlying ionotropic and metabotropic receptor kinetics. However, finding optimal theoretical frameworks to model them is very challenging. Most ionotropic receptors are accurately represented by the H-H theoretical framework for voltage-gated channels. Like for NMDA receptor (Mg^{2+}-blocker), some gating kinetic models may incorporate blocking mechanisms. For GPCRs, an enzymatic cascade is required to open the ion channel, including second-messenger signaling ($IP_3 \rightarrow Ca^{2+}$, DAG, and cAMP) and/or activation of voltage-gated channels by G_{β_γ} subunits. In general, three aspects must be considered while modeling receptor gating kinetics: **1)** activation/binding, **2)** deactivation/unbinding and **3)** desensitization. A *desensitized* state is analogous to the *inactivated* states of voltage-gated channels. The process that occurs during a sequence of rapid activations comprise priming, desensitization and saturation. A simple kinetic model for the most abundant excitatory and inhibitory neurotransmitter receptor is based on two variables (Table 7.3, Destexhe et al., 1998): the fraction of receptor in the open state r (7.7) for ionotropic/metabotropic receptors and the concentration of activated G-protein s (7.8) only for metabotropic receptors.

$$\frac{dr}{dt} = \alpha[T](1-r) - \beta r \tag{7.7}$$

$$\frac{ds}{dt} = K_r r - K_s s \tag{7.8}$$

TABLE 7.3
Receptor kinetics models for the most abundant excitatory (Glu) and inhibitory (GABA)

Receptor	γ_R	Channel Open Probability $P(r,s,V_m^{post})$	E_R
AMPA/k	$0.35-1.0\,nS$	$P_{AMPA/k}(r,s,V_m^{post})=r$ $\alpha=1.1\times10^6\,M^{-1}s^{-1},\ \beta=190\,s^{-1}$	$0\,mV$
NMDA	$0.01-0.6\,nS$	$P_{NMDA}(r,s,V_m^{post})=r\,B(V_m^{post})$ $\alpha=7.2\times10^4\,M^{-1}s^{-1},\ \beta=6.6\,s^{-1}$ *Magnesium Blocker* $B(V_m^{post})=\frac{1}{1+e^{-0.062\,V_m^{post}}[Mg^{2+}]/3.57}$	$0\,mV$
$GABA_A$	$0.25-1.2\,nS$	$P_{GABA_A}(r,s,V_m^{post})=r$ $\alpha=5\times10^6\,M^{-1}s^{-1},\ \beta=180\,s^{-1}$	$-80\,mV$
$GABA_B$	$1\,nS$	$P_{GABA_B}(r,s,V_m^{post})=\frac{s^n}{s^n+K_d}$ $\alpha=9\times10^4\,M^{-1}s^{-1},\ \beta=1.2\,s^{-1}$ $K_r=180\,s^{-1},\ K_s=34\,s^{-1}.\ K_d=100\,\mu M^4$	E_K

Under this general theoretical formalism, the current flowing through the receptor's channels can be modeled as:

$$I_R=\bar{g}_R\,P_R(r,s,V_m^{post})\,(V_m^{post}-E_R) \tag{7.9}$$

$P(r,s,V_m^{post})$ is the probability of the channel to be open. The maximum conductance and reversal potential for the channel is $\bar{g}_R$ and E_R, respectively.

The maximum conductivity is modeled as $\bar{g}_R=\gamma_R\rho_R$, with $\gamma_R\,(nS)$ and $\rho_R\,(cm^{-2})$ representing the single channel conductance and membrane channel density, respectively. The AMPA/k receptor is responsible for most part of the inward current depolarizing excitatory synapses. Therefore, changes in $\bar{g}_{AMPA/k}$ have been linked to LTP and LTD. Although changes in the probability of Glu release from presynaptic terminal have been linked to STDP as well, the two most accepted hypotheses are: **1**) changes in receptor density ρ_R by the insertion/removal of AMPA/k receptors to/from the PSD, and **2**) changes in receptor conductance γ_R by subunit phosphorylation/dephosphorylation. Two major mechanisms have been proposed for AMPA/k receptor trafficking: **1-a)** exocytic/endocytic exchange with intracellular pools of AMPA/k receptor subunits, and **1-b)** lateral diffusion to/from extra-synaptic membranes. Trafficking can take place by either protein–protein interactions or posttranslational modifications. These two hypotheses involve the GluR1 subunit, which is target of many enzymes at three serine sites, in particular S831 and S845. Intracellular Ca^{2+} constitutes the major trigger mechanisms underlying the action of three enzymes (PKC, CaMKII, and PKA) on the GluR1 subunit. There is substantial evidence for the existence of two sources of $[Ca^{2+}]_i$ that modulate separately the activity of these enzymes, one that is NMDA receptor dependent and another that is NMDA receptor independent. NMDA receptor expression/conductance can also change during synaptic plasticity, a phenomenon named meta-plasticity. Two models have been proposed to stabilize neuronal population with dynamically changing synaptic strengths: the sliding threshold model and the homeostatic scaling. Several biophysical models for synaptic plasticity have been proposed (Shouval et al. 2002; Rubin et al., 2005; Earnshaw and Bressloff, 2006). The are currently several software for modeling multiscale brain circuits (e.g., NetPyNE, http://netpyne.org/; BMTK, https://alleninstitute.github.io/bmtk/).

7.9 PROBLEMS AND EXERCISES

Problem 7.1 – Calculate the time it takes glutamate and GABA to diffuse from the presynaptic terminal to the postsynaptic site. The diffusion coefficients for these amino acids in the synaptic cleft is $D_{Glu} = 0.32\ \mu m^2 / ms$ and $D_{GABA} = 0.37\ \mu m^2 / ms$ (Rusakov et al., 2011). The gap between the pre and post -synaptic membrane is 30 nm. For this calculation, assume diffusion only in one dimension.

Problem 7.2 – You are studying synaptic transmission between two neurons (A and B). Using a whole-cell patch-clamp, a constant electric current of ~ 200 pA was injected (current clamp) into the soma of neuron A to generate an AP. An EPSP were concurrently recorded from a dendrite of neuron B using a dendritic patch-clamp protocol (Davie et al., 2006). Combining the voltage clamp technique with pharmacological manipulations, design an experimental paradigm to characterize the particular neurotransmitter and receptor underlying the EPSP. It is known from a previous study that neuron B express AMPA/k receptor, NMDA receptor and α_3-ACh receptor, with reversal potentials $0\ mV$, $0\ mV$ and $-15\ mV$, respectively.

REFERENCES

Blakely RD, Edwards RH. Vesicular and plasma membrane transporters for neurotransmitters. *Cold Spring Harb Perspect Biol* 4, a005595, 1–24, 2012.

Bliss TV, Lomo T. Long-lasting potentiation of synaptic transmission in the dentate area of the anaesthetized rabbit following stimulation of the perforant path. *J Physiol* 232, 331–356, 1973.

Breit M, Kessler K, Stepniewski M, Vlachos A, Queisser G. Spine-to-dendrite calcium modeling discloses relevance for precise positioning of ryanodine receptor-containing spine endoplasmic reticulum. *Sci Rep* 8, 15624, 2018.

Davie JT, Kole MHP, Letzkus JJ, Rancz EA, Spruston N, Stuart GJ, Häusser M. Dendritic patch-clamp recording. *Nat Protoc* 1(3), 1235–1247, 2006.

Destexhe A, Mainen Z, Sejnowski TJ. Synthesis of models for excitable membranes, synaptic transmission and neuromodulation using a common kinetic formalism. *J Comput Neurosci* 1, 195–230, 1994.

Destexhe A, Mainen ZF, Sejnowski TJ. Kinetic models of synaptic transmission. In: Koch C, Segev, I (Eds.), *Methods in Neuronal Modeling* (2nd Edition). MIT Press, Cambridge, pp. 1–25, 1998.

Earnshaw BA, Bressloff PC. Biophysical model of AMPA receptor trafficking and its regulation during long-term potentiation/long-term depression. *J Neurosci* 26(47), 12362–12373, 2006.

Goddard GV, McIntyre DC, Leech CK. A permanent change in brain function resulting from daily electrical stimulation. *Exp Neurol* 25(3), 295–330, 1969.

Johenning FW, Theis A-K, Pannasch U, Rückl M, Rüdiger S, Schmitz D. Ryanodine receptor activation induces long-term plasticity of spine calcium dynamics. *PLoS Biol* 13(6), e1002181, 2015. doi:10.1371/journal.pbio.1002181

Keizer J, Levine L. Ryanodine receptor adaptation and Ca2+-induced Ca2+ release-dependent Ca2+ oscillations. *Biophys J* 71, 3477–3487, 1996.

Koch C, Zador A, Brown TH. Dendritic spines: Convergence of theory and experiment. *Science* 256(5059), 973–974, 1992.

Li L, Bischofberger J, Jonas P. Differential gating and recruitment of P/Q-, N-, and R-type Ca2+ channels in hippocampal mossy fiber boutons. *J Neurosci* 27(49), 13420–13429, 2007.

Lisman J. The CaM kinase II hypothesis for the storage of synaptic memory. *Trends Neurosci* 17, 406–412, 1994.

Lytton WW, Sejnowski TJ. Simulations of cortical pyramidal neurons synchronized by inhibitory interneurons. *J Neurophysiol* 66(3), 1059–1079, 1991.

Perea G, Araque A. Properties of synaptically evoked astrocyte calcium signal reveal synaptic information processing by astrocytes. *J Neurosci* 25, 2192–2203, 2005.

Reuveni I, Friedman A, Amitai Y, Gutnick MJ. Stepwise repolarization from Ca2+ plateaus in neocortical pyramidal cells: Evidence for nonhomogeneous distribution of HVA Ca2+ channels in dendrites. *J Neurosci* 13, 4609–4621, 1993.

Rubin JE, Gerkin RC, Bi G-Q, Chow CC. Calcium time course as a signal for spike-timing–dependent plasticity. *J Neurophysiol* 93, 2600–2613, 2005.

Rusakov DA, Savtchenko LP, Zheng K, Henley JM. Shaping the synaptic signal: Molecular mobility inside and outside the cleft. *Trends Neurosci* 34(7), 359–369, 2011.

Saheki Y, De Camilli P. Synaptic vesicle endocytosis. *Cold Spring Harb Perspect Biol* 4, a005645, 1–30, 2012.

Shouval HZ, Castellani GC, Blais BS, Yeung LC, Cooper LN. Converging evidence for a simplified biophysical model of synaptic plasticity. *Biol Cybern* 87, 383–391, 2002.

Sjöström PJ, Rancz EA, Roth A, Häusser M. Dendritic excitability and synaptic plasticity. *Physiol Rev* 88, 769–840, 2008.

Südhof TC. Neurotransmitter release: The last millisecond in the life of a synaptic vesicle. *Neuron* 80, 675–690, 2013.

Volterra A, Bezzi P. Chapter 13: Release of transmitters from glial cells. In: Volterra A, et al. (Eds.), *The Tripartite Synapse: Glia in Synaptic Transmission.* Oxford University Press, Oxford, UK, pp. 164–184, 2002.

8 Astrocyte Physiology

8.1 OVERVIEW

Electrically silent non-neuronal cells were discovered by Rudolf Virchow (1821–1902) in 1856. They, named glia cells, can be found in the peripheral and central nervous systems. Satellite and Schwann cells are two important glial cells of the peripheral nervous system. The latter forms the myelin sheath in peripheral axons. Four types of glial cells have been identified in the central nervous system: **a)** olygodendrocytes, **b)** ependymal cells, **c)** microglia, and **d)** astroglial cells. As referred in Chapter 6, the myelin sheath for neuronal axons in the central nervous system are created by olygodendrocytes. Ependymal cells have been linked to the production of CSF and neuroregeneration. Microglia are involved in the immunite response. They activate upon the binding of pathogen associate molecular patterns (PAMS) on their pattern recognition receptor (PRR), which initiates the production of host-defense peptides (e.g., cathelin-related antimicrobial peptides "CRAMP"), cytokines (e.g., tumor necrosis factor "TNF", transforming growth factor beta "TGF-β", interleukin "IL-18/12/23"), chemokines (e.g., C-C motif ligand 2 "CCL2"), reactive oxygen species (ROS) and NO. These molecules have an important role in brain innate immunity. In addition, activated microglia upregulate the expression of major histocompatibility complex (MHC) class II molecules, which through T cell receptors (TCR) present antigens to T cells. Pro-inflammatory cytokines (e.g., IL) produced by an activated microglia skew CD4+ T cells into T helper 1 (T_H1) cells and differentiate/activate T_H17 cells. The last two mechanisms play a crucial role in the adaptive immune response. Astroglial cells in the brain are divided into two types: protoplasmic (grey matter) and fibrous (white matter) astrocytes. These cells were initially assumed to hold neurons in place. In fact, the name glia in Greek means glue. However, it was not until the groundbreaking demonstrations of dynamic Ca^{2+} fluctuations in cultured protoplasmic astrocytes (Cornell-Bell et al., 1990) that our simplified view of the role of these cells in brain functions was extended. This chapter is dedicated to discussing several aspects of astrocyte physiology as these cells are the most relevant glia cells in brain imaging. The physiology of this particular subtype of glial cells has not been extensively covered by classic neuroscience books; hence, some data and a large number of references are provided in this section. Henceforth, the term protoplasmic will be eluded.

DOI: 10.1201/9780367901509-8

8.2 ASTROCYTIC PROCESSES: VASCULAR AND SYNAPTIC END-FEET

In addition to provide cytoarchitectonic support to active neuronal structures, astrocytes play important role in trophic support (in particular for synapses), neuroprotection, neurogenesis (parenchymal astrocytes), cellular metabolism, housekeeping/homeostasis, information processing through modulating neuronal activity, neurotransmitter regulation, and blood flow control. Through a network of well-connected astrocytes with slightly overlapping processes, neuronal information is conveyed over long distances, adding a new dimension to the already complex brain. Although still highly debated, it is believed that there is a 1:1 glia-to-neuron ratio in the human brain, with astrocytes representing ~ 20–30% of the glial cells. Astrocytes comprises a cell body (or soma) that is intracellularly connected with symmetrically radiating astrocytic processes (Figure 8.1). The soma area has been estimated ($A = 2.5 \times 10^{-6}\ cm^2$) using prolate spheroid ($10.4 \pm 0.2\,\mu m$ –major cell axes x and $6.2 \pm 0.1\,\mu m$ – minor cell axes). The length and number of main processes vary with the species. The length ranges from ~ $37\,\mu m$ (rodents) to ~ $98\,\mu m$ (humans), and the number of main processes from 4 (rodents) to 37 (humans). Therefore, the average diameter of a human astrocyte is about $143\,\mu m$, while it drops to $56\,\mu m$ in rodents. Neighboring astrocytes only overlap at the end of their processes. Similarly, the overall length of the overlapping process is larger in humans (~ $53\,\mu m$) than in rodents (~ $5\,\mu m$), for a total average area of $205\,\mu m^2$ and $12\,\mu m^2$, respectively.

Astrocytes control water homeostasis through aquaporin channels (AQP4), crucial for their response to brain insults. Astrocytes have been classified on the basis of: **a)** ion channel and GPCR expression, **b)** K^+ uptake capabilities, **c)** morphological features, and **d)** progenitor cells (Type 1 and Type 2). According to the role in brain imaging, this book classifies them into neuronal and vascular-associated astrocytes (Figure 8.2). Neuronal-associated astrocytes embedded into neuronal networks are symmetrically distributed forming a reticulum. They project the end-feet of astrocytic into the neuronal synapses to form a tripartite unit (Araque et al., 1999; Chapter 7). There, astrocytes provide different types of support to synapses, e.g., Glu and GABA

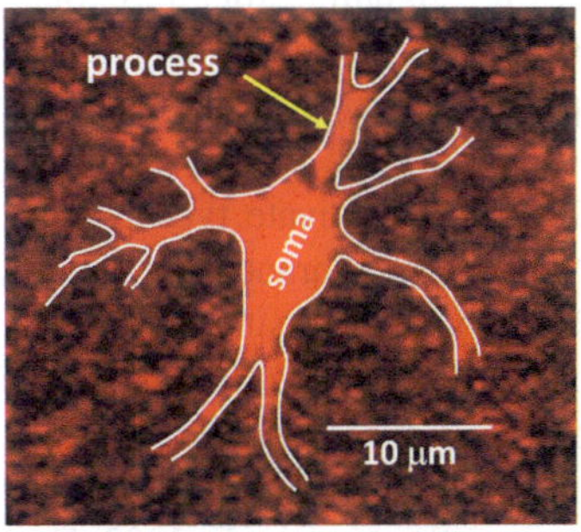

FIGURE 8.1 Anatomical features of a single hippocampal astrocyte, as stained with sulforhodamine 101.

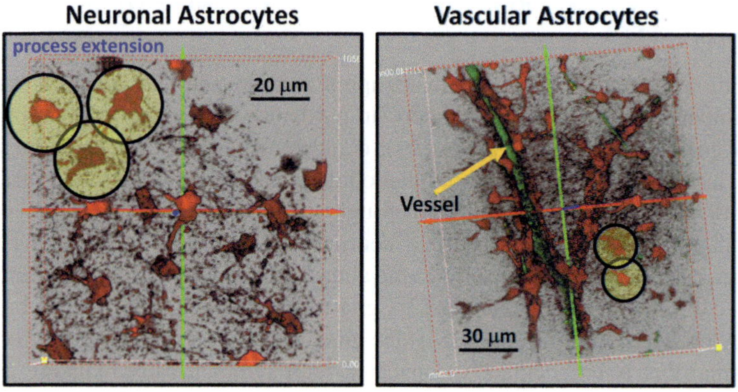

FIGURE 8.2 Classification of astrocytes according to their functional roles in connection to brain imaging (sulforhodamine 101).

Note: For this image, vessels were loaded with FITC-dextran via tail injection.

recycling, formation/elimination of synapses, and metabolic support preferably in the form of lactate. Vascular-associated astrocytes organize around vessels forming grape-like clusters. There, astrocytic end-feet interact with smooth muscle cells (SMC), pericytes and endothelial cells (EC) to regulate blood flow and participate in metabolite exchange. This multicellular system is referred to as the neurovascular unit (Lecrux and Hamel, 2011; Chapter 10).

8.3 LIGAND-GATED METABOTROPIC RECEPTORS IN ASTROCYTES

Ca^{2+} signaling constitutes the hallmark of astrocytic activity and it underlies the multipurpose role of these cells. Increases in $[Ca^{2+}]_i$ helps regulate/modulate a variety of biochemical processes via enzyme phosphorylation/dephosphorylation. It is also a crucial intracellular messenger in signal transduction. Cornell-Bell et al. (1990) reported different types of astrocytic Ca^{2+} signaling after the exogenous application of glutamate to a cultured astrocytic syncytium, from sustained/damped oscillations to complex biphasic responses. Group I metabotropic glutamate receptors (mGluR1 and mGluR5) underlie the Glu-evoked Ca^{2+} response, which are $G_{q/11}$-type GPCR. As discussed in Chapter 7, GTP-bound G-proteins activate type β-phospholipase C (PLCβ), which causes the cleavage of PIP_2 into membrane DAG and cytosolic IP_3. Astrocytes also express $G_{q/11}$-type GPCRs with binding sites for other neurotransmitters: serotonin ($5\text{-}HT_{2B}$), ACh (M_1 muscarinic), histamine (H_1) and NE/E (α_1 adrenergic). Therefore, they can sense/quantify changes in presynaptic neuronal activity via Ca^{2+} signaling. Ca^{2+} signaling is mediated by the activation of IP_3 receptors (IP_3R) on the membrane of the endoplasmic reticulum (ER) (Nett et al., 2002; Zur Nieden and Deitmer, 2006). IP_3R are arranged in clusters. These clusters organize spatially along the ER membrane to form microdomains for Ca^{2+} signaling enhancement. Depending on the observation technique, Ca^{2+} responses could be recorded at the level of a single IP_3R channel (i.e., blips), a cluster of channels forming a microdomain (i.e., puffs), or a global synchronized response of the entire astrocytic soma (i.e., spikes). Ca^{2+} microdomains are distributed along the processes of astrocytes forming a reticular meshwork of nodes/shafts with loop-like structures. These nodes comprise the Ca^{2+} microdomains. Synchronized activations of these microdomains due to extended synaptic activity produce Ca^{2+} puffs that could propagate to the astrocyte soma to generate a global Ca^{2+} spike. Spikes occur mainly in the soma of astrocytes, and their characteristics are determined by the degree of cluster coupling inside the astrocyte soma. By using the total internal reflection fluorescence (TIRF) on neuron-astrocytes co-cultures expressing a modified genetically-encoded Ca^{2+} sensor (Lck-GCaMP2), Shigetomi et al. (2010) reported the full-duration half-maximum ($T_{0.5} = 3.9 \pm 0.3\,s$) and full-width half-maximum ($FWHM = 5.1\,\mu m$) of single puffs in hippocampal astrocytes. The mean number of these clusters (puff sites) in the soma of hippocampal astrocytes was $n_{cluster} \approx 4$. By fitting a biophysical model to experimental data, Riera, Hatanaka, Uchida et al. (2011) estimated the approximately number of IP_3R channels in a single cluster $N_{channel} \approx 131$. Due to variations in the number of somatic clusters recruited and their activation timing, global Ca^{2+} events in astrocytes exhibit different temporal characteristics. In general, a few large clusters could become strongly interconnected in the astrocytic soma, giving rise to longer global spike-like events ($T_{0.5} \approx 5\,s$) with a large variation in peak amplitudes ($5\% \leq \Delta F / F \leq 30\%$). Because of this large variability, analysis of somatic Ca^{2+} signaling in astrocytes have been oversimplified to two basic indices: **(a)** cell type (e.g., hippocampal and neocortical astrocytes), **(b)** level of activation (i.e., inactive, active, and hyperactive), and **(c)** spike frequency, or the equivalent inter-spike-interval (ISI). A very interesting stochastic property of the somatic Ca^{2+} spikes is that the ISI variability increases linearly with the mean ISI (Skupin et al., 2008). However, the dynamics of somatic spike-like events is ultimately determined by the characteristics of the interactions between the IP_3R channels inside a puff-generating cluster. Complex Ca^{2+} spike-like transients in cortical astrocytes have been also reported under external sensory stimulations in vivo (Wang et al., 2006). Somatic Ca^{2+} spikes could happen spontaneously in situ (Nett et al., 2002) and in vivo (Hirase et al., 2004) without the application of either external stimuli or exogenous agonists. These spontaneous Ca^{2+} oscillations (SCOs) have been linked to different astrocytic functions. SCOs happens even in absence of neuronal activity (Nett et al., 2002; Zur Nieden and Deitmer, 2006; Shigetomi et al., 2010) and its frequency is related to the basal level of cytosolic Ca^{2+} and IP_3 at the soma (Riera, Hatanaka, Uchida et al., 2011). SCOs seems to play an important role in neuroprotection, as astrocytic Ca^{2+} signal shows hyperactive levels whenever astrocytes become reactive to a brain insult.

It has been also demonstrated that astrocytes respond to the application of adenosine and ATP through the activation of purinergic receptors. Adenosine causes changes in cAMP by binding to $G_{i/o}$-type GPCRs (P1 purinergic receptors $A_{1,3}$). By binding to P2Y purinergic receptors ($P2Y_{1,2,4,14}$), ATP also causes changes in astrocytic Ca^{2+} signaling via a variety of mechanisms. The role of ATP on inter-astrocytic Ca^{2+} signal propagation in a form of waves is discussed later in this chapter. Astrocytes also have voltage-gated ionic channels (VGCCs) and ligand-gated ionotropic receptors (P2X, AMPA, $GABA_{A/B}$), which can modulate $[Ca^{2+}]_i$ by means of Ca^{2+} and Na^+ / K^+ transmembrane fluxes. Group II metabotropic glutamate receptor (mGluR3) has also been proposed to cause changes in cytosolic IP_3 in astrocytes (Zur Nieden and Deitmer, 2006), but the effect may be negligible as group II/III mGluR antagonist RS-α-cyclopropyl-4-phosphonophenylglycine produce no effect on SCOs (Nett et al., 2002).

8.4 ASTROCYTIC CALCIUM MODELS

Multiple mechanisms can potentially cause changes in the cytosolic Ca^{2+} concentration in astrocytes (Figure 8.3). They can be grouped into those related to: **1)** Ca^{2+} ion fluxes through the plasma membrane and **2)** Ca^{2+} ion exchanges with intracellular stores/buffers.

i) The molar concentration of Ca^{2+} ions in the extracellular space is 10^4 higher than that in the intracellular space. Ca^{2+} ion entry from the extracellular space through the plasma membrane powered by this gradient is known to play a protagonist role in the genesis of SCOs. This Ca^{2+} gradient is maintained by active transport. Amount the most relevant mechanisms for Ca^{2+} ion in/out flux in astrocytes are:
   a) A passive leakage $j_{PM_{leak}}$ has been proposed by several groups. Using healthy hippocampal astrocytes in acute slice preparations from adult mice, Riera, Hatanaka, Uchida et al. (2011) estimated a value of $0.036\,\mu M / s$ for the total leakage into these astrocyte somas. Based on the estimated surface area of their soma, the leakage flow density must be approximately $14.4\,mM / cm^2 s$.

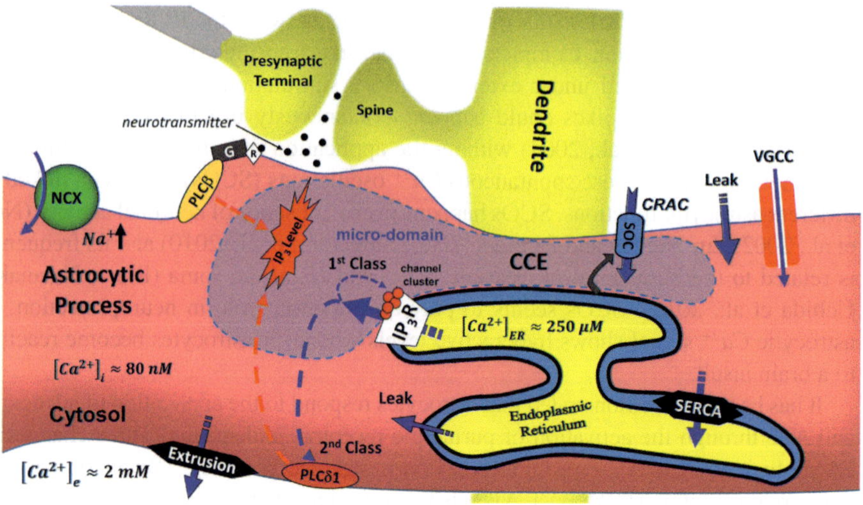

FIGURE 8.3 Mechanisms involved in astrocytic Ca^{2+} signaling.

Note: Pathways for Ca^{2+} influx/outflux are represented with blue arrows. Pathways involving IP_3 genesis are denoted by red arrows.

b) Several VGCCs (Cav1, Cav2.2 and Cav2.3) have been identified in rat cortical astrocytes (D'Ascenzo et al., 2004). However, their role in Ca^{2+} signal genesis is still under debate (Carmignoto et al., 1998). Astrocytes show a very stable V_r at $\sim -85\,mV$.

c) A capacitative Ca^{2+} entry (CCE) from the extracellular space occurs gradually while the ER is being depleted of Ca^{2+} due to the action of IP_3 on IP_3Rs. The CCE is facilitated by store-operated Ca^{2+} channels (SOC). The electric current through these channels is called Ca^{2+} release-activated Ca^{2+} current (CRAC). Four mechanisms have been proposed for the signaling between the ER store and the plasma membrane underlying the CCE: (1) the exocytosis model, (2) the existence of diffusible messengers (e.g., Ca^{2+} influx factor – nitric oxide), (3) a spatially-confined Ca^{2+} regulation system, and (4) a conformational coupling. The conformational coupling hypothesis has been supported by several studies. Two types of mechanisms have been proposed for the CRAC pore-forming subunit in astrocytes: **a)** the activation of the type-1 *canonical* transient receptor potential (TRPC1), and **b)** the activation of the calcium release-activated calcium channel protein 1 (orai1). The PLC_β-based hydrolysis of IP_3R-tethered PIP_2 has been suggested as the main working principle for SOC activation, leading to the production of an important intracellular messengers (i.e., IP_3). IP_3 acts on IP_3R receptors on the ER membrane causing Ca^{2+} depletion, which is sensed by the stromal interaction molecule 1 (stim1) and led to the activation of orai1. Stim1 also activates TRPC1 through its C-terminal polybasic domain. A more detailed discussion, with specific citations, can be found in Riera, Hatanaka et al. (2011).

d) Ca^{2+} is extruded across the plasma membrane by primary ($PMCA_{1,2,4}$ isoforms, $K_D \approx 100\,nM$, uniformly distributed) and secondary (NCX, NCX1 B-containing isoforms, $K_D \approx 1\,\mu M$, confined to microdomains of sub-plasma-membrane "junctional" ER) active transport. PMCA and NCX are highly expressed in the plasma membrane of astrocytes. These two active transport mechanisms seem to operate on two different cytosolic sub-compartments for Ca^{2+} ions. Extrusion of Ca^{2+} across the plasma membrane is required for steady-state behaviors. NCXs can translate Na^+-transients into intracellular Ca^{2+} signals. A transient Na^+ influx into astrocytic cytosol can happen by either the effect of AMPA receptors or by voltage-gated Na^+ channels. Details about these mechanisms for Na^+ influx in astrocytes are provided in Verkhratsky and Steinhäuser (2000). An influx of Na^+ ions also takes place through high-affinity transmembrane glutamate transporters EAAT1 (GLAST) and EAAT2 (GLT-1) expressed in astrocytes (Danbolt, 2001).

Therefore, the total Ca^{2+} in/out flux through the plasma membrane j_{PM} can be modeled as:

$$j_{PM} = j_{PM_{leak}} + j_{VGCC} + j_{CCE} + j_{NCX}^{PM} - j_{PMCA} \tag{8.1}$$

ii) The basal concentration of Ca^{2+} in the ER is relatively high, i.e., $[Ca^{2+}]_{ER} \approx 250\,\mu M$. As the cytosolic basal concentration is very low $[Ca^{2+}]_i \approx 80\,nM$, the ER represents the primary intracellular source for Ca^{2+} signaling in astrocytes. Ca^{2+} ions exchanges with the ER occur also by means of multiple mechanisms:
   a) Astrocytes in rat brains expressed majorly the type-2 IP_3R (Sharp et al., 1999). A flux of Ca^{2+} from the ER to the cytosol j_{IP_3R} is generated by the opening of these IP_3R channels after binding intracellular IP_3. The IP_3R has also an inactivation binding site for Ca^{2+} ions that could be either closed (occupied) or open (non-occupied).
   b) Similar to the plasma membrane, the ER membrane in astrocytes is also leaky. Therefore, a leak of Ca^{2+} ions from the ER $j_{ER_{leak}}$ has been included in many biophysical models.
   c) The SERCA represents the major mechanism for Ca^{2+} to be resequestered by the ER (active transport).

Hence, Ca^{2+} release from and sequestration by the ER j_{ER} can be modeled as:

$$j_{ER} = j_{IP_3R} + j_{ER_{leak}} - j_{SERCA} \tag{8.2}$$

Free Ca^{2+} within the astrocytic cytosol can diffuse and be subject to each of all reactions included in fluxes j_{PM} and j_{ER}. If mitochondrion Ca^{2+} sequestration is neglected and buffering mechanisms are very fast, Ca^{2+} signaling in astrocytes can be modeled by a reaction-diffusion equation.

$$\frac{\partial [Ca^{2+}]_i}{\partial t} = D_{Ca} \nabla^2 [Ca^{2+}]_i + \overbrace{j_{PM} + j_{ER}}^{reactions} \tag{8.3}$$

Biophysical models have been proposed to represent quantitatively each of the reactions in equation (8.3). In Chapters 6 and 7, H-H channel gating models for different types VGCCs were presented. Kinetic models for active transporters (*primary* – PMCA and SERCA; *secondary* – NCX) were also discussed in Chapter 6. Therefore, the discussion in this sub-section is focused on specific models for the IP_3R and CCE components.

<u>The IP_3R model</u>: A general equation (8.4) for the flow j_{IP_3R} of Ca^{2+} ions from the ER to the cytosol through open IP_3R channels and the ER membrane leakage $j_{ER_{leak}}$ was initially proposed by De Young and Keizer (1992) based on microscopic kinetics for IP_3R channels.

$$j_{ER} = j_{IP_3R} + j_{ER_{leak}} = c_1 (v_1 P_{open} + v_2)\left([Ca^{2+}]_{ER} - [Ca^{2+}]_i\right) \tag{8.4}$$

The proposed flux constants for these mechanisms were: maximum Ca^{2+} channel flux $v_1 = 6\,s^{-1}$ and Ca^{2+} leak flux constant $v_2 = 0.11\,s^{-1}$. P_{open} represented the open probability of an IP_3R channel. The volume rate between ER and cytosol $c_1 = V_{ER} / V_c \approx 0.185$ (dim/less, Alberts et al., 2002) was introduced to implicitly account for a rapid buffering capacity of the ER and cytosol. Under the assumption of a closed cell, the total free Ca^{2+} per cytosolic volume $c_o = c_1[Ca^{2+}]_{ER} + [Ca^{2+}]_i$ is conserved ($V_c \gg V_{ER}$); hence, $[Ca^{2+}]_{ER} = (c_o - [Ca^{2+}]_i) / c_1$ in (8.4). Data recorded from Purkinje cells of cerebellum showed that, in the physiological ranges, IP_3Rs display a bell-shaped dependence with $log_{10}[Ca^{2+}]_i$ and a sigmoidal dependence with $log_{10}[IP_3]_i$. These findings inspired De Young and Keizer (1992) to formulate the most universally used model for the kinetic of the IP_3R, i.e., $P_{open} = (X^e_{110})^3$, where X^e_{110} constitutes the fraction of the IP_3R in the state S_{110} of eight possible ones. A phenomenological equation was obtained (8.5) by fitting the eight-state model to the Purkinje cells data. In practice, this model is rarely used to describe Ca^{2+} signaling in astrocytes.

$$X^e_{110} = \frac{[Ca^{2+}]_i [IP_3]_i d_2}{\left([Ca^{2+}]_i [IP_3]_i + [IP_3]_i d_2 + d_1 d_2 + [Ca^{2+}]_i d_3\right)\left([Ca^{2+}]_i + d_5\right)} \quad (8.5)$$

With receptor dissociation constants for: IP_3 ($d_1 = 0.13\,\mu M$, $d_3 = 943.4\,nM$) and Ca^{2+} ((*inhibition*) $d_2 = 1.049\,\mu M$, (*activation*) $d_5 = 82.34\,nM$). Under the validity of a series of thermodynamic and protein kinetic assumptions, Li and Rinzel (1994) reduced the De Young-Keizer model for P_{open} to a two-variable model (8.6), which resembles the H-H model for voltage-gated ion channels. This simplified model has been widely used to explain the kinetics of the IP_3R in astrocytes. Ca^{2+} ions flow through the channel with a magnitude proportional to the product of the opening probabilities for an activation m and an inactivation h gate. Three independent particles are required to approach to the channels for each gate to open. The open probability for the activation gate m was approached by its values at the equilibrium (8.7). The temporal course of the inactivation gate is governed by equation (8.8).

$$P_{open} = \left(m_\infty([Ca^{2+}]_i, [IP_3]_i)\right)^3 h^3 \quad (8.6)$$

$$m_\infty\left([Ca^{2+}]_i, [IP_3]_i\right) = \frac{[IP_3]_i [Ca^{2+}]_i}{\left([IP_3]_i + d_{IP_3}\right)\left([Ca^{2+}]_i + d_{act}\right)} \quad (8.7)$$

$$\frac{dh}{dt} = \alpha_h\left([IP_3]_i\right)(1-h) - \beta_h\left([Ca^{2+}]_i\right)h \quad (8.8)$$

$$\alpha_h\left([IP_3]_i\right) = a d_2 \frac{[IP_3]_i + d_1}{[IP_3]_i + d_3} \quad \text{and} \quad \beta_h\left([Ca^{2+}]_i\right) = a[Ca^{2+}]_i$$

The cytosolic Ca^{2+} concentration and the total Ca^{2+} concentration in the astrocyte per cytosolic volume satisfy (8.9) and (8.10), respectively.

$$\frac{d[Ca^{2+}]_i}{dt} = \left(v_1 P_{open} + v_2\right)\left([Ca^{2+}]_{Total} - \left(1 + c_1\right)[Ca^{2+}]_i\right) - j_{SERCA} + \epsilon\, j_{PM} \quad (8.9)$$

$$\frac{d[Ca^{2+}]_{Total}}{dt} = \epsilon\, j_{PM} \quad (8.10)$$

The parameter $\epsilon = A_{PM} / A_{ER}$ is the ration between the areas of the plasma membrane A_{PM} and the ER membrane A_{ER}. Under the fast-buffering assumption, it was proposed $\epsilon \approx 0.01$. Li and Rinzel (1994) discussed several sets of adjusted parameters (d_{IP_3}, d_{act}, d_1,d_2 d_3, a, v_1 and v_2) for the model.

The CCE model: There have been two kinetics models proposed for the CRAC current j_{CCE} in astrocytes: **1)** an *exhaustive* model proposed by Höfer et al. (2002), $j_{CCE} = X_{CCE}([IP_3]_i)^2 / (([IP_3]_i)^2 + (K_{CCE})^2)$, with a maximal rate of activation-dependent Ca^{2+} influx $X_{CCE} = 0.2\mu M / s$ and its half-saturation constant $K_{CCE} = 1\mu M$, and **2)** a *phenomenological* model proposed by Di Garbo et al. (2007), $j_{CCE} = k_{CCE}(H_{CCE})^2 / (([Ca^{2+}]_{ER})^2 + (H_{CCE})^2)$, with the maximum CRAC channel flux $k_{CCE} = 0.01\mu M / s$ and a half-saturation constant $H_{CCE} = 10\mu M$.

8.4.1 Models for the Mitochondrion Sequestration and Protein Buffers

In the original Li-Rinzel model (8.6-8.10), mitochondrion sequestration j_M was not included. Also, it was assumed Ca^{2+} ions bind very quickly buffer reservoir. Ca^{2+} ion sequestration by mitochondria has been suggested as a crucial mechanism in shaping Ca^{2+} signal dynamics. It has been also demonstrated that mitochondria limit the rate/extent of Ca^{2+} wave propagation in astrocytes (Boitier et al., 1999). In astrocytes, mitochondria are located in both the soma and processes. They are efficient Ca^{2+} ion sequesters, in particular due to the existence of a mitochondrial Ca^{2+} uniporter (MCU) and NCXs. The MCU can transport Ca^{2+} into the mitochondrial matrix due to the existence of a negative membrane potential across its inner membrane. This negative potential is generated by the electron-transport chain (ETC). The MCU is only activated at very large $[Ca^{2+}]_i$ values ($\sim 0.5 - 10\mu M$). Such a high concentration of Ca^{2+} ions are only possible within microdomains in the astrocytic processes. In these microdomains, the mitochondria are juxtaposed to the IP_3Rs. Although it has been poorly study, mobile/stationary endogenous proteins (e.g., CaM, Parvalbumin, Calbindin, Troponin-C, phospholipids), exogenous (e.g., EGTA, BAPTA), and genetically-encoded/AM-ester Ca^{2+} indicators can act as Ca^{2+} buffers with different dynamics. Buffering can reduce Ca^{2+} ion diffusion range, giving rise to an effective diffusion coefficient (D_{Ca}^{eff}), which limits propagation within an astrocytic network.

Although a more complex model for the MCU mechanism has been developed recently (e.g., Wacquier et al., 2016), a simple theoretical approach was early proposed for cardiac cells (Williams et al., 2013): $j_{MCU} = P_O M_{mito} N_{MCU} i_{MCU} / (zFV_c)$, with $i_{MCU} = g_{MCU}(\Delta\psi_m - E_{Ca})$. The number of mitochondria per astrocytes M_{mito}, the number of MCUs per mitochondrion N_{MCU}, and the single-channel MCU current i_{MCU} must be adjusted for astrocytes. Parameters $\Delta\psi_m$, P_O and g_{MCU} represent the mitochondrial inner membrane potential, the open probability, and the single-channel MCU conductance, respectively. In principle, the MCU and mitochondrion NCX mechanisms could be added to the Li-Rinzel model. It can also be generalized if the fast-buffering assumption is not satisfied. In this case, the value for parameter ϵ must be revisited. Recently, Moshkforoush et al. (2020) proposed a simple model to capture different types of fast-buffering mechanisms in astrocytes, which consists of the substitution of $d[Ca^{2+}]_i / dt$ by $(1+\theta)d[Ca^{2+}]_i / dt$ in equation (8.9). The buffering parameter $\theta = b_t K_b / (K_b + [Ca^{2+}]_i)^2$ was defined using a reduced reaction-diffusion model (Smith et al., 1996). Parameters b_t and K_b are the total buffer protein concentration ($160\mu M \leq b_t \leq 240\mu M$) and the buffer rate constant ratio ($16\mu M \leq K_b \leq 24\mu M$, respectively.

8.4.2 A Model for IP_3 Dynamics

The main production of IP_3, quantified by flux $j_{GPCR} \sim \mu M / s$, originates from the action of a ligand L (e.g., neurotransmitters, purines) on $G_{q/11}$-type GPCR receptors (activation of PLC_β), which can be represented by a simple kinetic model:

$$\frac{d[G]}{dt} = k_a^G (\delta+\rho)\left(\left[[G_T]-[G]\right]\right) - k_d^G [G] \tag{8.11}$$

k_a^G and k_d^G are the G-protein activation and degradation rates, respectively. $[G]$ and $[G_T]$ are the concentration of activated and total G-protein, respectively. δ is the ratio of unbound/bound receptor at rest, useful to explain background IP_3 levels. The basal production of IP_3 for healthy hippocampal astrocytes in acute slices is $\sim 0.32\mu M / s$ (Riera, Hatanaka, Uchida et al., 2011). During agonist-related activation, the production of IP_3 changes with time according to $j_{GPCR} = \alpha_G [G]$, with G changing dynamically as described by equation (8.11). The variable ρ represents the ratio of bound of ligand L to total receptors. For fast binding kinetics, the following relationship is satisfied $\rho = [L]/([L]+K_{RG})$, with the dissociation for ligand binding receptor K_{RG} and ligand concentration $[L]$. IP_3 degrades with a relatively short lifetime $\sim 12.5 s$ by the action of the 3-kinase and 5-phosphatase. At high concentration, IP_3 can be also degraded by dephosphorylation. IP_3 degradation has been modeled by a negative flux directly proportional to its concentration: $j_{Deg} = K_{IP_3} [IP_3]_i$, with $K_{IP_3} \approx 0.08 s^{-1}$. IP_3 could be also produced by the subtype $PLC_{\delta 1}$, highly expressed in astrocytes. The $PLC_{\delta 1}$ protein is regulated by cytosolic Ca^{2+}, introducing positive feedback $j_{PLC_{\delta 1}}$ in IP_3 production. De Young and Keizer (1992) were the first in recognizing this mechanism, which was modeled by a normalized single binding Michaelis-Menten kinetics: $j_{PLC_{\delta 1}} = v_\delta([Ca^{2+}]_i + (1-\alpha)K_{Ca\delta}) / ([Ca^{2+}]_i + K_{Ca\delta})$, with half-saturation constant $K_{Ca\delta} = 1.1\mu M$ and maximum rate for IP_3 production $v_\delta = 1.2 - 2.8 s^{-1}$. However, a Hill equation with coefficient $n = 2$ has been used consistently by many others to reproduce the $PLC_{\delta 1}$-related positive feedback: $j_{PLC_{\delta 1}} = v_\delta([Ca^{2+}]_i)^2 / (([Ca^{2+}]_i)^2 + (K_{Ca\delta})^2)$, with dissociation constant $K_{Ca\delta} = 0.3\mu M$ and maximum rate $v_\delta = 0 - 0.05\mu M / s$.

Using digitalized experimental data for $PLC_{\delta1}$, Riera et al. (2010) provided new estimators for $K_{Ca\delta} = 0.55\mu M$ and $v_\delta = 0.152\mu M / s$. IP_3 can also diffuse inside astrocytes ($D_{IP_3} = 280\mu m / s^2$). Therefore, an equivalent reaction-diffusion equation for IP_3 is defined:

$$\frac{d[IP_3]_i}{dt} = D_{IP_3}\nabla^2[IP_3]_i + \overbrace{j_{GPCR} + j_{PLC_{\delta1}} - j_{Deg}}^{reactions} \tag{8.12}$$

According to the underlying mechanism, Ca^{2+} oscillations are classified into two classes (Sneyd et al., 2006). The positive Ca^{2+} feedback in the IP_3R is the major contributor to oscillations in the 1st class, referred by some researcher as a type of CICR. This type of oscillation occurs at constant IP_3. In the 2nd class, Ca^{2+} signal modulates the IP_3 levels through a PLC-based production and a kinase-3 degradation ($IP_3 \rightarrow IP_4$). For astrocytes, both mechanisms have been identified and might operate simultaneously.

8.5 STOCHASTICITY IN IP_3R CLUSTERS

Ca^{2+} signaling in astrocytes occurs at different physical scales (blips, puffs, and spikes). As IP_3Rs are organized in densely-packaged clusters, Ca^{2+} blips due to the opening of single IP_3R channels create steep concentration gradients inside certain cytosolic microdomains that couple channels strongly inside the cluster. A single puff is generated as the results of such strong intra-cluster reaction-diffusion phenomenon. Ca^{2+} ions from multiple puffs in astrocytic processes diffuse (weak coupling) to the soma, where depending on the spatiotemporal characteristics of the puffs they could create a global spike-like event. This hierarchical organization in multiple scales underlie the random character of astrocytic Ca^{2+} signaling from local puffs to global spikes. Stochastic models have been proposed to explain Ca^{2+} signal randomness in astrocytic processes and somas. These models range from those spatially extended that requires heavy computation to very simplified versions. The spatially extended models, resulting from realistic reaction-diffusion systems, rely on the solution of complex partial differential equations using finite element solver with a fine discretization. Fine discretization in time and space are needed because of the steep gradient and dynamic stiffness. Savtchenko et al. (2018) proposed one of the most ambitious astrocytic models up today by combining a morphology-generating tool (ASTRO, https://github.com/LeonidSavtchenko/Astro/tree/master/neuronSims) and the NEURON platform. However, simplified models are also very useful. In what follow, two simplified models, the Green's cell model (Skupin et al., 2010) and the Langevin approximation (Shuai and Jung, 2002), are described in more details.

The Green's cell model was introduced to simulate stochastic hierarchical Ca^{2+} signaling systems at three different physical levels: the IP_3R channel (blips), the cluster (puffs) and whole cell (spikes). Under this theoretical framework, active molecules are treated individually as Markov chains, but their coupling take place in a form of a deterministic diffusion. IP_3R channels are modeled using the original De Young-Keizer model. The model assumes that $n_{cluster}$ IP_3R clusters are distributed with random positions $\vec{r}_j = (r_j, \theta_j, \varphi_j)$ inside the cytoplasm, of a spherical-like astrocyte of radius R. The Ca^{2+} concentration at any particular position $\vec{r} = (r, \theta, \varphi)$ inside the cytoplasm is represented by 3-component Green's functions: the spherical Bessel functions of the first kind ($\sqrt{\pi/2}J_{l+1/2}(\lambda_{lp}r)r^{-1/2}$) and the Legendre polynomial $P(cos\varnothing_j)$, with $cos\varnothing_j = cos(\theta)cos(\theta_j) + sin(\theta)sin(\theta_j)cos(\varphi - \varphi_j)$. If each cluster "$j$" produces a Ca^{2+} current $J_j(t)$, it can be demonstrated that the average Ca^{2+} concentration in the cell cytosol can be described by (8.13).

$$\left\langle [Ca^{2+}]_i(t) \right\rangle = \sum_{j=1}^{n_{cluster}} \left[\frac{\sqrt{2/\pi}}{R^3} \sum_p^{\infty} \frac{sin(\lambda_p R) - R\lambda_p cos(\lambda_p R)}{\lambda_p^{5/2}} X_j^{(p)}(t, r_j) + X_j^{(0)}(t) \right] \tag{8.13}$$

The three-component response functions $X_j^{(p)}(t, r_j)$ and $X_j^{(0)}(t)$ are defined as:

$$X_j^{(p)}(t, r_j) = \frac{J_{1/2}(\lambda_p r_j)}{R^2(J_{1/2}^2(\lambda_p R) - J_{-1/2}(\lambda_p R)J_{3/2}(\lambda_p R))\sqrt{r_j}} \sum_{i=1}^{3} \int_0^t \sigma_j(\tau) e^{s_i(t-\tau)} \frac{adj(\mathbf{M}_p)\mathbf{e}_1 d\tau}{\partial |\mathbf{M}_p| / \partial s \big|_{s=s_i}} \tag{8.14}$$

$$X_j^{(0)}(t) = \frac{3}{4\pi b_c^3} \sum_{i=1}^{3} \int_0^t \sigma_j(\tau) e^{s_i(t-\tau)} \frac{adj(\mathbf{M}_0)\mathbf{e}_1 d\tau}{\partial |\mathbf{M}_0| / \partial s \big|_{s=s_i}} \tag{8.15}$$

By definition, $(\mathbf{M}_p)^{-1} = adj(\mathbf{M}_p)/|\mathbf{M}_p|$. The coefficients $\lambda_p = \lambda_{0p}$ for the spherical Bessel functions $J_l(\lambda_{lp} r)$ are calculated using the boundary conditions at the plasma membrane $r = R$. Defining $c_0 = [Ca^{2+}]_i(0)/K_B$, $b_0 = 1/(c_0 + 1)$, $b_{i0} = 1/(c_0\kappa + 1)$, $\sigma_m = (1 + c_0)$, $\sigma_{im} = \epsilon_R(1 + \kappa c_0)$, the dissociation constant ratio $k = K_B / K_{B_i}$, the ratio of diffusion coefficient $d = D_B / D_{Ca}$, the time separation for the mobile $\epsilon_\tau = [B]_T / K_B$ and immobile $\epsilon_\tau^i = [B_i]_T / K_B$ buffers, the ratio of buffer influence $\epsilon_R = [B_i]_T k_{B_i}^- / [B]_T k_B^-$, the dimensionless cell radius b_c, $\sigma_c = (b_0 + b_{i0}\epsilon_R\kappa + \sigma_p - \sigma_l)$, and $\sigma_{l,p} = P_{l,p} / k_B^+[B]_T$, the matrix M_p can be written as:

$$\mathbf{M}_p = \begin{bmatrix} \lambda_p^2 + \sigma_c s & \sigma_m & \sigma_{im} \\ b_0 & d\epsilon_\tau \lambda_p^2 + \epsilon_\tau s + \sigma_m & 0 \\ b_{i0}\epsilon_R \kappa & 0 & \epsilon_\tau^i s + \sigma_{im} \end{bmatrix}$$

$[B]_T = 50\,\mu M$ and $[B_i]_T = 30\,\mu M$ are the total mobile and immobile buffer, respectively. $K_B = k_B^- / k_B^+$, with dissociation rate $k_B^- = 100\,s^{-1}$ ($k_B^- = 0.3\,s^{-1}$) and capture rate $k_B^+ = 600(\mu Ms)^{-1}$ ($k_B^+ = 1.5(\mu Ms)^{-1}$) of mobile BAPTA (EGTA). $K_{B_i} = k_{B_i}^- / k_{B_i}^+$, with dissociation rate $k_{B_i}^- = 2\,s^{-1}$ and capture rate $k_{B_i}^+ = 1(\mu Ms)^{-1}$ of immobile buffers. $D_B = 95\,\mu m^2/s$ is the diffusion coefficient of mobile buffers. Symbol $\boldsymbol{e}_1 = (1,0,0)'$ stands for the unit vector. The time constants s_i of the response functions must be determined based on both the Ca^{2+}-buffer interaction and the $\sigma_{l,p}$ scaled fluxes. P_p and

P_l are the estimated pump and leak rate constants, respectively. The scaled channel cluster current is defined as:

$$\sigma_j(t) = \frac{J_j(t)}{k_B^-[B]_T}\left(\frac{k_B^+[B]_T}{D_{Ca}}\right)^{3/2} \tag{8.16}$$

The Langevin approximation is a stochastic version of the Li-Rinzel model for a cluster of IP_3Rs in which the inactivation gate h for each IP_3R is modeled as a stochastic process. Each inactivation gate is assumed to be a two-state Markov process, with the unbinding (open) rate α_h and the binding (close) rate β_h. However, each IP_3R channels can exist in four different sub-states (0, 1, 2, or 3 sites occupied) and only if all three-sites of the h gate are bound, the channel opens. The kinetic scheme describing the channels dynamic transition from these states is:

$$[n_0] \xleftrightarrow{(\beta_h, 3\alpha_h)} [n_1] \xleftrightarrow{(2\beta_h, 2\alpha_h)} [n_2] \xleftrightarrow{(3\beta_h, \alpha_h)} [n_3]$$

Under this construct, the analysis could be simplified to determine the number of IP_3R channels $[n_i]$ in the cluster with i open gates $(\forall i = \{1,2,3\})$. Finally, the total number of fully-open (i.e., the three gates are open) channels in the cluster $n^h_{open} = [n_3]$. Therefore, P_{open} (8.6) can be redefined as:

$$P_{open} = \left(m_\infty\left([Ca^{2+}]_i, [IP_3]_i\right)\right)^3 n^h_{open} / N_{channel} \tag{8.17}$$

Assuming a slow Ca^{2+} dynamics and an instantaneous diffusion inside the cluster, a master equation for the sub-state statistics (i.e., $P_h(n_i,t)$ – probability to have n_i channels with i open h-gate at time t) in the cluster is obtained:

$$\begin{aligned}\dot{P}_h(n_i,t) = {} & (N_{channel} - n_i + 1)\alpha_h P_h(n_i - 1, t) + (n_i + 1)\beta_h P_h(n_i + 1, t) \\ & -\left(n_i\beta_h + (N_{channel} - n_i)\alpha_h\right)P_h(n_i,t)\end{aligned}$$

For the particular case of a cluster with a large number of IP_3R channels, the master equation can be approximated by a Fokker-Planck equation, a linear PDE for the probability of fraction of h-open gates $h = n^h_{open} / N_{channel}$. Each Fokker-Planck equation has associated an equivalent Langevin equation.

$$dh = \alpha_h(1-h) - \beta_h h + g_h d\omega \tag{8.18}$$

With $g_h d\omega$ defined as a multiplicative Wiener process, with $\langle (g_h) \rangle^2 = (\alpha_h(1-h) - \beta_h h) / N_{channel}$. α_h and β_h depend on $[Ca^{2+}]_i$ and $[IP_3]_i$ (8.8). The Langevin model was able to reproduce $\langle [Ca^{2+}]_i \rangle$ with < 5 % errors for clusters with $N_{channel} > 15$.

Astrocytic Ca^{2+} signal models have been also extended to describe complex functions of these cells: **a)** astrocyte-neuron interactions and **b)** astrocyte-vascular interactions. Reviews discussing Ca^{2+} signal models in astrocytes and their implications in astrocytic function are recommended for further readings (Manninen et al., 2018; Oschmann et al., 2018; Guttenplan and Liddelow, 2018).

8.6 INTERCELLULAR CALCIUM WAVES – EXTRACELLULAR SIGNALING VS. GAP-JUNCTIONS

Despite astrocytes are not as electrically excitable as neurons, they communicate rapidly and coordinately through a syncytium. Intercellular Ca^{2+} waves constitute the most useful experimental tool to study astrocytic communication (Figure 8.4). Typically, these waves are non-regenerative and can propagate up to ~ $350\,\mu m$ from their point of origin.

Two mechanisms have been suggested to underlie communication between astrocytes (Figure 8.5): **1)** molecules/ion (e.g., IP_3 and Ca^{2+}) diffusion through gap-junctions expressed in their overlapping processes and **2)** gliotransmission (mostly ATP). As astrocytes are organized around neurons and vessels, communication between them has impact on a multicellular level, helping modulate/control neuronal excitability and CBF, respectively. In the neocortex, astrocytes hold an organization tightly linked to the concept of functional columns. For example, gap-junctions between astrocytes from two contiguous columns in the somateosensory cortex of rodents are restricted.

Gap junction channels are formed by two apposed hemichannels with an inner pore diameter of ~ $11\,A°$. A *reflexive* gap junction connects processes in the same

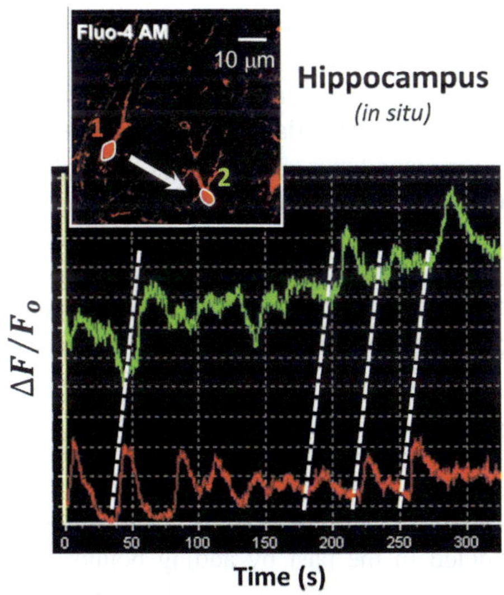

FIGURE 8.4 A Ca^{2+} wave propagating between two astrocytes in a hippocampal acute slice.

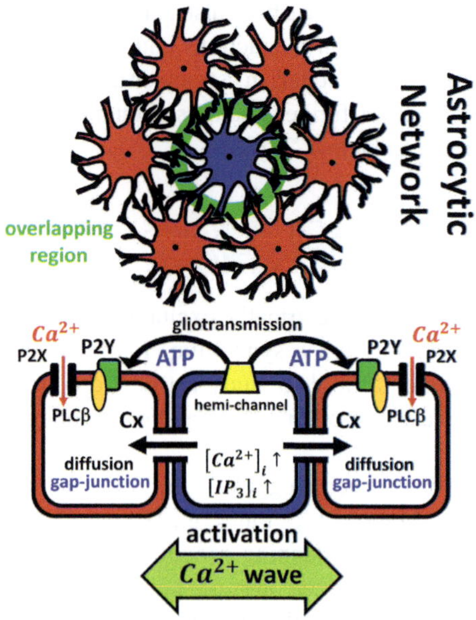

FIGURE 8.5 An illustration of the two mechanisms by mean of which an astrocyte can communicate with their neighbors in a syncytium.

astrocyte. Reflexive gap junctions are homogeneously distributed within the processes. However, intercellular gap junctions seem to be located in the narrow overlapping interface between astrocytes. Via ionic/biochemical coupling in the central pore of hemichannels, gap junctions show selectively to certain small molecules/ions ($\sim 1.1 kDa$). The gap junctional permeability can be dynamically regulated through different mechanisms (Giaume et al., 2010). Two types of gap-junctions have been identified: **a)** one via connexin Cx26 and **b)** another involving connexins Cx43 and Cx30. Cx43 is one of the most abundant gap junctions in astrocytes. This particular connexin is selective for second messengers (IP_3, Ca^{2+} and cAMP), aminoacids (Glu/aspartate), nucleotides (ATP/ADP, NAD), metabolites (glucose, lactate) and RNA. Some connexins operate as hemichannels (functional unpaired connexons), facilitating the exchanges of molecules/ions between the cytosol and the ECS. Secretion of ATP, most likely via these hemichannels, underlie the other form of communication between astrocytes. Extracellular ATP binds a variety of purinergic receptors expressed by astrocytes ($P2Y_{1,2,4,14}$ and $P2X_7$). Through the activation of PLCβ by an ATP-binding conformation change ($G_{q/11}$), P2Y receptors produce intracellular IP_3, which indirectly causes increases in cytosolic Ca^{2+} concentration. Direct increases in $[Ca^{2+}]_i$ might occur after ATP binds P2X receptors, which opens channels for Na^+ and Ca^{2+}. The effect of gap junctions in the intercellular astrocytic Ca^{2+} signaling has been modeled in the past by adding boundary conditions (Fick's law) at the connexin sites: $D_{Ca}\nabla[Ca^{2+}]_i \cdot \vec{n} = P_{Ca}^{l,m}\left([Ca^{2+}]_i^l - [Ca^{2+}]_i^k\right)$ for Ca^{2+} and/or $D_{IP_3}\nabla[IP_3]_i \cdot \vec{n} = P_{IP_3}^{l,m}\left([IP_3]_i^l - [IP_3]_i^k\right)$ for IP_3. Vector $\vec{n}$ represents the orientation of the hemichannels. $P_{Ca}^{l,m}$ and $P_{IP_3}^{l,m}$ are the permeability to Ca^{2+} and IP_3 of the gap-junction separating the l-th and m-th astrocytes. $[Ca^{2+}]_i^{l,k}$ and $[IP_3]_i^{l,k}$ represent the cytosolic concentrations in the astrocyte (l,k) for Ca^{2+} and IP_3, respectively. Modeling the effect of ATP extracellular signaling in Ca^{2+} wave propagation is a little more complicated. Bennett et al. (2005) proposed the use of equation (8.11) to model the P2Y metabotropic receptor for ligand ATP, with $k_a^G = 0.017\,s^{-1}$ and $k_d^G = 0.15\,s^{-1}$. They modeled extracellular ATP using a reaction-diffusion model $\partial[ATP]_e / \partial t = D_{ATP}\nabla^2[ATP]_e + J_{prod} - J_{deg}$, with $D_{ATP} = 300\,\mu m^2 / s$. The breakdown of ATP by an ectonucleotidase was model by a single binding Michaelis-Menten kinetics $J_{deg} = V_{deg}[ATP]/([ATP]+K_{deg})$. The mechanisms underlying the production of ATP (J_{prod}) are still elusive. Models that include a P2X component in the Ca^{2+} signal genesis have also been proposed; they are discussed in Manninen et al. (2018).

8.7 METABOLIC PATHWAYS AND CONTROL POINTS IN ASTROCYTES

As for most living cells, neurons and astrocytes in the brain extract energy stored in nutrients by a process involving three main components: the glycolysis, the tricarboxylic acid cycle (TCA-cycle) and the oxidative-phosphorylation. These three components are typically regulated by a network of enzymes interacting dynamically with each other. To minimize toxic residuals from the metabolism of complex nutrients (e.g., glycerol, amino acids, and fatty acids), these two brain cells preferentially use a monosaccharide (glucose "Glc") as the main source of energy. As astrocytes forms a syncytium connecting neurons and the surrounding vasculature, they are ideal cells for the local regulation of Glc supply/metabolism in a way that satisfy an optimal offer-demand principle, also protecting the brain from both the accumulation and shortage of metabolites. To accomplish it, metabolite synthesis/degradation must take place in these cells in a compartmentalized way at both multi and single-cellular levels. Crucial findings about the neuronal-astrocytic interplay for metabolite interactions have been obtained from the isolated nerve ending particles, i.e., the synaptosomes. In this section, we provide elements in the Glc metabolism by astrocytes needed to explain their most relevant functioning principles (Chapter 9).

8.7.1 Regulation Mechanisms of Glycolysis

Glycolysis, also known as the Embdem-Meyerhoff pathway, comprises two phases (Figure 8.6): preparatory and pay-off. During the preparatory phase, one Glc is converted into two D-glyceraldehyde 3-phosphate (GADP) molecules, a process expending two ATPs. Each GADP molecule is used to produce two ATPs and one reduced nicotinamide adenine dinucleotide (NADH) cofactor in the pay-off phase, a process that runs twice. In the context of brain imaging, only three enzymes are considered to play regulatory roles: the hexokinase (HK), the phosphofructokinase-1 (PFK-1) and the pyruvate kinase (PK). First of all, a family of HK enzymes regulate the influx of Glc into cells. Taking extracellular Glu into the cells occurs in a concentration-dependent manner via glucose transporters (GLUT); i.e., a type of

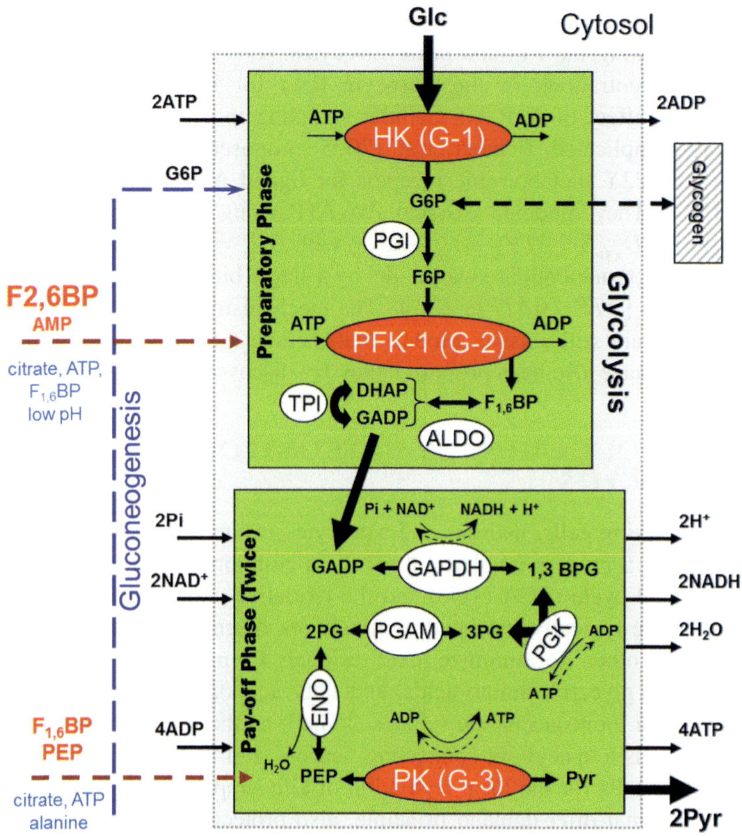

FIGURE 8.6 The three checkpoints in the glycolysis pathways (G-1, G-2, and G-3). (Adapted from Riera et al. (2008).)

uniporter. Glc phosphorylation into G6P by HK in the cytoplasm is an irreversible chemical reaction with Mg^{2+} as cofactor. G6P cannot leak out from brain cells as they lack transmembrane transporters for G6P. With the exception of the glucokinase found in glucose-sensing neurons, most HK isozymes in the brain (I, II and III) have high affinity for Glc and are inhibited by G6P. The brain has no glucose-6-phosphatase; hence, G6P cannot be converted back into Glc (last step in gluconeogenesis). Therefore, HK is considered the first checkpoint in the glycolytic pathway (G-1) as it facilitates taking up extracellular Glc by the brain cells. Glc is delivered from blood to the extracellular milieu via the GLUT-1 (55-kDa subtype), highly expressed by brain endothelial cells (ECs). Astrocytes and neurons express preferentially isoforms GLUT-1 (45-kDa subtype) and GLUT-3, respectively. Depending on cells nutrient demand, GP6 could be stored in the form of glycogen polymers, a mechanism, named the glycogen-shunt (Chapter 9). Isomerization of G6P into β-D-fructose 6-phosphate (F6P) is carried out by the phosphoglucose isomerase (PGI). The transformation of F6P into β-D-fructose 1,6-bisphosphate ($F_{1,6}BP$) is performed by the allosteric enzyme PFK-1. For many reasons, PFK-1 is considered a key checkpoint (G-2) in the glycolytic pathway. First, the PFK-1 enzymatic reaction is energetically very favorable; hence, almost irreversibly $F_{1,6}BP$ proceed down the glycolytic pathway after this step. Second, PFK-1 shows allosteric inhibition by citrate, the first sub-product of the TCA-cycle in the matrix of the mitochondrion; hence, it helps drop-off the glycolytic pathway if the TCA-cycle is overrunning. Mitochondrial tricarboxylate carriers (or citrate transporters) have been found in the neurons and glial cells. PFK-1 is also inhibited by a low pH level, which may prevent the accumulation of cytosolic H^+ produced in the pay-off phase. ATP is not only a substrate for PFK-1, but also an allosteric inhibitor; hence, the glycolytic pathway is downregulated at the PFK-1 level in the presence of high ATP concentrations in the cytosol. Lastly, PFK-1 is inhibited by its own product $F_{1,6}BP$. By increases the affinity for F6P, β-D-fructose 2,6-bisphosphate ($F_{2,6}BP$) constitutes one of the most important regulators of PFK-1. The binding of $F_{2,6}BP$ also reduces the inhibitory effect of ATP. $F_{2,6}BP$ is reversibly produced in the brain from F6P by the 6-phosphofructo-2-kinase/fructose-2,6-bisphosphatase (PFK-2). This reaction consumes one ATP. In case of a local increase in Glc concentration, glycolysis will be accelerated by this feedforward mechanism. The expression of liver (L) and muscle (M) isoforms are similar in neurons and astrocytes. However, the brain isoform (C) shows fourfold greater expression in astrocytes than in neurons. Among the four PFK-2.(1-4) isoforms, PFK-2.(3) shows the highest kinase-bisphosphatase activity ratio, producing more $F_{2,6}BP$ at a given F6P concentration. In comparison to neurons, PFK-2.(3) is also highly expressed in astrocytes. Almeida et al. (2004) demonstrated a rapid up-regulation of the glycolytic pathway in astrocytes after the inhibition of mitochondrial oxidative-phosphorylation (cytochrome c) by NO, which was independent of cyclic guanosine monophosphate (GMP) and associated with an enhancement of PFK-1 (C) activity. Therefore, NO might not only down-regulates oxygen consumption by mitochondria, but could also trigger the transfer of lactate from astrocytes to neurons by activating the glycolytic pathways in astrocytes (Chapter 9). PFK-1 is also activated by high-levels of adenosine monophosphate (AMP). Almeida et al. (2004) also demonstrated that the phosphorylation of the PFK-2.(3) isoform was mediated by the AMP-activated protein kinase (AMPK). AMP signaling is crucial in maintaining energetic equilibrium, as it can switch cells from ATP consumption to ATP production (short-term regulation). The preparatory phase ended with an aldol reaction by the fructose 1,6 bisphosphate aldolase (ALDO), converting $F_{1,6}BP$ into dihydroxyacetone phosphate (DHAP) and glyceraldehyde 3-phosphate (GADP), which are rapidly and reversibly interconverted by the triose-phosphate isomerase (TPI). There are five steps in the pay-off phase: **1)** a redox reaction by the glyceraldehyde phosphate dehydrogenase (GAPDH; <u>substrates</u> – GADP, oxidized nicotinamide adenine dinucleotide cofactor "NAD^+", P_i and <u>products</u> – 1,3-bisphosphoglycerate "1,3BPG", reduced nicotinamide adenine dinucleotide cofactor "NADH", H^+), **2)** a first substrate-level phosphorylation by the phosphoglycerate kinase (PGK, <u>substrates</u> – 1,3BPG, ADP and <u>products</u> – 3-phosphoglycerate "3PG", ATP), **3)** a mutase reaction by the phosphoglycerate mutase (PGAM, <u>substrate</u> – 3PG and <u>product</u> – 2-phosphoglycerate "2PG", **4)** a hydration by the enolase (ENO, <u>substrates</u> – 2PG and <u>product</u> – phosphoenolpyruvate "PEP", H_2O), and **5)** a second substrate-level phosphorylation by the allosteric pyruvate kinase (PK), converting irreversibly PEP and ADP into pyruvate (Pyr) and ATP with Mg^{2+} as a cofactor. PK constitutes the last checkpoint (G-3) in the glycolytic pathway. PK is also inhibited by citrate and ATP for cells energetically charged. High levels of $F_{1,6}BP$ and PEP activate PK, which might speed up the pay-off phase. PK is inhibited by alanine, an important aspect for maintaining the ammonia homeostasis in the brain (Chapter 9). Low ADP concentration inhibits PGK, suggesting a regulatory role for this enzyme. However, low ADP causes an equivalent effect in glycolysis via PK.

8.7.2 The Pyruvate Breakdown Process

Depending on the mechanism used to shuttle NADH + H^+ from the cytosol to the mitochondrion, a single Glc yield either 30 (glycerol 3-phosphate shuttle) or 32 (malate-aspartate shuttle, MAS) ATP molecules. Twenty-five of these ATP molecules are produced by breaking down the carbon skeletons in the end-product of glycolysis (2 Pyr molecules), which occur in three sequential steps: **(i)** Pyr decarboxylation, **(ii)** TCA-cycle and **(iii)** oxidative-phosphorylation. The main catabolic reactions and checkpoints in this three-step process are discussed henceforth. Details about these three steps can be found in Riera et al. (2008).

Pyr decarboxylation: Pyr, a polar molecule, is transported through the impermeable inner mitochondrial membrane by the mitochondrial pyruvate carriers (MPC1 and MPC2). After entering the mitochondrion by this carrier-mediated and regulated process, Pyr can be metabolized by either: **1)** an irreversible carboxylation of Pyr into oxaloacetate "OAA" by the Pyr carboxylase, or **2)** an oxidative decarboxylation of Pyr into acetylcoenzyme A (Acetyl-CoA) by the Pyr dehydrogenase (PDH)-complex. The first pathway is essential for gluconeogenesis (OAA→PEP, by cytosolic PEP carboxykinase) in case of accumulated Pyr in the mitochondrion, a process requiring the transfer of OAA to the cytosol by the MAS. It is also crucial to incorporate extra OAAs in the TCA-cycle if needed. The second pathway prepares the Pyr for its oxidation in the TCA-cycle. In case of energy demand by the cell, the PDH-complex produces Acetyl-CoA from Pyr (Figure 8.7, top). Acetyl-CoA has a large size; hence, it cannot leave the mitochondrion, which ensure a positive flow of carbon toward the TCA-cycle (Figure 8.7, bottom). In a desphosphorylated state, the PDH-complex is active. It becomes inactive if phosphorylated by the PDH-kinase, which requires ATP consumption. The presence of energy-rich molecules (e.g. NADH, ATP, Acetyl-CoA) enhances PDH-kinase activity, which help reduce the flow of carbon into the TCA-cycle. The concentration of NAD^+, coenzyme A (CoA-SH) and ADP increases as the energetic charge of the cell drops, which promote carbon flow into the TCA-cycle by reducing the activity of the PDH-kinase. In addition, high levels of Pyr and Ca^{2+} inhibit this enzyme. NADH and Acetyl-CoA reduce the affinity of the PDH-complex complex for Pyr, a negative allosteric effect. The PDH phosphatase uninterruptedly dephosphorylated inactive PDH complexes, a process that can be boosted by the presence of Ca^{2+} and Mg^{2+}. For all these reasons, the PDH-complex constitutes the first checkpoint in mitochondrial respiration (H-1).

TCA-cycle: The PDH-complex pathway is followed by the TCA-cycle (also known as the citric acid cycle or the Krebs cycle) (Figure 8.7, bottom). In the TCA-cycle,

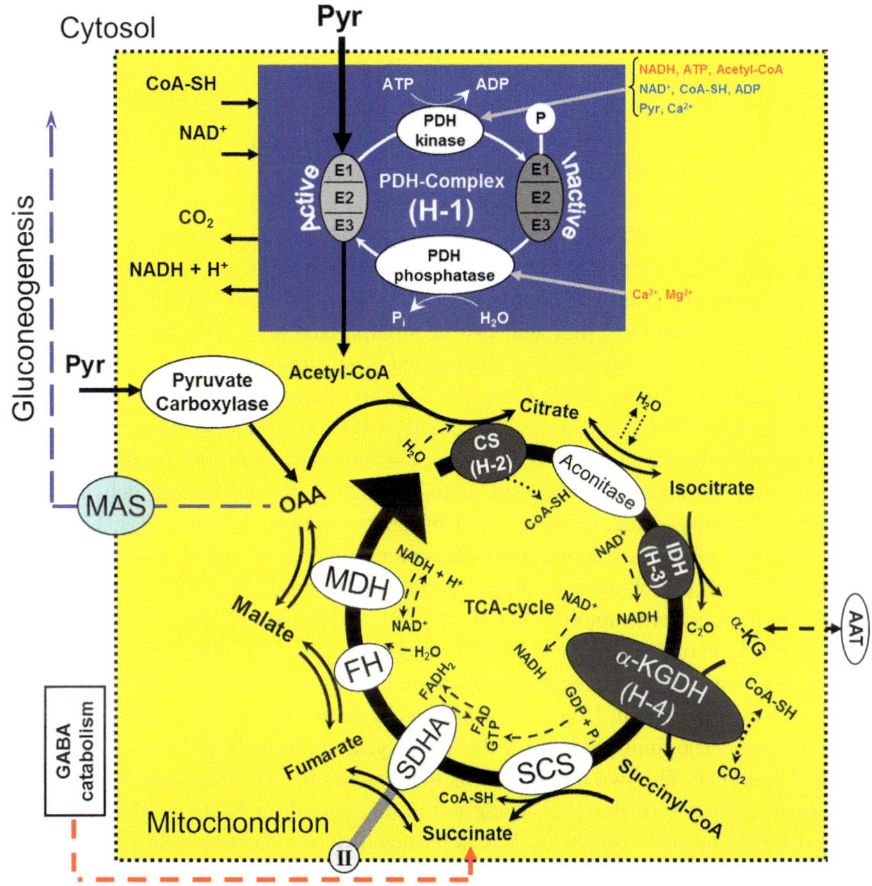

FIGURE 8.7 The four checkpoints in the PHD-complex (H-1) and TCA-cycle (H-2, H-3, H-4). (Adapted from Riera et al. (2008).)

a series of endergonic and exergonic chemical reactions takes place, resulting in the catabolism of energy-rich molecules. These reactions provide important precursors for different cellular functions. In summary, two carbons are oxidized to CO_2 during the TCA-cycle. The energy from the involved chemical reactions is stored in two energy-rich components: **1)** guanosine triphosphate (GTP) and **2)** two electron donors, i.e. NADH and reduced flavin adenine dinucleotide ($FADH_2$). Three checkpoints in the TCA-cycle are relevance to brain imaging. The condensation of OAA to citrate by the citrate synthase (CS) constitutes the first checkpoint (H-2). Mitochondrial citrate inhibits the activity of CS. Note that the final fate of the glucose sub-products is determined by citrate inhibitory feedback, both in the glycolysis (PFK-1 and PK) and TCA-cycle (CS). The activity of CS is also allosterically regulated as a function of the mitochondrion energy charge. It is inhibited by the presence of energy-rich molecules (ATP and NADH) and its activity enhanced by increases in ADP levels. This step is followed by an isomerization (aconitinase) of citrate into isocitrate by means of a sequential dehydration/hydration. The isocitrate dehydrogenase (IDH) is the rate-limiting step in the cycle. There are negative (e.g. NADH, ATP) and positive (isocitrate, ADP, AMP, Ca^{2+}) allosteric effectors on IDH; hence it constitutes an important checkpoint in the TCA-cycle (H-3). A product of IDH, α-ketoglutarate (α-KG), drives the MAS through the malate/α-KG carrier. The α-ketoglutarate dehydrogenase (α-KGDH) is the last checkpoint (H-4) in the TCA-cycle. α-KGDH activity is inhibited by ATP and activated by NAD^+, both allosterically. Succinyl-CoA and NADH are negative product effectors for this enzyme. The presence of Ca^{2+} increases the activity of α-KGDH. The TCA-cycle continues from this point with four sequential reactions that are not much relevant for brain imaging. Therefore, they will just be summarized here: **1)** substrate-level phosphorylation by the succinyl-CoA synthetase (SCS, substrates – Succinyl-CoA, GDP, P_i) and products – succinate, GTP, CoA-SH, **2)** a first dehydrogenation by the succinate dehydrogenase (SDHA, substrates – succinate, FAD and products – fumarate, FADH2), **3)** a hydration by fumarase (FH, substrates – fumarate, H_2O and product – malate, and **4)** a highly endergonic second dehydrogenation by the malate dehydrogenase (MDH, substrates – malate, NAD^+ and products – OAA, NADH, H^+). During the SCS reaction, GTP is used in a trans-phosphorylation catalyzed by the mitochondrial nucleoside diphosphokinase to phosphorylate ADP, producing ATP, and generating GDP. Through protein subunits (SDHB, SDHC, and SDHD), SDHA is coupled to the inner membrane of the mitochondrion, forming the complex II of the electron-transport chain. The thermodynamic equilibrium in the last seven reversible reaction is in favor of the product formation. Hence, the TCA-cycle always flows in a clockwise direction (black arrow). The final products of the TCA-cycle are three NADH, one $FADH_2$, two H^+ and one GTP. The two carbon atoms that entered the TCA-cycle leaves it as CO_2. The main regulator of the TCA-cycle is the mitochondrial NAD^+/NADH ratio. This ratio depends on the oxygen level as determined by the electron-transport/oxidative-phosphorylation pathway.

Oxidative-phosphorylation: Through the electron-transport/oxidative-phosphorylation pathway, ATP is created from high-energy electron donors located inside the mitochondrial matrix. First, electrons enter the electron-transport chain (ETC) from NADH and $FADH_2$ donors in its complexes I and II, respectively. These electrons are then transferred to coenzyme Q (CoQ). They are carried through the mitochondrial inner membrane to complex III by the CoQ reduced form (ubiquinol, $CoQH_2$). From complex III, electrons are transferred to a peripheral membrane protein, cytochrome c. In synaptosomes, NO is a potent reversible inhibitor of cytochrome c at the nM level. Cytochrome c transfers electrons into complex IV, which are used in there to reduce O_2 to H_2O. The ETC is a multi-step redox process involving several enzymes, many of them crossing the mitochondrial inner membrane to deliver H^+ to the outer chamber like ferryboats, named proton pumps. As discussed in Chapter 6, the energy stored in the H^+ ion gradient is used by the ATP synthase to phosphorylate ADP, creating ATP – the final product of respiration. The ETC is very efficient. However, some electrons can prematurely leak to O_2, resulting in the formation of ROS. In addition, the potential involvement of vascular signaling molecules (e.g. endothelium-derived factors) on electron-transport chain enzymes is still under investigation, which might have further implications in brain imaging.

8.8 PROBLEMS AND EXERCISES

Problem 8.1 – If its basal Ca^{2+} concentration is 80 nM, calculate the amount of Ca^{2+} ions in the cytosol of an average human astrocyte soma at rest. Calculate the PMCA rate $\overline{v}_{max}^{PMCA}$ if the leak across the plasma membrane is $j_{PM_{leak}} = 14.4\ mM / cm^2 s$. Ignore any other type of fluxes across the plasma membrane of the astrocyte.

Problem 8.2 – Calculate the total energy (enthalpy) in the form of Glc in the circulation system of a normal average 20 years old American man ($weighs = 197.9\,pounds$). Consider that: a) the average concentration of glucose in blood is $5.5\,mmol / L$, b) the average blood volume for males is $75\,ml / kg$, and c) the enthalpy energy releases at standard temperature/pressure due to the decomposition of ATP into hydrated ADP is $30.5\,kJ / mol$.

REFERENCES

Alberts BD, Lewis BJ, Raff M, Roberts K, Watson JD. *Molecular Biology of the Cell*. Garland Publishing Inc., New York, pp. 405–408, 2002.

Almeida A, Moncada S, Bolanos JP. Nitric oxide switches on glycolysis through the AMP protein kinase and 6-phosphofructo-2-kinase pathway. *Nat Cell Biol* 6, 45–51, 2004.

Araque A, Parpura V, Sanzgiri RP, Haydon PG. 1999. Tripartite synapses: Glia, the unacknowledged partner. *Trends Neurosci* 22: 208–215.

Bennett MR, Farnell L, Gibson WG. A quantitative model of purinergic junctional transmission of calcium waves in astrocyte networks. *Biophys J* 89, 2235–2250, 2005.

Boitier E, Rea R, Duchen MR. Mitochondria exert a negative feedback on the propagation of intracellular Ca2+ waves in rat cortical astrocytes. *J Cell Biol*, 145(4), 795–808, 1999.

Carmignoto G, Pasti GL, Pozzan T. On the role of voltage-dependent calcium channels in calcium signaling of astrocytes in situ. *J Neurosci* 18(12), 4637–4645, 1998.

Cornell-Bell AH, Finkbeiner SM, Cooper MS, Smith SJ. Glutamate induces calcium waves in cultured astrocytes: Long-range glial signaling. *Science* 247, 470–473, 1990.

Danbolt NC. Glutamate uptake. *Progr Neurobiol* 65, 1–105, 2001.

D'Ascenzo M, Vairano M, Andreassi C, Navarra P, Battista Azzena G, Grassi C. Electrophysiological and molecular evidence of L-(Cav1), N- (Cav2.2), and R- (Cav2.3) type Ca2+ channels in rat cortical astrocytes. *Glia* 45, 354–363, 2004.

De Young GW, Keizer J. A single-pool inositol 1,4,5-trisphosphate-receptor based model for agonist-stimulated oscillations in Ca2+ concentration. *Proc Natl Acad Sci* 89, 9895–9899, 1992.

Di Garbo A, Barbi M, Chillemi S, Alloisio S, Nobile M. Calcium signalling in astrocytes and modulation of neural activity. *BioSystems* 89, 74–83, 2007.

Giaume C, Koulakoff A, Roux L, Holcman D, Rouach N. Astroglial networks: A step further in neuroglial and gliovascular interactions. *Nat Rev Neurosci* 11, 87–99, 2010.

Guttenplan KA, Liddelow SA. Astrocytes and microglia: Models and tools. *J Exp Med* 216(1), 71–83, 2018.

Hirase H, Qian L, Barthó, P, Buzsáki G. Calcium dynamics of cortical astrocytic networks in vivo. *PLoS Biol* 2, 0494–0499, 2004.

Höfer T, Venance L, Giaume C. Control and plasticity of intercellular calcium waves in astrocytes: A modeling approach. *J Neurosci* 22, 4850–4859, 2002.

Lecrux C, Hamel E. The neurovascular unit in brain function and disease. *Acta Physiologica* 203(1), 47–59, 2011.

Li YX, Rinzel J. Equations for InsP3 receptor-mediated [Ca2+]i oscillations derived from a detailed kinetic model: A Hodgkin and Huxley like formalism. *J Theor Biol* 166, 461–473, 1994.

Manninen T, Havela R, Linne M-L. Computational models for calcium-mediated astrocyte functions. *Front Comput Neurosci* 12(article 14), 1–27, 2018.

Moshkforoush A, Balachandar L, Moncion C, Montejo KA, Riera Diaz J. Unraveling ChR2-driven stochastic Ca2+ dynamics in astrocytes: A call for new interventional paradigms. *PLoS Comp Biol*, submitted, 2020. https://doi.org/10.1371/journal.pcbi.1008648

Nett WJ, Oloff SH, McCarthy KD. Hippocampal astrocytes in situ exhibit calcium oscillations that occur independent of neuronal activity. *J Neurophysiol* 87, 528–537, 2002.

Oschmann F, Berry H, Obermayer K, Lenk K. From in silico astrocyte cell models to neuron-astrocyte network models: A review. *Brain Res Bull* 136, 76–84, 2018.

Riera J, Hatanaka R, Ozaki T, Kawashima R. Modeling the spontaneous Ca2+ oscillations in astrocytes: Inconsistencies and usefulness. *J Integr Neurosci* 10(4), 439–473, 2011.

Riera J, Hatanaka R, Uchida T, Ozaki T, Kawashima R. Quantifying the uncertainty of spontaneous Ca2+ oscillations in astrocytes: Particulars of Alzheimer's disease. *Biophys J* 101, 554–564, 2011.

Riera JJ, Schousboe A, Waagepetersen HS, Howarth C, Hyder F. The micro-architecture of the cerebral cortex: Functional neuroimaging models and metabolism. *NeuroImage* 40, 1436–1459, 2008.

Savtchenko LP, Bard L, Jensen TP, Reynolds JP, Kraev I, Medvedev N, Stewart MG, Henneberger C, Rusakov DA. Disentangling astroglial physiology with a realistic cell model in silico. *Nat Commun* 9(3554), 1–15, 2018.

Sharp AH, Nucifora Jr. FC, Blondel O, Sheppard CA, Zhang C, Snyder SH, Russel JT, Ryugo DK, Ross CA. Differential cellular expression of isoforms of inositol 1,4,5-triphosphate receptors in neurons and Glia in Brain. *J Comp Neurol* 406, 207–220, 1999.

Shigetomi E, Kracun S, Sofroniew MV, Khakh BS. A genetically targeted optical sensor to monitor calcium signals in astrocyte processes. *Nat Neurosci* 13, 759–766, 2010.

Shuai J, Jung P. Stochastic properties of Ca2+ release of inositol 1,4,5-trisphosphate receptor clusters. *Biophys J* 83, 87–97, 2002.

Skupin A, Kettenmann H, Falcke M. Calcium signals driven by single channel noise. *PLoS Comput Biol* 6 , e1000870, 1–13, 2010.

Skupin A, Kettenmann H, Winkler U, Wartenberg M, Sauer H, Tovey SC, Taylor CW, Falcke M. How does intracellular Ca2+ oscillate: By chance or by the clock? *Biophys J* 94, 2404–2411, 2008.

Smith GD, Wagner J, Keizer J. Validity of the rapid buffering approximation near a point source of calcium ions. *Biophys J* 70(6), 2527–2539, 1996.

Sneyd J, Tsaneva-Atanasova K, Reznikov V, Bai Y, Sanderson MJ, Yule DI. A method for determining the dependence of calcium oscillations on inositol trisphosphate oscillations. *Proc Natl Acad Sci USA* 103(6), 1675–1680, 2006.

Verkhratsky A, Steinhäuser C. Ion channels in glial cells. *Brain Res Rev* 32, 380–412, 2000.

Wacquier B, Combettes L, Tran Van Nhieu G, Dupont G. Interplay between intracellular Ca2+ oscillations and Ca2+-stimulated mitochondrial metabolism. *Sci Rep* 6, 19316, 2016.

Wang X, Lou N, Xu Q, Tian GF, Peng WG, Han X, Kang J, Takano T, Nedergaard M. Astrocytic Ca2+ signaling evoked by sensory stimulation in vivo. *Nat Neurosci* 9, 816–823, 2006.

Williams GSB, Boyman L, Chikando AC. Mitochondrial calcium uptake. *Proc Natl Acad Sci USA* 110 (26), 10479–10486. https://doi.org/10.1073/pnas.1300410110

Zur Nieden R, Deitmer JW. The role of metabotropic glutamate receptors for the generation of calcium oscillations in rat hippocampal astrocytes in situ. *Cereb Cortex* 16, 676–687, 2006.

9 Astrocyte Functional Roles

9.1 OVERVIEW

Internal energy available in brain cells is used to do chemical, electric, and mechanical work crucial for their functions. During cellular work, a small portion of this energy is irreversibly lost in the form of heat. Energy, majorly in the form of D-glucose (Glc), is provided to the brain through a Glc carrier (GLUT-1) densely expressed in the membranes of endothelial cells (EC) in brain capillaries. To a minor degree, energy can be also delivered via amino acid carriers (e.g., LAT1) and transporters for nucleosides/nucleobases also present in ECs. Therefore, the continuous restoration of internal energy in brain cells occurs mostly by the delivery of Gibb's free energy stored in Glc. As discussed in Chapter 8, Glc consumption (CMR_{Glc}) includes oxidative ($CMR_{Glc}(ox)$) and non-oxidative ($CMR_{Glc}(non-ox)$) components. The oxygen-to-glucose index ($OGI = CMR_{O_2} / CMR_{Glc}$) is used to quantify the degree of oxidative vs. non-oxidative Glc breakdown. The cerebral metabolic rate of oxygen (CMR_{O_2}) depends on the cellular activity, mostly by neurons. The oxidative breakdown of Glc is represented by the reaction $C_6H_{12}O_6 + 6O_2 \rightarrow 6CO_2 + 6H_2O$. The total ATP regeneration rate is defined by the generic equation $V_{ATP} = \varphi CMR_{O_2} + \gamma CMR_{Glc}$, with $\varphi = 6\,ATP / oxygen$ and $\gamma = 2\,ATP / Glc$. The OGI can be estimated using two main methods: **1)** the cerebral arterio-venous differences of oxygen and glucose (gold standard), and **2)** PET imaging with radiotracers (i.e., ^{18}F-fluorodeoxyglucose "FDG" and ^{15}O-O_2). FDG is used to estimate CMR_{Glc} and ^{15}O-O_2 to estimate CMR_{O_2}. These two PET-based magnitudes can be also used to estimate V_{ATP}. However, absolute values of V_{ATP} are typically obtained with MRS (^{31}P and ^{13}C). Under awake resting conditions, measurements from these two methods in humans have shown $OGI \in (5.5-5.8)$ (Siesjo, 1978; Sokoloff, 1991). The OGI deviation from the theoretical stoichiometric value of 6 is due to biosynthetic reactions and efflux of lactate from the brain. Hence, $CMR_{Glc}(non-ox)$ at rest is quite negligible (about 9 %). However, increases in neuronal activity (e.g., sensory stimulation) correlate with decreases in OGI, which have been linked to quick energy mobilization for glutamate/GABA neurotransmitter recycling by astrocytes via glycolysis, resulting in lactate overproduction by the lactate dehydrogenase (LDH) in these cells, and hence, increased lactate efflux from them to the extracellular space. A glycolytic index ($GI = CMR_{Glc} - s\,CMR_{O_2}$) was introduced to quantify glycolysis in brain

DOI: 10.1201/9780367901509-9

tissue, with $s = 0.19$ for gray matter (Hyder et al., 2016). The *OGI* depends also on the balance between the glycogenolysis and glycogenesis rates in astrocytes by phosphoglucose mutase (PGM) activity, a process referred to as the glycogen-shunt. These glycogenolysis and glycogenesis rates vary depending on the energetic charge of astrocytes. There are two main processes dictating astrocytic energetic charges: **a)** the rate of the glutamate/GABA-glutamine recycling and **b)** the rate of astrocytic calcium oscillations. Energy charges associated with GABA-glutamine recycling has been considered negligible, but exact quantification is still elusive. However, it has been suggested that the glutamate-glutamine recycling consumes 2 ATPs per glutamate (Magistretti et al., 1999); one of them is used to convert glutamate into glutamine by the glutamine synthetase (GS) and the other by the Na-K ATPase pump to reestablish the Na^+ gradient created during glutamate uptake by astrocytes. Glutamate release from presynaptic terminals is also known to trigger increases in astrocytic calcium excitability by its action on metabotropic receptors (mGluR), a mechanism underlying the main form of communication between neuron and the surrounding vasculature to alter energy supply according to demand. The latter leads to energy consumption via Ca^{2+}-ATPase pumping to reestablish gradients across the plasma and endoplasmic reticulum (ER) membranes. During surges in neuronal activity, Glc influx from vessels to brain tissue via GLUT-1 is very limited. Therefore, neuronal activation causes temporary increases in the glycogenolysis rates, facilitating the use of glycogen-stored energy in astrocytes. There are two theoretical models to explain how increases in neuronal energy demands are met by metabolizing astrocytic glycogen: **1)** the astrocyte-neuron lactate shuttle (ANLS) and **2)** the glucose sparing by glycogenolysis (GSG). For the ANLS model, ATP produced from glycogenolysis and Glc uptake by astrocytes is used to satisfy increases in energy demand due to glutamate/GABA-glutamine recycling and a by-product of the associated glycolysis (pyruvate) is used to produce and export lactate to neurons for their energetic demand. For the GSG, astrocytes also use energy from glycogenolysis to satisfy increases in their energy consumption due to the amino acid neurotransmitters recycling, but they reduce Glc uptake to enable their use by nearby neurons.

In turn, astrocytic calcium oscillations cause regional increases in cerebral blood flow (CBF) via a variety of mechanisms, leading to an increase in the oxygen delivery from the microvasculature to the brain tissue. Increases in CBF must be such that compensate for an expected reduction in the oxygen extraction fraction (OEF, $E = (C_B^a - C_B^v) / C_B^a$) from capillary to brain tissue due to a shortening in the hemoglobin transition time inside the capillary. According to the Fick's principle $E = CMR_{O_2} / (C_B^a f)$. Zheng et al. (2002) proposed a general equation to describe the dependency of E with CBF (f):

$$\frac{\varphi}{f}\frac{dE}{dt} + E = (1-g)\left[1-\left(1-\frac{E_0}{1-g_0}\right)^{1/f}\right] \tag{9.1}$$

In this general model, variable g changes dynamically as described by a set of differential equations. The well-accepted model for oxygen transport to tissue from capillaries proposed by Buxton and Frank (1997) is obtained by assuming $\varphi = 0$

and $g = g_0 = 0$. Beyond isoelectricity, neuronal oxidative demand is linearly related to neuronal activity. These linear relationships seem to be conserved across species (Hyder, 2009). According to the ANLS model, evoked surges in neuronal activity will lead to increases in CMR_{O_2} due to enhanced oxidative metabolism of the exported lactate from astrocytes to neurons. Unnecessary lactate in the extracellular space will be drained out through the microvasculature, which has been reported during increases of neuronal activity. Disproportional increases in partial pressure of oxygen (PaO_2) in brain tissue could lead to accumulation of reactive oxygen species (ROS) and, hence, oxidative cellular stress. ROS toxicity causes oxidation of membrane lipids, and conformational changes in proteins, which might have a very adverse impact in brain functions. In $100\,g$ of brain tissue at rest, f and CMR_{O_2} are approximately equal to $19\,ml\,/\,min$ and $1.3\,ml\,/\,min$, respectively (Powers et al.,1985). The average arterial oxygen content for young healthy subjects (age~ 26 y/o) is $C_a^B = 9.4 \pm 0.5\,\mu mol\,ml^{-1}$ (Hyder et al., 2016). To deliver just the necessary oxygen during increases in neuronal activity, regulation of regional CBF must be performed in a very precise way. In this chapter, we will discuss mechanisms present in astrocytes to estimate the regional metabolic cost and to accordingly adjust local blood supply by the release of vasoactive substances to the nearby vasculature. The material is mainly based on recently published results; therefore, the number of cited references is substantially larger than for other chapters. Most of these references are also cited in two reviews (Riera et al., 2008; Riera and Sumiyoshi, 2010).

9.2 GLUTAMATE/GABA-GLUTAMINE RECYCLING

To maintain chemical neurotransmission, presynaptic terminals must continuously create and refill synaptic vesicles (SV) for neurotransmitter release. In Chapter 7, we presented the four main steps underlying this process: **1)** SV pool establishment, **2)** SV clustering, **3)** SV refilling, and **4)** docking/priming SVs. SV refilling depends on available neurotransmitters inside the synaptic terminal. The concentration of a neurotransmitter inside the synaptic terminals could increase in two ways: **a)** its de-novo synthesis from substrates and **b)** its reuptake from the synaptic cleft. The enzymatic machinery necessary for de-novo synthesis is available in the neuron ending terminals.

As the action of the released neurotransmitters should be terminated, reuptake is also a form of neurotransmitter clearance from the synaptic cleft with minimum energetic cost. Although it is expected that a small portion of the released neurotransmitters diffuse passively out of the synaptic cleft contributing to volume transmission, most of the released neurotransmitters is either *directly* or *indirectly* recycled back into the terminal. Indirect recycling occurs if neurotransmitter clearance is accelerated by enzymatic catabolism in the cleft, followed by the reuptake of the inactive residuals, and the resynthesis of the neurotransmitter inside the terminal from these residuals. For example, acetylcholinesterase (AChE) catalyzes the breakdown of acetylcholine (ACh) into choline in the cleft, which is transported back to the presynaptic terminal by the high-affinity choline transporter (ChT). Inside a cholinergic synapse, biosynthesis of ACh takes place by the action of the choline acetyl transferase (CAT) on

TABLE 9.1
Mechanisms for the synthesis and reuptake of monoaminergic neurotransmitters

<table>
<tr><th>Neurotransmitters</th><th>Precursor</th><th>Synthesis mechanism</th><th>Reuptake mechanism</th></tr>
<tr><td>Epinephrine</td><td>Tyrosine</td><td>Tyrosine Hydroxylase</td><td>Epinephrine Transporter (ET)</td></tr>
<tr><td>Norepinephrine</td><td rowspan="3">Tryptophan</td><td rowspan="3">Tryptophan Hydroxylase</td><td>Norepinephrine Transporter (NET)</td></tr>
<tr><td>Dopamine (DA)</td><td>Dopamine Transporter (DAT)</td></tr>
<tr><td>Serotonin (5-HT)</td><td>Serotonin Transporter (SERT)</td></tr>
<tr><td>Histamine</td><td>Histidine</td><td>Histidine Decarboxylase</td><td>Polyspecific Transporters: Organic Cation Transporter (OCT) and Plasma Membrane Monoamine Transporter (PMAT)</td></tr>
</table>

two substrates (i.e., choline and acetyl-CoA). The vesicle ACh transporter (VAChT) moves ACh into the synaptic vesicles. CAT and VAChT are biomarkers for ACh presynaptic nerve endings. However, direct reuptake is the most common form of neurotransmitter recycling. Reuptakes is performed by the quick action of co-transporters that use Na^+-K^+-Cl^- gradients to move the respective neurotransmitter back to the presynaptic terminals. For example, released monoamines are transported back into the synaptic terminals by a variety of co-transporters, (Table 9.1).

Although each monoamine transporter has preferential affinity for a given neurotransmitter, it could in some cases also transport multiple monoamines. Monoamines can also be directly synthesized from precursors inside the presynaptic terminals by a variety of enzymes. Once inside the terminal, monoamines are packed into the vesicles by vesicle monoamine transporters (VMAT). Inside the terminal and in the cleft, monoamines can also be catabolized to inactive compounds by the action of several enzymes. The monoamine oxidase (MAO) and the catechol-O-methyl transferase (COMT) are responsible for the catabolism of catecholamines (DA, E, and NE). The MAO-A isoform catabolizes indolamines (5-HT and melatonin). The histamine N-methyltransferase (HNMT) metabolizes histamine.

GABA and glycine are preferentially taken up directly into presynaptic GABAergic terminals by a variety of co-transporters, which leads to a significant reduction in the energy cost, and in the speed for the recycling of these inhibitory neurotransmitters in presynaptic terminals. In particular, GABA recycling into the neuronal terminals has been well-documented in the past (Hertz and Schousboe, 1987; Schousboe, Sarup, Bak et al., 2004; Schousboe, Sarup et al., 2004; Schousboe, Larsson et al., 2004). In cortical GABAergic interneurons, three types of GABA transporters (GAT-1&3) and a betaine-GABA transporter (BGT-1) are the most reported co-transporters for synaptic GABA clearance. BGT-1 has a lower affinity for GABA. The aminoacid glycine is transported back into the presynaptic terminals by the glycine transporters

($GlyT_1$ and $GlyT_2$), but it could be also synthesized in the cytosol from serine by the phosphoserine. Different types of vesicular inhibitory amino acid transporters (VIAAT) have been identified in GABAergic (e.g., VGAT) and glycinergic neuronal terminals.

However, the existence of high-affinity glutamate transporter in presynaptic terminals has been questioned, although if present, they might help maintain glutamatergic homeostasis (Waagepetersen et al., 2005). For example, a second isoform of EAAT2 (GLT-1b) has been only identified in cultured neurons (Chen et al., 2002), and glutamate transporters EAAT3 (or EAAC-1) have been primarily found in the postsynaptic contacts of the neuronal synapses (Danbolt, 2001).

Due to the lack of the enzyme pyruvate carboxylase, neurons cannot perform de-novo synthesis of glutamate and GABA from Glc. The pyruvate carboxylase and GS are exclusively expressed in astrocytes (Norenberg and Martinez-Hernandez, 1979; Yu et al., 1983; Shank et al., 1985). Protuberances in astrocytic processes, in the form of end-feed, surround synapses (Chapter 8). These perisynaptic end-feet together with presynaptic terminals and postsynaptic sites form a tripartite unit.

Bergles and Jahr (1997) provided initial evidence for the existence of specific carriers in the membrane of hippocampal astrocytes that use the Na^+ electrochemical potential as a driving force to cotransport glutamate. Nowadays we know that high-affinity glutamate transporters EAAT1 (or GLAST) and EAAT2 (or GLT-1) are abundantly expressed in these perisynaptic astrocytic end-feet (Gegelashvili and Schousboe, 1997; Danbolt, 2001), suggesting that astrocytes are the most important cells in the clearance of glutamate from excitatory synapses. A small portion of the released glutamate into the synaptic cleft could also be moved back into the terminal through the transporter GLT-1b (Chen et al., 2002). EAAT1&2 are symporters that move glutamate out from the synaptic cleft into astrocytes using Na^+ and K^+ gradients. Moving in each glutamate consumes indirectly an ATP molecule by the action of the Na-K ATPase pump to replace the equilibrium gradients for these two cations. Notice that the entire process will help buffer any excess of K^+ in the ECS due to neuronal activation and will also increases cytosolic *Pi* and H^+ concentrations. Interestingly, a very small portion of released GABA at inhibitory synapses can also be transported into perisynaptic astrocytic end-feet by means of BGT-1 and GAT-3 (Schousboe and Kanner, 2002), which might indirectly affect the observed rate of glutamate-glutamine cycling (Schousboe, Sarup, Bak et al., 2004). Inside astrocytes, GABA is catabolized to succinate to enter the TCA-cycle. Therefore, exogenous GABA could be used as energy source for astrocytes (McKenna and Sonnewald, 2005). In cultured astrocytes, GABA supports oxidative metabolism (Yu and Hertz, 1983). In fact, metabolism in the cerebral cortex could be changed (i.e., stimulated, dampened) by local inhibitions (Nasrallah et al., 2007). The transferred carbon from GABA could flow out of the TCA-cycle as α-KG. On the other hand, glutamate can enter the TCA-cycle as α-KG by either transamination via the aspartate aminotransferase (AAT, or transaminase), an enzymatic reaction producing aspartate from OAA, or by oxidative deamination catalyzed by the glutamate dehydrogenase (GDH, reductive amination), an enzymatic reaction producing ammonia while reducing NAD^+ to NADH. These two enzymes can also reversibly convert α-KG into glutamate. Astrocytic glutamate could be converted to glutamine (Gln) by GS, a chemical reaction consuming another ATP.

System A transporter family mediates the uptake of glutamine in presynaptic neurons, while release from astrocytes is mediated by the system N and L transporter families (Bak, Schousboe, and Waagepetersen, 2006). The system N can mediate glutamine transport in both the outward and the inward directions (Bröer et al., 2002). The efflux of glutamine from the astrocytic cytosol (N family) takes place through SNAT3 and SNAT5 (Bröer and Brookes, 2001), which act as 1 Na^+ symporters and 1 H^+ antiporters for glutamine export. SNAT3 and EAAT1 may form a complex (Martínez-Lozada et al., 2013), suggesting that one of the three Na^+ ions imported to the cytosol through EAATs are exported via SNAT3 (Todd et al., 2017). In this case, the final energy cost for the astrocytic Na-K ATPase involved in the glutamate-glutamine recycling might be reduced. As discussed later, energy cost associated with the GABA-glutamine recycling is negligible. Extracellular glutamine is then imported into surrounding neurons through SNAT1, SNAT2 and SNAT7, which uses Na^+ natural gradient (A family). Back in the neurons, glutamate and ammonia are regenerated from glutamine via phosphate-activated glutaminase (PAG), an amidohydrolase enzyme. PAG is ubiquitously expressed (Waagepetersen et al., 2007), but with more prominence in glutamatergic structures (Laake et al., 1999). A small portion of GABA being recycled via the astrocytic network might be exported to inhibitory neurons in the form of glutamine. In the GABAergic terminals, glutamate and ammonia are regenerated from glutamine via PAG. This glutamate is finally used to synthesize GABA in presynaptic terminals by the glutamic acid decarboxylases (GAD), a well-known biomarker for GABAergic neurons (Saito et al., 1974).

We will refer to the entire process as the glutamate/GABA-glutamine recycling mechanism (Figure 9.1). The cycle can be divided into two parts: **1)** the transfer of the carbon skeletons of these amino acids and **2)** the concomitant transfer of the ammonia nitrogen. These two parts will be considered separately henceforth. For this cycle to work properly, specific transporters for glutamate, GABA and glutamine must be differentially expressed in neuronal and astrocytic plasma membranes (Schousboe, Sarup, Bak et al., 2004; Bak, Schousboe, and Waagepetersen, 2006). Note that several ions (e.g. Na^+, K^+, H^+, Cl^-) are co/anti transported with glutamate, glutamine, and GABA during the neurotransmitter recycling process. In order to reestablish the ionic concentrations to their equilibrium, ATP is required by different ATPases.

The version of the glutamate-glutamine cycle formulated by Van den Berg and Garfinkel (1971) was initially concerned merely on the fate of the carbon skeletons of the amino acids, operating in a stoichiometrical manner (Cotman et al., 1981). The cycle was suggested based on the existence of at least two distinct metabolic pools for glutamate and glutamine representing primarily neurons and astrocytes, respectively (Van den Berg and Garfinkel, 1971; Benjamin and Quastel, 1972; Berl and Clarke, 1983; Ottersen et al., 1992). Demonstrations of considerable astrocytic oxidative metabolism of glutamate through the TCA-cycle challenge the initial proposal (Yu et al., 1982; McKenna et al., 1996). Actually, glutamate could in principle contribute to the oxidative energy metabolism (Hertz et al., 2007). Nowadays, the cycle has been extended to include GABA and a TCA-principle for carbon interplay between the two main aminoacid neurotransmitters. The latter is seen as crucial for establishing an

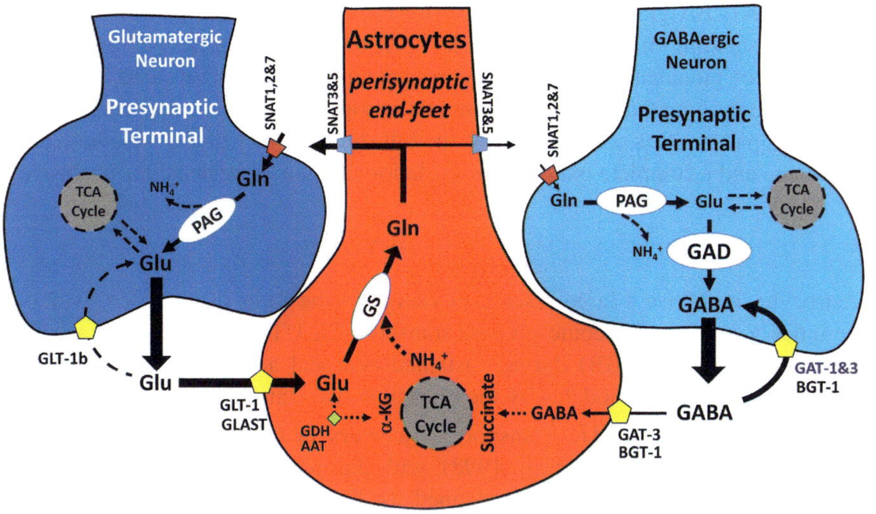

FIGURE 9.1 Illustration of the glutamate/GABA-glutamine cycle in the tripartite synaptic units.

Note: The more relevant a pathway is, the thicker its arrow.

optimal balance between excitation and inhibition that can adjust dynamically during brain functions.

Production of glutamate from glutamine in neuronal synaptic terminals by PAG produce ammonia (NH_4^+). Conversion of glutamate into glutamine in astrocytes by GS requires ammonia. Ammonia transfer seems to happen mainly by facilitated diffusion between these two cellular substrates. However, nitrogen could also be transferred as the amino group in an amino acid, and to that end, two intercellular compartmental shuttles have been proposed, i.e., the branched chain amino acid (BCAA) (Yudkoff, 1997; Lieth et al., 2001) and the alanine-lactate (Waagepetersen et al., 2000; Zwingmann et al., 2000). The alanine-lactate shuttle is stimulated by an increase in neuronal activity requiring glutamate-glutamine recycling (Bak et al., 2005). With regard to the BCAA shuttle, it has been demonstrated that [^{15}N] valine metabolism in cultured astrocytes is activity-dependent (Bak et al., 2007). For nitrogen transfer to occur properly via these shuttles, GDH-catalyzed reaction should occur as a reductive amination. GDH, which is located in the mitochondria, is an important branch-point enzyme for carbon and nitrogen metabolism. As a result of the high K_m value of GDH for ammonia (Zaganas et al., 2001), an oxidative deamination (opposite reaction) is the most favorable reaction in the brain. However, GDH has a higher activity and a lower K_m in glutamatergic neurons (Zaganas et al., 2001), which together with the high ammonia concentration in the mitochondrion created by the high PAG activity in these neurons facilitate reductive amination (Bak, Schousboe, and Waagepetersen, 2006). In summary, the overall carbon and ammonia homeostasis in the nerve ending, which is crucial for its well-functioning, is controlled during the glutamate/GABA-glutamine recycling. However, there are still unraveled issues on our understanding of this cycle. For example, the extent of astrocytic and neuronal clearance of glutamate and GABA neurotransmitter is yet to be quantified. Further clarifications about the transfer of ammonia nitrogen between the neuronal and astrocytic compartments is needed. Finally, the full stoichiometry of the cycle with carbon exchange and a complete oxidative degradation of the neurotransmitter is elusive.

9.3 THE GLYCOGEN-SHUNT PROPOSITION

Astrocytes comprise a large reserve of energy in the form of glycogen, which can be quickly used to overcome glucose transport (GLUT-1, 55kDa) limit across the blood-brain barrier (BBB) during sustained neuronal activity. The maximum rate for GLUT-1 has been modeled by the reversible two-state Michaelis-Menten model, $V_{GT}^{max} = T_{GT}^{max} / \left(K_{GT} / G_0 + 1\right)$. Transport parameters in this model have been fitted using MRS data (Gruetter et al., 1998), with the maximum glucose transport capacity and the transporter affinity defined as $T_{GT}^{max} = 0.69 \pm 0.09\,\mu mol\,g^{-1} min^{-1}$ and $K_{GT} = 0.6 \pm 2.0\,mM$, respectively. Note that the steady-state plasma glucose content can vary substantially, $4.6\,mM \leq G_0 \leq 29\,mM$. Hence, the maximum sustained uptake of glucose is defined $T_{GT}^{max} G_0 / (K_{GT} + G_0) \cong T_{GT}^{max}$. As mentioned in Chapter 8, cytosolic G6P can be either converted into F6P by PGI or stored as glycogen polymers. Although the pentose phosphate pathway is known to be irrelevant for the glucose metabolism in the brain, G6P could be also oxidatively metabolized in the brain by the G6P dehydrogenase. Glycogen polymers in the brain (Figure 9.2), created progressively from excessive GP6, are used as an emergency fuel during intense energy demand associated with high neuronal activity in a healthy brain (Gibbs et al., 2006). Glycogen can also be used during energetic stress in pathological conditions (e.g., cerebral ischemia and hypoglycemia). Therefore, glycogen polymers are considered as the largest energy reservoirs in the brain. These polymers have been observed *in vivo* in the brain (Öz et al., 2003; Gruetter et al., 2003). The glycogenolysis consists of two parts: **1)** the phosphorolytic cleavage (catalysis) of a bond in orthophosphate

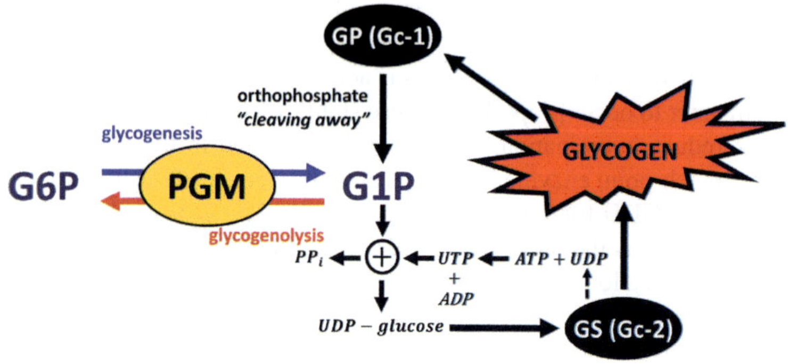

FIGURE 9.2 The glycogen-shunt hypothesis.

by the glycogen phosphorylase (GP), degrading glycogen polymers to glucose-1-phosphate (G1P), and **2)** the reversibly isomerization of G1P to G6P by the phosphoglucose mutase (PGM). GP is regulated by phosphorylation and allosteric factors. There are three known GP isoforms. The brain contains two of them, the brain, and the muscle types. GP is reversibly phosphorylated by phophorylase kinase, converting the enzyme from a less activated (*allosterically regulated*) **form b** to a more activated (*allosterically unregulated*) **form a**. On the other hand, phosphoprotein phosphatase 1 dephosphorylates GP. Intense astrocytic Ca^{2+} signaling causes the activation of the phosphorylase kinase, converting GP into its activated **form a**. Any cyclic AMP (cAMP) cascade of hormonal origin could also lead to the conversion of GP into its **form a**. G6P overloading and ATP produces an inhibitory effect on GP.

However, the phosphorylated GP is less sensitive to allosteric inhibitors. Note that cAMP (negative) and ATP (positive) are two allosteric effectors reflecting cellular energy charge. For all these reasons, GP is considered a checkpoint (Gc-1) in the glycogen-shunt pathway. The glycogenesis comprises three parts: **1)** reversible isomerization (PGM) of G6P for the G1P synthesis, **2)** the conversion of G1P into UDP-glucose, the activated Glc donor, and **3)** the formation of glycogen polymers from G1P by the allosteric enzyme glycogen synthase (GS). Note that to synthesize a glycogen polymer from G1P, 1 ATP equivalent is required, making this pathway energetically expensive. GS is allosterically inhibited by ATP, ADP, and inorganic phosphate (Pi). GS is activated by G6P overloading. GS is phosphorylated by phosphorylase kinase and dephosphorylated by phosphoprotein phosphatase 1. Under low glucose levels or high energetic stress, GS is also inhibited by protein kinase A (PKA) phosphorylation. As for GP, allosteric effects on GS are also overridden by reversible phosphorylation. For all these reasons, GS is also considered a key checkpoint in the glycogen-shunt (Gc-2).

Experimental data in conditions with different energetic stress (i.e., sensory stimulation, seizures, and ischemia) indicated that changes in neocortical glycogen production V_G (*glycogenolysis*) with respect to that for resting neuronal activity (RA) ($V_{G(RA)}$) are linearly proportional to the associated increase in glucose utilization:$\Delta V_G = V_G - V_{G(RA)} = 0.962\,\Delta CMR_{Glc}$, with $\Delta CMR_{Glc} = CMR_{Glc} - CMR_{Glc(RA)}$. The Glc consumption at resting neuronal activity is defined as $CMR_{Glc(RA)}$. In principle, the glycogenolysis rate includes two components $V_G = V_G^B - V_G^S$, with V_G^B and V_G^S representing the rate of breakdown and synthesis of glycogen polymers. Storages of glycogen in the brain are found predominantly in astrocytes. Astrocytes express both GP and GS (Cataldo and Broadwell, 1986; Ignacio et al., 1990). It has been demonstrated that expression of GP in the brain is high in astrocytes and very low in neurons (Cataldo and Broadwell, 1986), although GS expression in neurons is large (Vilchez et al., 2007). Interestingly, phosphorylase kinase subunits (Phka1, Phkg2) are found only in astrocytes (Pfeiffer et al., 1992). As OGI is smaller than the theoretical stoichiometric value of 6, it has been assumed that a fraction of Glc used by the brain might be originated from glycogenolysis in astrocytes, which could drastically increase with an enhancement of energetic demand during neuronal activity (Shulman et al., 2001). If the astrocytes are energetically charged and the levels of mitochondrial Pyr are high, Pyr carboxylation into OAA will take place. OAA is then transported back into the cytosol by the malate-aspartate shuttle (MAS) to initiate the first step in gluconeogenesis. In this step, cytosolic OAA is decarboxylated and simultaneously phosphorylated to produce PEP, a reaction catalyzed by the PEP carboxykinase. As the brain lacks glucose-6-phosphatase, the final product of this upstream pathway is G6P, which can be converted into glycogen polymers. The interplay between the glycolytic pathway and glycogen-shunt can be assessed by means of OGI measurements (Shulman et al., 2001).

9.4 ENERGETICS FOR EXCITATION AND INHIBITION

The mammalian brain has phylogenetically developed a mechanism for glutamate neurotransmitter recycling that is energetically expensive, involving a non-neuronal intermediator, i.e., the astrocytes. Apparently, this is not an evolution error, but an efficient way to optimize energy usage with minimal metabolite residuals and toxicity. Let us first discuss mechanisms used by astrocytes to estimate the energetic cost caused by increases in neuronal activity. According to measurements by PET and fMRI, Glc usage in the cerebral cortex is mostly related to glutamatergic synaptic activity in neurons (Magistretti et al., 1999). Contributions from other types of neurotransmissions during neuronal activity (e.g., GABA) are associated with a less metabolic cost. Although our understanding improved in the early 2000s through the introduction of two alternative techniques: **a)** fMRI and transcranial magnetic stimulation (Waldvogel et al., 2000) and **b)** microspectrofluorimetry (Chatton et al., 2003), exact quantifications of the energetic costs associated with different types of synaptic activity remained controversial for several years. Many in-vivo observations were obtained under anesthesia protocols that enhanced GABAergic activity (e.g., benzodiazepines and barbiturates). Energetic rates/fluxes associated with anesthesia were expected to be different, complicating data interpretation. Challenges have been also associated with the use of a variety of species under different energetic demands above isoelectricity. Henceforth, we present in this book results from a meta-analysis (Rothman et al., 2022) of MRS measurements using ^{13}C and ^{14}C tracers in the brains of different species (e.g., rats, mice, tree-shrews, and humans) under three energetic conditions (e.g., physiological activation, sustained seizure, and ischemia). Data revealed a linear relationship (Eq. 9.2) between the rates for glutamate/GABA – glutamine recycling V_{cycle} and neuronal glucose oxidation $CMR_{Glc-N}(ox)$. Please note that $CMR_{Glc-N}(ox)$ and V_{cycle} are given in $\mu mol\, g^{-1}\, min^{-1}$.

$$CMR_{Glc-N}(ox) = 0.784\, V_{cycle} + 0.110 \tag{9.2}$$

The intercept of $0.110\,\mu mol\, g^{-1}\, min^{-1}$ is related to a necessary oxidative neuronal energy for non-recycling activities at isoelectricity. The total neurotransmitter recycling rate comprises both glutamatergic and GABAergic components: $V_{cycle} = V_{cycle}^{Glu} + V_{cycle}^{GABA}$. Data from rats have been used to separate contributions to neuronal Glc oxidation from the recycling of these two amino acid neurotransmitters.

$$CMR_{Glc-N}^{Glu}(ox) = 0.812\, V_{cycle}^{Glu} + 0.077 \tag{9.3}$$

$$CMR^{GABA}_{Glc-N}(ox) = 0.763V^{GABA}_{cycle} + 0.011 \quad (9.4)$$

Direct measurements of ^{13}C turnover from Glc and acetate into glutamate, glutamine, and GABA can be used to make distinctions between glutamatergic and GABAergic fluxes (Pfeuffer et al., 1999). Henry et al. (2006) reviewed the ingredients necessary for a dynamic metabolic modeling based on such type of NMR data. By using ^{13}C NMR spectroscopy in anesthetized (halothane/N_20) rats combined with a three-compartment metabolic modeling, it was shown that above isoelectricity V_{cycle} can vary between $0-0.58\,\mu mol\,g^{-1}\,min^{-1}$, from which V^{Glu}_{cycle} and V^{GABA}_{cycle} account for the 77% and 23%, respectively (Patel et al., 2005). The authors used a combination of [1-^{13}C] glucose and [2-^{13}C] acetate infusion. A recent meta-analysis shows that the ranges of variation for V^{Glu}_{cycle} and V^{GABA}_{cycle} could be anywhere between $0-0.45\,\mu mol\,g^{-1}\,min^{-1}$ and $0-0.14\,\mu mol\,g^{-1}\,min^{-1}$, respectively (Rothman et al., 2022).

On the other hand, Glc breakdown depends on its uptake into cells by different types of Glc carrier. Glc continuously delivered to the extracellular milieu from capillaries through GLUT-1 is taken up into nearby cells by different Glc transmembrane carrier subtypes: neurons (GLUT-3) and astrocytes (GLUT-1, 45 kDa, less glycosylated). This Glc uptake occurs in a concentration-dependent manner. Simpson et al. (2007) provided transport properties (e.g., capacity and affinity for substrate) for these facilitative and bidirectional transmembrane carriers. Compared to GLUT-1, GLUT-3-mediated Glc transport exhibits high-affinity and high-capacity. Therefore, there is a heterogeneous degree of cell specific Glc uptake, with neurons having the capacity to take up more Glc than astrocytes during periods of intense neuronal activity. All these findings suggest that Glc usage is highly compartmented in the brain with cell-dependent distributions.

$$CMR_{Glc}(ox) = CMR_{Glc-N}(ox) + CMR_{Glc-A}(ox) \quad (9.5)$$

$$CMR_{Glc-N}(ox) = CMR^{Glu}_{Glc-N}(ox) + CMR^{GABA}_{Glc-N}(ox) \quad (9.6)$$

Glc oxidation in glutamatergic and GABAergic neurons is 82% ($0.495 \pm 0.015\,\mu mol\,g^{-1}\,min^{-1}$) and 18 % ($0.107 \pm 0.012\,\mu mol\,g^{-1}\,min^{-1}$), respectively (Hyder et al., 2006; Patel et al., 2005). As pyruvate carboxylase is only expressed in astrocytes, the following flux relationships must be hold: $V_{PDH-N} = V_{TCA-N}$ and $V_{PDH-A} = V_{TCA-A} + V_{PC}$, with V_{PDH-N} and V_{PDH-A} defining the pyruvate flux into the PHD-complex in neurons and astrocytes, respectively. The pyruvate fluxes into the TCA-cycle in neurons and astrocytes are respectively V_{TCA-N} and V_{TCA-A}. Finally, V_{PC} corresponds to the flux of pyruvate from the astrocytic pyruvate carboxylase. Using ^{13}C MRS data from multiple studies, Hyder et al. (2006) provided estimates for these fluxes ($\mu mol\ g^{-1}\,min^{-1}$) above isoelectricity: $V_{PDH-N} = 0.71 \pm 0.19$, $V_{PDH-A} = 0.13 \pm 0.04$, and $V_{PC} = 0.06 \pm 0.03$. These authors proposed the following theoretical model for Glc oxidation in neurons $CMR_{Glc-N}(ox) = V_{TCA-N}/2$ and astrocytes $CMR_{Glc-A}(ox) = V_{TCA-A}/2 + \delta V_{PC}$. Parameter δ reflects the percentage of pyruvate from the V_{PC} flux used to create ATP. δ can take values from 0 to 1, with 1 indicating that 100 % of pyruvate from V_{PC} is oxidated in the TCA-cycle. Because of the incompleteness of the TCA-cycle, $\delta = 0.4$ when all pyruvate from V_{PC} is used to produce glutamine, and indirectly ammonia detoxification. By using this model with $\delta = 0.4$, and ^{13}C NMR data from multiple studies, they estimated that neurons and astrocytes are responsible for at least 92 % and 8 % of the total Glc oxidation, respectively.

An initial study reported that $100g$ of human brain consumes $5.6mg$ of Glc per minute (Erbsloh et al., 1958). In a more recent work using ^{18}F-fluoro-2-deoxy-d-glucose (PET), MRI for partial volume correction (i.e., removing CSF contribution) and blood samples from the radial artery (Ibáñez et al., 2004), it has been demonstrated that brain Glc consumption at rest ($CMR_{Glc(RA)}$) it higher than initially predicted, with not significant differences between young ($0.534 \pm 0.084\,\mu mol\, g^{-1} min^{-1}$) and old ($0.516 \pm 0.073\,\mu mol\, g^{-1} min^{-1}$) human subjects. In this late study, the brain resting state was ensured by keeping subject's eyes covered and ears plugged with cotton while their head rested in a thermoplastic mask. During sensory stimulation, changes in total Glc consumption ΔCMR_{Glc} with respect to resting neuronal activity can be as large as $1.7\,\mu mol\, g^{-1} min^{-1}$, a result obtained by blocking glycogenolysis (i.e., $\Delta V_G = 0$, Figure 3, Rothman et al., 2022). Therefore, the brain needs to use glycogen energy reservoirs to compensate for changes in Glc consumption during neuronal activation as the maximum rate for GLUT-1 is limited to $T_{GT}^{max} = 0.69 \pm 0.09\,\mu mol\, g^{-1} min^{-1}$. At that maximum Glc consumption during sensory stimulation, glycogen will contribute with the needed Glc to meet brain energy demand $\Delta V_G = 0.962(1.7\,\mu mol\, g^{-1} min^{-1}) = 1.63\,\mu mol\, g^{-1} min^{-1}$. Therefore, the glycogen-shunt constitutes a fight-or-flight kind of response. In fact, a variety of peptide-hormones (e.g., glucagon, insulin, and vasoactive intestinal peptide) and neurotransmitters (e.g., adrenaline, noradrenaline, serotonin, and histamine) regulates storage/mobilization of glycogen polymers. Evidence suggests that glycogen turnover is essential to maintain healthy brain function, and dysregulation in the mechanisms underlying glycogenesis/glycogenolysis might results in dysfunctions of the cerebral cortex. For example, it has been demonstrated that glycogen turnover is necessary for normal glutamate uptake in cultured astrocytes (Sickmann et al., 2007). Similarly, Gibbs et al. (2006) showed a crucial role of glycogen turnover for memory consolidation in young chickens *in vivo*. These studies used 1,4-dideoxy-1,4-imino-D-arabinitol (DAB), which is a specific GP inhibitor. Other iminosugars, such as isofagomine, can also prevent the catalysis of glycogen polymer. The breakdown of glycogen polymer into G6P occurs preferentially in astrocytes, which contains the largest glycogen polymer storage.

9.5 INCREASES IN ENERGY DEMAND: THE ASTROCYTE-NEURON INTERPLAY

During increases in energy demand, glycogenolysis is activated, resulting in the overproduction of pyruvate. The production of pyruvate in astrocytes can result in increases in the synthesis of three subproducts: **a)** lactate by LDH to be exported to the extracellular space, **b)** glutamine to be used in the neurotransmitter recycling process, and **c)** ATP due to its oxidation in the TCA-cycle to satisfy energetic deficits.

Therefore, the fluxes of these metabolites (e.g., Glc, lactate, and pyruvate) across different cellular compartments is driven by glycogen polymer turnover in astrocytes, neurotransmitter recycling demands in neurons, and energy charge in both brain cells (Sickmann et al., 2005). Henceforth, we will provide details of two theoretical frameworks proposed to explain inter-compartmental fluxes of these metabolites and the action of shuttles to maintain neurotransmitter, ammonia, and carbon homeostasis among domains in the cell. For both frameworks, Glc is continuously delivered from capillaries (red) to the extracellular milieu through GLUT-1 at a limited rate.

9.5.1 The Astrocyte-Neuron Lactate Shuttle (ANLS)

Cytosolic Pyr could in principle be reduced to lactate by the enzyme LDH. This phenomenon occurs mainly whenever oxygen is depleted during intense cellular activity; hence, it is referred sometimes as to anaerobic glycolysis. In mammals, multiple LDH isozymes have been characterized in the past. Interestingly, most of these isoenzymes result from combining LDH-1 (heart type) and LDH-5 (muscle type) subunits. The reversible REDOX reaction by the LDH enzyme is: Lactate + $NAD^+ \rightleftharpoons$(Oxidation/Reduction) Pyr + NADH. Using polyclonal antibodies against these two subunits, Bittar et al. (1996) showed that LDH-1 was present in both neurons and astrocytes. However, immunoreactivity against LDH-5 was restricted to astrocytes. The kinetic characteristics of the LDH-1 and LDH-5 subunits are very distinctive. The inhibition by lactate of Pyr reduction in terms of the dissociation constant is stronger in the LDH-1 subtypes. The inhibition by Pyr of lactate oxidation was very similar in both subunits. However, pyruvate binding affinity for LDH-5 is greater than for LDH-1. These LDH cellular distributions and kinetics suggested a favorable flow of lactate. At physiological pH, lactate is negatively charged. Therefore, they flow across cellular membranes via monocarboxylate transporters (MCT). MCT expression is also different in brain cells, with MCT-2 and MCT-1&4 subtypes found mostly in neurons and astrocytes, respectively. The MCT-2 subtype has the highest affinity for extracellular lactate, which might help the flow of this metabolite into the neurons rather than into astrocytes. The glutamate-glutamine cycle requires the consumption of two ATP molecules: one ATP is used by the GS enzyme to convert glutamate into glutamine and another ATP is used by the Na-K ATPase pump to reestablish the Na^+ gradient. Based on all these observations, Magistretti et al. (1999) suggested that the glutamate-glutamine cycle might trigger glycolysis in astrocytes to meet ATP deficit during neuronal activity, resulting in the overproduction of lactate. Lactate produced by astrocytes is then released for oxidation in adjacent neurons, a mechanism known as the ANLS (Figure 9.3). An excessive production of lactate not used by neurons during activation could efflux into the circulating blood. Activation of astrocytic glycolysis might also stimulate Glc uptake by astrocytes from the extracellular milieu (Pellerin and Magistretti, 2004), which was consistent with a partitioned Glc uptake of ~26 % by neurons and of ~74 % by glia reported later by Hyder et al. (2006). However, it was not clear whether the oxidation of lactate during neuronal activity was derived directly from Glc uptaken by astrocytes and/or from increases in glycogenolysis (Hertz et al., 2007). Also, glutamatergic activity during activation of neurons was better preserved

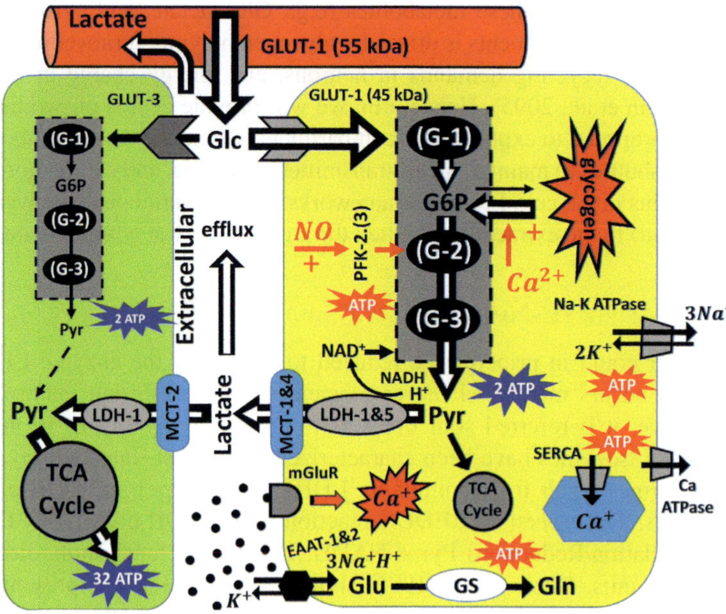

FIGURE 9.3 The schematics of the ANLS model.

by Glc metabolism than by lactate metabolism (Bak, Schousboe et al., 2006). The ANLS is theoretically defined by the following equations:

$$\Delta CMR_{Glc-A} = \Delta V_{cycle} \tag{9.7}$$

With $\Delta CMR_{Glc-A} = CMR_{Glc-A} - CMR_{Glc-A(RA)}$ and $\Delta V_{cycle} = V_{cycle} - V_{cycle(RA)}$. Note that $CMR_{Glc-A(RA)}$ and $V_{cycle(RA)}$ are Glc consumption in astrocytes and neurotransmitter recycling rate at resting neuronal activity. The model proposes a 1:1 stoichiometry for Glc and Glu. Above isoelectricity during neuronal activation, there is no Glc uptake in neurons coupled to ΔV_{cycle}.

$$\Delta CMR_{Glc-N} = 0 \tag{9.8}$$

As mentioned before, the lactate generated by glycolysis in astrocytes is shuttled to nearby neurons. There is a 3/2 greater yield of ATP production using glycogen; hence, it can be assumed that:

$$\Delta V_G = {}^{2}\!/_{3}\, \Delta V_{cycle} \tag{9.9}$$

The 2/3 factor was estimated by inhibiting glycogenolysis (Dienel and Cruz, 2015). ATP consumption by astrocytes per variation in the recycling is $V_{ATP-A} = 2\,\Delta V_{cycle}$ and a lactate is shuttled to neurons for a final production of ATP in the TCA-cycle at a rate $V_{ATP-N} = 30\,\Delta V_{cycle}$.

In Figure 9.3, the more relevant a pathway is, the thicker its arrow. White/black arrows indicate metabolic pathways increases during neuronal activation (black) with respect to isoelectricity (white).

The impact of each metabolic pathway is represented by the size of the rectangles/circles.

Experimental studies provide evidence for and against the ANLS. The model is strongly supported by special astrocytic features: **a)** they ideally positioned between neurons and vessels with processes on intraparenchymal capillaries for efficient Glc uptake via GLUT-1 and perisynaptic end-feet endowed with high-affinity transporters EAAT1&2 for a quick glutamate clearance, **b)** their perisynaptic end-feet are enriched with mGluR, crucial for the initiation of Ca^{2+} signaling, thus converting GP into its activated **form a** to stimulate glycogenolysis, **c)** astrocytic end-feet on inhibitory synapses possess GABA transporters BGT-1 and GAT-3 (Schousboe and Kanner, 2002), and **d)** they are rich in lactate MCT-1&4 monocarboxylate transporters. Waagepetersen et al. (2000) extended the astrocyte-neuron lactate shuttle to include a mechanism for ammonia homeostasis in glutamatergic neurons and astrocytes, i.e., the lactate-alanine shuttle. In cerebellar cocultures, it has been shown that the lactate-alanine shuttle and the associated glutamate-glutamine cycle can coexist. However, these two shuttles were uncoupled and only the latter was activity-dependent (Bak et al., 2005).

9.5.2 The Glucose Sparing by Glycogenolysis (GSG)

The ANLS model ignores the energetics in astrocytes of: **1)** glutamate/GABA oxidation, **2)** completed stoichiometric for changes in transmembrane ionic concentrations for $Na^+/K^+/H^+/Cl^-$, and **3)** Ca^{2+} ion pumping across plasma/ER membranes. Therefore, the ANLS model has been debated in the past by multiple authors (Dienel and Hertz, 2001). For example, using a theoretical framework for Glc and lactate transport, Simpson et al. (2007) suggested that Glc diffuses directly to neurons through the basal lamina and interstitium. These authors have also claimed that neurons are responsible for the generation of the lactate transients observed during neuronal activity. Rotman et al. (2022) proposed an alternative model to account for some of the missing energetic components (Figure 9.4). The model is based on five main assumptions:

i) Glc transport from capillaries to tissue is limited to V_{GT}^{max} (GLUT-1).
ii) There is no glycogenolysis at rest ($V_{G(RA)} = 0$), i.e., $\Delta V_G = V_G$.
iii) Astrocytes fully support energy deficits above isoelectricity by using glycogenolysis, i.e., $\Delta CMR_{Glc-A} = 0$ whenever $\Delta V_{cycle} > 0$.
iv) In the absence of glycogen polymers (e.g., depletion or inhibition of V_G^B), $\Delta CMR_{Glc-A} = (3/2)\overline{\Delta V_G}$, with $\overline{\Delta V_G}$ indicating the expected supply from glycogen polymers.
v) A linear relation between increase in glucose consumption and Glu/GABA recycling rate above isoelectricity; i.e., $\Delta CMR_{Glc-N}(ox) = C_{cycle}\,\Delta V_{cycle}$.

Neurons have minimal to no glycogen reserves. Hence, the maximum rate for neuronal signaling must be restricted by their speed of metabolizing glucose; and the latter is limited by the maximum Glc transport rate from capillaries V_{GT}^{max}. The GSG model operates under two modes: **a)** complete Glc sparing mode, and **b)** astrocytic Glc stealing mode.

Under complete Glc sparing, $\Delta CMR_{Glc} = 2C_{cycle}\,\Delta V_{cycle}$, as $\Delta CMR_{Glc-A} = 0$. The model implicitly assumes that $CMR_{Glc-N}(ox) = CMR_{Glc-N}(non-ox)$. The maximum Glc consumption in neurons is defined by the equation: $\Delta CMR_{Glc-N-max} = V_{GT}^{max} - CMR_{Glc(RA)}$. Based on experimental data from rat cerebral cortex at normoglycemia and theoretical models, $V_{GT}^{max} / CMR_{Glc-A(RA)} = 1.57$; hence,

$$\Delta CMR_{Glc-N-max} = 0.57\,CMR_{Glc(RA)} \tag{9.10}$$

The GSG model also predicts the maximum rate for neuronal signaling as $\Delta V_{cycle-max} = (0.29 / C_{cycle})CMR_{Glc(RA)}$. A ^{13}C MRS meta-analysis was performed to determine $C_{cycle} = 0.784$ in the rat cerebral cortex (Rotman et al., 2022), leading to:

$$\Delta V_{cycle-max} = 0.57\,V_{cycle(RA)} \tag{9.11}$$

The sparing of Glc for neuron takes place until the limit V_{GT}^{max} is reached.

Under the astrocytic Glc stealing mode, further increases in the rate of glucose consumption has to be balanced by a reduction of Glc uptake by astrocytes (i.e., $\Delta CMR_{Glc-A} < 0$). The stolen Glc from astrocytes has to be replaced by additional glycogen breakdown ($\Delta V_G \uparrow$). The GSG model predicts $\Delta CMR_{Glc} = \Delta V_G$ and $\Delta V_G = 2C_{cycle}\,\Delta V_{cycle}$, which must be held above isoelectricity to maintain energy cost

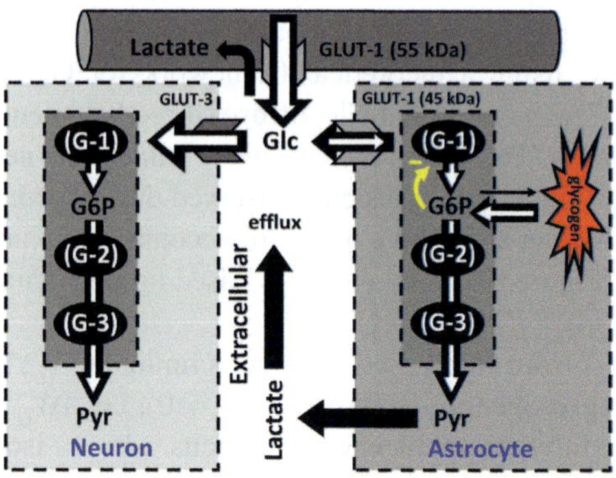

FIGURE 9.4 The schematics of the GSG model.

Note: Similar to Figure 9.3, white/black arrows are used to indicate metabolic pathways increasing with neuronal activation (black) with respect to isoelectricity (white). Whether the GSG model is working under complete sparing (inflow) or stealing (outflow) mode is determined by the direction of the Glc flow through the astrocytic GLUT-1 (45 KDa).

due to neuronal signaling and astrocytic functioning. The total Glc usage can be estimated by the following relationship:

$$\Delta CMR_{Glc-N-max} = 0.94\, CMR_{Glc(RA)} \tag{9.12}$$

$$\Delta V_{cycle-max} = 0.92\, V_{cycle(RA)} \tag{9.13}$$

Equations (9.12)–(9.13) were based on values of $CMR_{Glc-N(RA)} = 0.56\,\mu mol\, g^{-1} min^{-1}$ and $CMR_{Glc-A(RA)} = 0.32\,\mu mol\, g^{-1} min^{-1}$, which were calculated using a meta-analysis of ^{13}C MRS data for an average $V_{cycle(RA)} = 0.57\,\mu mol\, g^{-1} min^{-1}$ (Rotman et al., 2022). If glycogenolysis is depleted/inhibited, the GSG model predicts $\Delta CMR_{Glc-N-max} = 0.24\, CMR_{Glc(RA)}$ and $\Delta V_{cycle-max} = 0.23\, V_{cycle(RA)}$. Note that overproduction of G6P from glycogenolysis inhibit Glc degradation by hexokinase (HK), part of the first checkpoint in the glycolytic pathway (G-1) (Chapter 8). This inhibition will create a reduction of Glc update by astrocytes with respect to that for isoelectricity.

According to the GSG model, total ATP production in neurons under stimulation is represented by $\Delta V_{ATP-N} = 32\Delta CMR_{Glc-N}(ox) + 2\Delta CMR_{Glc-N}(non-ox) \approx 34\Delta CMR_{Glc-N}(ox) = 26.65\Delta V_{cycle}$.

The ATP production by oxidative Glc metabolism is $V_{ATP-N}(ox) = 30\Delta\, CMR_{Glc-N}(ox) = 23.52\Delta V_{cycle}$. For a non-oxidative glycogenolysis in astrocytes, the stoichiometric relationship is 3 ATP per glucosyl moiety. However, astrocytes produce ATP from glycogen-derived pyruvate participating in the oxidation and resynthesis of Glu/GABA. Therefore, $\Delta V_{ATP-A} = 3\,\Delta V_G + 4.8\,\Delta V_{cycle} = 9.48\,\Delta V_{cycle}$.

9.6 THE NEUROVASCULAR/METABOLIC COUPLING

Residuals from metabolisms can create significant levels of toxicity, from acids (e.g., lactate) to reactive oxygen species (ROS) after the reduction of unused molecular oxygen to superoxide (O_2^-). Therefore, increases in Glc metabolism due to cellular activation must be accompanied by a precise control of regional cerebral blood flow (rCBF) to supply the necessary oxygen/Glc and to drain out residual metabolites (e.g., lactate, ROS), which is called the functional hyperemic response. To that end, brain cells are endowed with multiple phasic and tonic mechanisms (Figure 9.5) that compete/cooperate to ensure optimal energy supply and minimal tissue toxicity (Riera and Sumiyoshi, 2010).

9.6.1 The Role of Neuronal NO/PGE$_2$ Signaling – Phasic Mechanisms

Nitric oxide (NO) is one of the most important elements for this accurate rCBF regulation. For example, NO reacts quickly with O_2^-, producing peroxynitrite ($ONOO^-$). The balance between NO and $ONOO^-$ in brain tissue has an important impact on regional vascular control. NO diffuses quickly and isotropically through brain tissue, but it is annihilated by the oxyhemoglobin. Although NO could react with oxygen to form stronger oxidants (e.g., nitrogen dioxide), it is a process that requires a long time (~minutes) to occur. Therefore, NO is a perfect intercellular messenger to link long-term metabolic demand/usage and short-term requirement for changes in rCBF. During neuronal activation, NO is released to the extracellular space from terminals of a small populations of GABAergic interneurons by the activation of the type I (neuronal) nitric oxide synthase (nNOS). Frequently, these nitrergic GABAergic interneurons are also positive to neuropeptide Y (NPY) and somatostatin (SOM) immunostaining. However, they are rarely positive to vasoactive intestinal peptide (VIP) and cholecystokinin (CCK). They reveal some levels of colocalization with two calcium-binding proteins (calbindin and calretinin), but none with parvalbumin. Nitrergic GABAergic interneurons are not homogeneously distributed along all cortical layers. Some of these neurons could release NO perivascularly through vessel innervations (Estrada and DeFelipe, 1998). The role of nNOS in rCBF control has been previously demonstrated with different experimental approaches (Ma et al., 1996; Stefanovic et al., 2007). As discussed in Chapter 10, NO released by these inhibitory interneurons causes vascular dilations via cyclic GMP (cGMP)-dependent mechanisms in smooth muscle cells (SMC). In addition to strong dilations of SMCs, NO released by these interneurons could also enhance cellular glycolytic respiration in astrocytes through its action on PFK-2 to fuel nearby neurons (lactate), a sort of neurometabolic coupling (Almeida et al., 2004). In fact, NO signaling can shift cortical metabolism from oxidative to glycolytic mode by the simultaneous inhibition of mitochondrial cytochrome c oxidase and stimulation of the astrocytic glycolysis.

Generally, glutamatergic neurons do not express nNOS. Certain populations of pyramidal cells (PCs) expressing inducible prostanoid-synthesizing enzyme cyclooxygenase-2 (COX-2) participate in rCBF control by the release of prostaglandin E_2 (PGE_2). Released PGE_2 reaches receptors on either nearby glial processes or SMCs surrounding penetrating arterioles. Postsynaptic COX-2 expression in these PCs frequently shows robust synaptic antilocalization with presynaptic nNOS (Wang et al., 2005). In this way, NO released from nitrergic GABAergic interneurons might enhance vascular dilations by also activating COX-2 in these PCs, with the latter also targeting directly cortical microvessels through axonal varicosities. A substantial reduction of rCBF in rodents has been reported after application of selective inhibition of COX-2 (Niwa et al., 2000; Bakalova et al., 2002). The rCBF response was impaired in mice lacking the COX-2 gene.

9.6.2 The Role of Astrocytic-Mediated Signaling – Tonic Mechanisms

A large component in the functional hyperemia is mediated by Ca^{2+} signaling in astrocytes, which is evoked by the actions of neurotransmitters from nearby neurons on astrocytic perisynaptic end-feet during synaptic activation. As discussed in Chapter 8, group I mGluR has been proposed as the main GPCR underling Ca^{2+} responses in astrocytes evoked by released glutamate from neuronal synapses. As these mGluR are embedded together with glutamate transporters EAAT1&2 in the perisynaptic end-feet, astrocytic Ca^{2+} signaling might represent long-term energetic cost for glutamate/GABA-glutamine recycling during synaptic activation. Therefore, it is natural to hypothesize that an increase in rCBF must take place at the same time that astrocytic Ca^{2+} signaling stimulates glycogenolysis. We will discuss henceforth two astrocyte-medicated pathways involved in tonic control of rCBF: **a)** the ecosainoid pathway and **b)** the K^+ siphoning pathway.

Astrocytic Ca^{2+} signaling occurs in the form of either isolated global Ca^{2+} spikes or waves propagating along the astrocytic network. However, astrocytic Ca^{2+} waves have been demonstrated only in non-physiological situations; hence, we will refer henceforth only to global Ca^{2+} spikes. Ca^{2+} signaling in astrocytes can be observed either spontaneously or as a response to stimulation. Increases in the frequency of Ca^{2+} spikes cause the overproduction of arachidonic acid (AA) via the action of cytosolic phospholipase A2 ($cPLA_2$) enzymes on membrane phospholipids. A secretory (s) subtype of PLA_2 enzymes has also been found extracellularly on perivascular astrocytic end-feet. AA produced by the $sPLA_2$ can freely diffuse toward adjacent SMCs. In SMCs, AA serves as a substrate for a particular subtype (CYP4A) of the cytochrome P450 to produce a lipid acid, i.e., 20-hydroxyeicosatetraenoic (20-HETE) acid. Through both the direct blocking (phosphorylation) of the calcium-activated potassium (BK) channels and the PKC-dependent activation of the L-type calcium channels, 20-HETE has a contractive effect on the SMCs ($[^{20}HETE]$, Chapter 10). The contractive effect of 20-HETE on vessels may also be caused by the activation of the transient receptor potential canonical 6 (TRPC6) channels. Note that $sPLA_2$ produces continuously AA in the perivascular space, which is expected to keep SMC on a quasi-constricted state under physiological level of NO. NO increases cause inhibition of CYP4A, a mechanism that might amplify the vasodilatory effect of nitrergic GABAergic interneurons. However, NO signaling from neurons and, also, from endothelial cells (EC, Chapter 10) activates the soluble guanylate cyclase (sGC) in SMCs, creating cGMP-mediated relaxations.

On the other hand, AA produced intracellularly by $cPLA_2$ (Eq. 9.14) constitutes a substrate for a cytochrome P450 subtype highly expressed in astrocytes (CYP2C), increasing the production of epoxyeicosatrienoic (EET) acid (Eq. 9.15). EET is a lipophilic acid; hence, it can diffuse inside brain tissue, producing SMC hyperpolarization by the activation of BK channels. Similarly, L-type calcium channel deactivation caused by these hyperpolarized SMCs creates vessel dilations. Intracellular increases of AA in astrocytes also cause the production of prostaglandin E_2 (PGE_2) by the constitutive COX-1 enzyme (Eq. 9.16). PGE_2 is exported to the extracellular space through multidrug resistance proteins (MRPs). MRP1, MRP4 and MRP5 are expressed in the astrocytes mouse brains. Exported PGE_2 produces cyclic adenosine monophosphate (cAMP) formation in nearby SMCs by its action of PGE_2 receptors. cAMP activates BK channels via phosphorylation of the serine-869 residue by the protein kinase A. BK activation causes L-Type Ca^{2+} channel deactivation, and hence, vasodilation. Extracellular PGE_2 can be taken back into astrocytes by the prostaglandin transporter (PGT), a mechanism that in terms transports (antiport) a lactate to the neuropil. This constitutes a mechanism to bring vasodilation to an end whenever lactate is accumulated inside astrocytes, which might reflect high energy charge in the form of pyruvate and NADH due to hyperglycolytic activity, or alternatively due to TCA suppression. On the other hand, vasodilation is enhanced when extracellular lactate levels are substantially increased, which is critical to avoid tissue toxicity. MPR and PGT might ensure

TABLE 9.2
Model parameters for the eicosanoid astrocytic pathway

Parameter [units]	Symbol	Value
Relaxation time for AA [s]	τ_{AA}	1
Michaelis constant for AA [μM]	K_{AA}	0.8
Maximum rate of production of AA [μM/s]	O_{AA}	5
Relaxation time for EET [s]	τ_{EET}	0.606
Maximum rate of production of EET [1/s]	O_{EET}	6.15
Relaxation time for PGE_2 [s]	τ_{PGE_2}	1
Michaelis constant for PGE_2 [μM]	K_{PGE_2}	1
Maximum rate of production of PGE_2 [μM/s]	O_{PGE_2}	2

PGE_2 and lactate homeostasis (housekeeping) in the neocortex by dynamically regulating CBF and intercellular lactate flow via MCTs. A theoretical model for the eicosanoid pathway can be found in Suarez et al. (2025), which is summarized in (9.14), (9.16–9.23), and Table 9.2.

$$\frac{d[AA]_i}{dt} = -\frac{[AA]_i}{\tau_{AA}} + \frac{O_{AA}[Ca^{2+}]_i}{K_{AA} + [Ca^{2+}]_i} \tag{9.14}$$

$$\frac{d[EET]}{dt} = -\frac{[EET]}{\tau_{EET}} + O_{EET}[AA]_i \tag{9.15}$$

$$\frac{d[PGE_2]}{dt} = -\frac{[PGE_2]}{\tau_{PGE_2}} + \frac{O_{PGE_2}[AA]_i}{K_{PGE_2} + [AA]_i} \tag{9.16}$$

Hadfield et al. (2013) proposed equation (9.15) for EET. However, Farr and David (2011) proposed a direct relationship between Ca^{2+} and EET: $d[EET]/dt = -k_{EET}[EET] + v_{EET}([Ca^{2+}]_i - Ca_0^{2+})$, with rate constants for EET degradation $k_{EET} = 7.1\,s^{-1}$ and production $v_{EET} = 72\,s^{-1}$, respectively. Ca_0^{2+} represents the astrocytic basal Ca^{2+} level. This simplified model was adopted by Suarez et al. (2025)

Large-conductance BK channels are abundant in astrocytic perivascular end-feet. Therefore, global astrocytic Ca^{2+} signaling causes K^+ outflow into the perivascular space due to the activation of these BK channels. Depending on the perivascular concentration of K^+, brain vessels can either dilate or constrict. For concentrations below $26.3\,mM$, $K_{IR}2.1$ channels create hyperpolarization of SMCs, and consequently vasodilation. The effect is reverse for extracellular concentration of K^+ above $26.3\,mM$. Due to a more depolarized resting membrane potential in SMCs ($\sim -40\,mV$), a high concentration is needed for the inward effect to occur in these cell type. This mechanism has been sometime referred as the K^+ siphoning. The following general model is introduced to describe astrocytic membrane potential (Farr and David, 2011).

$$\frac{dV_m}{dt} = \frac{1}{C_m}\left(-I_{BK}^{(ast)} - I_L - \sum_k I_k + I_{gap}\right) \tag{9.17}$$

$$I_L = g_L\left(V_m - V_L\right) \tag{9.18}$$

$$I_{BK}^{(ast)} = g_{BK}n\left(V_m - V_{BK}\right) \tag{9.19}$$

$$\frac{dn}{dt} = \phi_{BK}(n_{\infty,BK} - n) \tag{9.20}$$

$$\phi_{BK} = \psi_n \cosh\left(\frac{V_m - v_{3,BK}}{2v_{4,BK}}\right) \tag{9.21}$$

$$n_{\infty,BK} = 0.5\left(1 + \tanh\left(\frac{V_m + EET_{shift}\left[EET\right] - v_{3,BK}}{v_{4,BK}}\right)\right) \tag{9.22}$$

$$v_{3,BK} = -\frac{v_{5,BK}}{2}\tanh\left(\frac{\left[Ca^{2+}\right] - Ca_{3,BK}}{Ca_{4,BK}}\right) + v_{6,BK} \tag{9.23}$$

The parameters suggested for this model are: $\psi_n = 2.664\, s^{-1}$, $v_{6,BK} = -15$ mV, $v_{5,BK} = 8\, mV$, $Ca_{3,BK} = 0.15\, \mu M$, $Ca_{4,BK} = 10^{-3}\, \mu M$, $v_{4,BK} = 14.5\, mV$, $EET_{shift} = 3\, mV / \mu M$, $g_{BK} = 195\, pS$, and $V_{BK} = -95\, mV$ (Suarez et al., 2025). In this model, the presence of multiple ionic channels/receptors could affect the astrocyte membrane potential by means of Ca^{2+} and Na^+ / K^+ transmembrane fluxes $\sum_k I_k$.

In particular, changes in the membrane potential will impact intracellular Ca^{2+} concentration via VGCCs. However, compared to the contribution of mGluR/IP_3R mechanisms, the effect of VGCCs can be in principle neglected. The model includes a leakage channel with parameters: $V_L = -70\, mV$ and $g_L = 78.54\, pS$. The model can be also generalized to include currents through the gap-junctions I_{gap}. It has been reported a total membrane capacitance of the astrocyte of $C_m = 40\, pF$. To improve accuracy of this model, adjustments of g_{BK} and C_m to account only for astrocytic end-feet must be done in the future. I_{gap} represents current leaking through the gap-junctions. For simplicity, membrane potential can be kept constant at $V_m = -70\, mV$ by setting the $C_m = \infty$.

In SMCs, contractions/dilations are determined by the state of the myosin light chain (MLC). There are three factors impacting the state of MLC: cAMP, cGMP and intracellular Ca^{2+}. The effect of cAMP on BK channels was discussed above, an effect related to PGE_2 production by astrocytes. cGMP is produced by soluble guanylate cyclase (sGC) through the activation by NO. On SMCs, cGMP activates serine/threonine protein kinases (PKGs) balanced by the inactivation of phosphodiesterases (PDEs), a mechanism discussed in Chapter 10. However, Ca^{2+}-mediated phosphorylation of MLC is the most important mechanism in vessel diameter regulation.

It has been suggested that a signal from vessels can also impact astrocytic Ca^{2+} dynamics by the activation of a type-4 transient receptor potential cation (TRPV4) channel expressed in their end-feet. In general, TRPV4 channels are activated by mechanical stretches, which is determined by the Ca^{2+} concentrations in the astrocyte cytosol $[Ca^{2+}]$ and perivascular space $[Ca^{2+}]_{PV}$. Slow and fast decays are associated with the presence of low and high extracellular Ca^{2+} in the perivascular space. The current through the TRPV4 channel has been modeled as $I_{TRPV4} = g_{TRPV4}s(V_m - V_{TRPV4})$; where, $g_{TRPV4} = 50\,pS$ and $Vg_{TRPV4} = 6\,mV$ are the maximum conductance and reversal potential for the TRPV4 channel, respectively. The channel open probability satisfies:

$$\frac{ds}{dt} = (s_\infty - s)[Ca^{2+}]_{PV} / \tau_{TRPV4}, \text{with } \tau_{TRPV4} = 0.9\,s^{-1}$$

s_∞ is the Ca^{2+}-dependent steady-state open probability of the TRPV4 channel, represented by the following equation: $s_\infty = \left(\frac{1}{1+e^{-(\epsilon-\epsilon_{1/2})/\kappa}}\right)\left[\left(\frac{1}{1+H_{Ca}}\right)\left(H_{Ca} + tanh\left(\frac{V_m - v_{1,TRPV4}}{v_{2,TRPV4}}\right)\right)\right]$, with $H_{Ca} = \frac{[Ca^{2+}]}{\gamma_i} + \frac{[Ca^{2+}]_{PV}}{\gamma_e}$. Model parameters are defined as: $\gamma_i = 0.01\,\mu M$, $\gamma_e = 0.2\,mM$, $\epsilon_{1/2} = 0.1$, $\kappa = 0.1$, $v_{1,TRPV4} = 120\,mV$, and $v_{2,TRPV4} = 13\,mM$. The TRPV4 signal will change cytosolic Ca^{2+} concentration according to the following Ca^{2+} influx $J_{Ca,TRPV4} = -\frac{1}{2}I_{TRPV4}/(C_A\gamma)$, with $C_A = 40\,pF$ and $\gamma = 834.3\,mV\,\mu M^{-1}$ the astrocyte membrane capacitance and a scaling factor relating ion flux to membrane potential, respectively. The variable ϵ represent the strain on the perivascular end-foot, which is considered proportional to the local relative radial strain r on the arteriole respect to the resting value r_0, $\epsilon = (r - r_0)/r_0$.

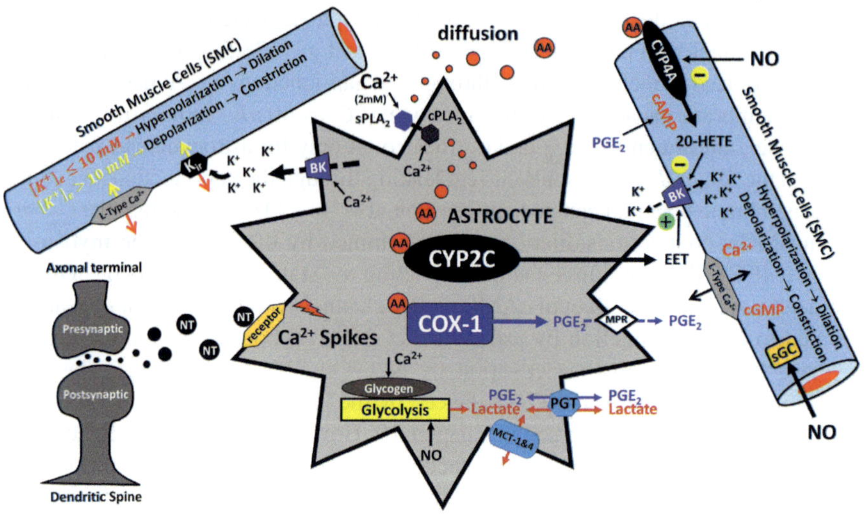

FIGURE 9.5 The neuro-vascular/metabolic pathways are illustrated.

9.7 NEURONAL MODULATION – GLIOTRANSMISSION

Using neuron and astrocytes co-culture from neocortical tissue, Parpura et al. (1994) demonstrated that astrocytic Ca^{2+} signal also influences excitatory synaptic activity through a NMDAR-dependent mechanism, a finding later demonstrated also in situ (Pasti et al., 1997). In addition to glutamate, astrocytes also release ATP and D-serine in a Ca^{2+}-dependent way. These gliotransmitters released by astrocytes reach extra-synaptic NMDAR impacting on both postsynaptic electric activity and existing mechanisms for synaptic plasticity. Through interaction with the PSD-95 multi-protein complex, ATP can pre-modulate excitatory synapses efficacy (Lalo et al., 2016). On the other hand, D-serine can modulate NMDAR activity by its action on glycine sites. Computational models describing astrocyte-neuron bidirectional communication are being proposed (Volman et al., 2012; Lenk et al., 2020). However, proofs for the existence of gliotransmission have been controversial (see companion papers: Fiacco and McCarthy, 2018; Savtchouk and Volterra, 2018) and, if it exists, an evaluation of its energetic cost needs to be performed to improve current biophysical models for brain imaging data analysis as well as its impact in neuronal computational models. Unfortunately, the lack of an experimental protocol to evaluate the existence of gliotransmission in vivo constitutes a major limitation.

9.8 PROBLEMS AND EXERCISES

Problem 9.1 – The Michaelis-Menten kinetic theory explains how reaction rates depend on the concentration of the substrate S and enzyme E. In this equation, k_1, k_2, and k_3 represent the binding affinity of the enzyme to the substrate creating the enzyme-substrate complex, the final catalysis reaction rate, and the dissociation of the complex. k_2 defines the rate limiting. The reverse reaction of catalysis for the most can be neglected $k_4 = 0$.

$$S + E \underset{k_3}{\overset{k_1}{\rightleftarrows}} [ES] \underset{k_4}{\overset{k_2}{\rightleftarrows}} E + P$$

Demonstrate the final Michaelis-Menten kinetic equation for the rate of catalysis of an enzyme under two approximations: **a)** the rapid equilibrium and **b)** steady-state. Discuss the differences between K_s and K_M. Discuss the limit cases of zero-order and first-order kinetics.

Problem 9.2 – Discuss how to calculate K_s and K_M parameters by using the Lineweaver-Burk plot.

Problem 9.3 – Demonstrate the Hill–Langmuir kinetic equation for enzymes with n binding sites.

REFERENCES

Almeida A, Moncada S, Bolaños JP. Nitric oxide switches on glycolysis through the AMP protein kinase and 6-phosphofructo-2-kinase pathway. *Nat Cell Biol* 6(1), 45–51, 2004.

Bak LK, Johansen ML, Schousboe A, Waagepetersen HS. Among the branched-chain amino acids only valine metabolism is upregulated in astrocytes during glutamate exposure. *J Neurosci Res* 85, 3465–3470, 2007.

Bak LK, Schousboe A, Sonnewald U, Waagepetersen HS. Glucose is necessary to maintain neurotransmitter homeostasis during synaptic activity in cultured glutamatergic neurons. *J Cereb Blood Flow Metab* 26, 1285–1297, 2006.

Bak LK, Schousboe A, Waagepetersen HS. The glutamate/ GABA-glutamine cycle: aspects of transport, neurotransmitter homeostasis and ammonia transfer. *J Neurochem* 98, 641–653, 2006.

Bak LK, Sickmann HM, Schousboe A, Waagepetersen HS. Activity of the lactate-alanine shuttle is independent of glutamate-glutamine cycle activity in cerebellar neuronal-astrocytic cultures. *J Neurosci Res* 79, 88–96, 2005.

Bakalova R, Matsuura T, Kanno I. The cyclooxygenase inhibitors indomethacin and rofecoxib reduce regional cerebral blood flow evoked by somatosensory stimulation in rats. *Exp Biol Med* 227, 465–473, 2002.

Benjamin AM, Quastel JH. Locations of amino acids in brain slices from the rat. Tetrodotoxin-sensitive release of amino acids. *Biochem J* 128, 631–646, 1972.

Bergles DE, Jahr CE. Synaptic activation of glutamate transporters in hippocampal astrocytes. *Neuron* 19, 1297–1308, 1997.

Berl S, Clarke DD. The metabolic compartmentation concept. In: Hertz L, Kvamme E, McGeer EG, Schousboe A (Eds.), *Glutamine, Glutamate and GABA in the Central Nervous System.* Alan R. Liss, Inc., New York, pp. 205–217, 1983. https://archive.org/details/glutamineglutama0007unse

Bittar PG, Charnay Y, Pellerin L, Bouras C, Magistretti PJ. Selective distribution of lactate dehydrogenase isoenzymes in neurons and astrocytes of human brain. *J Cereb Blood Flow Metab* 16(6), 1079–1089, 1996.

Bröer A, Albers A, Setiawan I, Edwards RH, Chaudhry FA, Lang F, Wagner CA, Bröer S. Regulation of the glutamine transporter SN1 by extracellular pH and intracellular sodium ions. *J Physiol* 539, 3–14, 2002.

Bröer S, Brookes N. Transfer of glutamine between astrocytes and neurons. *J Neurochem* 77, 705–719, 2001.

Buxton RB, Frank LR. A model for the coupling between cerebral blood flow and oxygen metabolism during neural stimulation. *J Cereb Blood Flow Metab* 17, 64–72, 1997.

Cataldo AM, Broadwell RD. Cytochemical identification of cerebral glycogen and glucose-6-phosphatase activity under normal and experimental conditions. II. Choroid plexus and ependymal epithelia, endothelia and pericytes. *J Neurocytol* 15, 511–524, 1986.

Chatton JY, Pellerin L, Magistretti PJ. GABA uptake into astrocytes is not associated with significant metabolic cost: Implications for brain imaging of inhibitory transmission. *PNAS* 100, 12456–12461, 2003.

Chen W, Aoki C, Mahadomrongkul V, Gruber CE, Wang GJ, Blitzblau R, Irwin N, Rosenberg PA. Expression of a variant form of the glutamate transporter GLT1 in neuronal cultures and in neurons and astrocytes in the rat brain. *J Neurosci* 22, 2142–2152, 2002.

Cotman CW, Foster A, Lanthorn T. An overview of glutamate as a neurotransmitter. *Adv Biochem Psychopharmacol* 27, 1–27, 1981.

Danbolt NC. Glutamate uptake. *Progr Neurobiol* 65, 1–105, 2001.

Dienel GA, Cruz NF. Contributions of glycogen to astrocytic energetics during brain activation. *Metab Brain Dis* 30, 281–298, 2015.

Dienel GA, Hertz, L. Glucose and lactate metabolism during brain activation. *J Neurosci Res* 66, 824–838, 2001.

Erbsloh F, Bernsmeier A, Hillesheim H. The glucose consumption of the brain & its dependence on the liver. *Arch Psychiatr Nervenkr Z Gesamte Neurol Psychiatr* 196, 611–626, 1958.

Estrada C, DeFelipe J. Nitric oxide-producing neurons in the neocortex: Morphological and functional relationship with intraparenchymal microvasculature. *Cereb Cortex* 8, 193–203, 1998.

Farr H, David T. Models of neurovascular coupling via potassium and EET signalling. *J Theor Biol* 286, 13–23, 2011.

Fiacco TA, McCarthy KD. Multiple lines of evidence indicate that gliotransmission does not occur under physiological conditions. *J Neurosci* 38(1), 3–13, 2018.

Gegelashvili G, Schousboe A. High-affinity glutamate transporters: Regulation of expression and activity. *Mol Pharmacol* 52, 6–15, 1997.

Gibbs ME, Anderson DG, Hertz L. Inhibition of glycogenolysis in astrocytes interrupts memory consolidation in young chickens. *Glia* 54, 214–222, 2006.

Gruetter R, Adriany G, Choi IY, Henry PG, Lei H, Öz G. Localized in vivo ^{13}C NMR spectroscopy of the brain. *NMR Biomed* 16, 313–338, 2003.

Gruetter R, Ugurbil K, Seaquist ER. Steady-state cerebral glucose concentrations and transport in the human brain. *J Neurochem* 70, 397–408, 1998.

Henry PG, Adriany G, Deelchand D, Gruetter R, Marjanska M, Öz G, Seaquist ER, Shestov A, Ugurbil K. In vivo ^{13}C NMR spectroscopy and metabolic modeling in the brain: A practical perspective. *Magn Reson Imaging* 24(4), 527–539, 2006.

Hertz L, Peng L, Dienel GA. Energy metabolism in astrocytes: high rate of oxidative metabolism and spatiotemporal dependence on glycolysis/glycogenolysis. *J Cereb Blood Flow Metab* 27, 219–249, 2007.

Hertz L, Schousboe A. Primary cultures of GABAergic and glutamatergic neurons as model systems to study neurotransmitter functions. I. Differentiated cells. In: Vernadakis A, Privat A, Lauder JM, Timiras PS, Giacobini E (Eds.), *Model Systems of Development and Aging of the Nervous System*. Nijhoff Publishing Company, Boston, pp. 19–31, 1987.

Hyder F. Dynamic imaging of brain function. *Methods Mol Biol* 489, 3–22, 2009.

Hyder F, Herman P, Bailey CJ, Møller A, Globinsky R, Fulbright RK, Rothman DL, Gjedde A. Uniform distributions of glucose oxidation and oxygen extraction in gray matter of normal human brain: No evidence of regional differences of aerobic glycolysis. *J Cereb Blood Flow Metab* 36(5), 903–916, 2016.

Hyder F, Patel AB, Gjedde A, Rothman DL, Behar KL, Shulman RG. Neuronal-glial glucose oxidation and glutamatergic-GABAergic function. *J Cereb Blood Flow Metab* 26, 865–877, 2006.

Ibáñez V, Pietrini P, Furey ML, Alexander GE, Millet P, Bokdea ALW, Teichberg D, Schapiro MB, Horwitz B, Rapoport SI. Resting state brain glucose metabolism is not reduced in normotensive healthy men during aging, after correction for brain atrophy. *Brain Res Bull* 63, 147–154, 2004.

Ignacio PC, Baldwin BA, Vijayan VK, Tait RC, Gorin FA. Brain isozyme of glycogen phosphorylase: immunohistological localization within the central nervous system. *Brain Res* 529, 42–49, 1990.

Laake JH, Takumi Y, Eidet J, Torgner IA, Roberg B, Kvamme E, Ottersen OP. Postembedding immunogold labelling reveals subcellular localization and pathway-specific enrichment of phosphate activated glutaminase in rat cerebellum. *Neuroscience* 88, 1137–1151, 1999.

Lalo U, Palygin O, Verkhratsky A, Grant SGN, Pankratov Y. ATP from synaptic terminals and astrocytes regulates NMDA receptors and synaptic plasticity through PSD-95 multi-protein complex. *Sci Rep* 6, 33609, 2016.

Lenk K, Satuvuori E, Lallouette J, Ladrón-de-Guevara A. Berry H, Hyttinen JAK. A computational model of interactions between neuronal and astrocytic networks: The role of astrocytes in the stability of the neuronal firing rate. *Front Comput Neuros* 13(92), 1–19, 2020.

Lieth E, LaNoue KF, Berkich DA, Xu B, Ratz M, Taylor C, Hutson SM. Nitrogen shuttling between neurons and glial cells during glutamate synthesis. *J Neurochem* 76, 1712–1723, 2001.

Ma J, Ayata C, Huang PL, Fishman MC, Moskowitz MA. Regional cerebral blood flow response to vibrissal stimulation in mice lacking type I NOS gene expression. *Am J Physiol Heart Circ Physiol* 270, H1085–H1090, 1996.

Magistretti PJ, Pellerin L, Rothman DL, Shulman RG. Energy on demand. *Science* 283, 496–497, 1999.

Martínez-Lozada Z, Guillem AM, Flores-Méndez M, Hernández-Kelly LC, Vela C, Meza E, Zepeda RC, Caba M, Rodríguez A, Ortega A. GLAST/EAAT1-induced glutamine release via SNAT3 in Bergmann glial cells: Evidence of a functional and physical coupling. *J Neurochem* 125, 545–554, 2013.

McKenna MC, Sonnewald U. GABA alters the metabolic fate of [U-13C]glutamate in cultured cortical astrocytes. *J Neurosci* Res 79, 81–87, 2005.

McKenna MC, Sonnewald U, Huang X, Stevenson J, Zielke HR. Exogenous glutamate concentration regulates the metabolic fate of glutamate in astrocytes. *J Neurochem* 66, 386–395, 1996.

Nasrallah FA, Griffin JL, Balcar VJ, Rae C. Understanding your inhibitions: Modulation of brain cortical metabolism by GABAB receptors. *J Cereb Blood Flow Metab* 27, 1510–1520, 2007.

Niwa K, Araki E, Morham SG, Ross ME, Ladecola, C. Cyclooxygenase-2 contributes to functional hyperemia in whisker-barrel cortex. *J Neurosci* 20, 763–770, 2000.

Norenberg MD, Martinez-Hernandez A. Fine structural localization of glutamine synthetase in astrocytes of rat brain. *Brain Res* 161, 303–310, 1979.

Ottersen OP, Zhang N, Walberg F. Metabolic compartmentation of glutamate and glutamine: Morphological evidence obtained by quantitative immunocytochemistry in rat cerebellum. *Neuroscience* 46, 519–534, 1992.

Öz G, Henry PG, Seaquist ER, Gruetter R. Direct, noninvasive measurement of brain glycogen metabolism in humans. *Neurochem Int* 43, 323–329, 2003.

Parpura V, Basarsky TA, Liu F, Jeftinija K, Jeftinija S, Haydon PG. Glutamate-mediated astrocyte-neuron signalling. *Nature* 369, 744–747, 1994.

Pasti L, Volterra A, Pozzan T, Carmignoto G. Intracellular calcium oscillations in astrocytes: A highly plastic, bidirectional form of communication between neurons and astrocytes in situ. *J Neurosci* 17(20), 7817–7830, 1997.

Patel AB, de Graaf RA, Mason GF, Rothman DL, Shulman RG, Behar KL. The contribution of GABA to glutamate-glutamine cycling and energy metabolism in the rat cortex in vivo. *PNAS* 102(15), 5588–5593, 2005.

Pellerin L, Magistretti PJ. Neuroenergetics: Calling upon astrocytes to satisfy hungry neurons. *Neuroscientist* 10(1), 53–62, 2004.

Pfeiffer B, Meyermann R, Hamprecht B. Immunohistochemical co-localization of glycogen phosphorylase with the astroglial markers glial fibrillary acidic protein and S-100 protein in rat brain sections. *Histochemistry* 97(5), 405–412, 1992.

Pfeuffer J, Tkac I, Choi IY, Merkle H, Ugurbil K, Garwood M, Gruetter R. Localized in vivo 1H NMR detection of neurotransmitter labeling in rat brain during infusion of [1-13C] D-glucose. *Magn Reson Med* 41, 1077–1083, 1999.

Powers WJ, Grubb Jr RL , Darriet D, Raichle ME. Cerebral blood flow and cerebral metabolic rate of oxygen requirements for cerebral function and viability in humans. *J Cereb Blood Flow Metab* 5, 600–608, 1985.

Riera J, Schousboe A, Waagepetersen HS, Howarth C, Hyder F. The micro-architecture of the cerebral cortex: Functional neuroimaging models and metabolism. *NeuroImage* 40, 1436–1459, 2008.

Riera J, Sumiyoshi A. Brain oscillations: Ideal scenery to understand the neurovascular coupling. *Curr Opin Neurol* 23, 374–381, 2010.

Rothman DL, Dienel GA, Behar KL, Hyder F, DiNuzzo M, Giove F, Mangia S. Glucose sparing by glycogenolysis (GSG) determines the relationship between brain metabolism and neurotransmission. *J Cereb Blood Flow Metab* 41(11), 1–17, 2022.

Saito K, Barber R, Wu J, Matsuda T, Roberts E, Vaughn JE. Immunohistochemical localizaation of glutamate decarboxylase in rat cerebellum. *PNAS* 71, 269–273, 1974.

Savtchouk I, Volterra A. Gliotransmission: Beyond Black-and-White. *J Neurosci* 38(1), 14–25, 2018.

Schousboe A, Kanner B. GABA transporters: Functional and pharmacological properties. In: Egebjerg J, Schousboe A, Krogsgaard-Larsen P (Eds.), *Glutamate and GABA Receptors and Transporters*. Taylor & Francis, London, UK, pp. 337–349, 2002.

Schousboe A, Larsson OM, Sarup A, White HS. Role of the betaine/GABA transporter (BGT-1/GAT2) for the control of epilepsy. *Eur J Pharmacol* 500, 281–287, 2004.

Schousboe A, Sarup A, Bak LK, Waagepetersen HS, Larsson OM. Role of astrocytic transport processes in glutamatergic and GABAergic neurotransmission. *Neurochem Int* 45, 512–527, 2004.

Schousboe A, Sarup A, Larsson OM, White HS. GABA transporters as drug targets for modulation of GABAergic activity. *Biochem Pharmacol* 68, 1557–1563, 2004.

Shank RP, Bennett GS, Freytag SO, Campbell GL. Pyruvate carboxylase: Astrocyte-specific enzyme implicated in the replenishment of amino acid neurotransmitter pools. *Brain Res* 329, 364–367, 1985.

Shulman RG, Hyder F, Rothman DL. Cerebral energetics and the glycogen shunt: Neurochemical basis of functional imaging. *PNAS* 98, 6417–6422, 2001.

Sickmann HM, Schousboe A, Bouman SD,Waagepetersen HS. A functioning glycolysis is important for maintenance of glutamate transport in cultured astrocytes. *J Neurochem* 102 (Suppl. 1), 285, 2007.

Sickmann HM, Schousboe A, Fosgerau K, Waagepetersen HS. Compartmentation of lactate originating from glycogen and glucose in cultured astrocytes. *Neurochem Res* 30(10), 1295–1304, 2005.

Siesjo B. *Brain Energy Metabolism*. Wiley, New York, 1978.

Simpson IA, Carruthers A, Vannucci SJ. Supply and demand in cerebral energy metabolism: The role of nutrient transporters. *J Cereb Blood Flow Metab* 27, 1766–1791, 2007.

Sokoloff L. Relationship between functional activity and energy metabolism in the nervous system: Whether, where and why? In: Lassen NA, Ingvar DH, Raichle ME, Friberg L (Eds.), *Brain Work and Mental Activity*. Munksgaard, Copenhagen, pp. 52–64, 1991.

Stefanovic B, Schwindt W, Hoehn M, Silva AC. Functional uncoupling of hemodynamic from neuronal response by inhibition of neuronal nitric oxide synthase. *J Cereb Blood Flow Metab* 27, 741–754, 2007.

Suarez A, Fernandez L, Riera J. Deciphering different aspects of the eicosanoid pathway in the neurovascular coupling. *J Cereb Blood Flow Metab*, 2025. https://doi.org/10.1177/0271678X241311010

Todd AC, Marx M-C, Hulme SR, Bröer S, Billups B. SNAT3-mediated glutamine transport in perisynaptic astrocytes in situ is regulated by intracellular sodium. *Glia* 65, 900–916, 2017.

Van den Berg CJ, Garfinkel D. A simulation study of brain compartments: Metabolism of glutamate and related substances in mouse brain. *Biochem J* 23, 211–218, 1971.

Vilchez D, Ros S, Cifuentes D, Pujadas L, Valles J, Garcia-Fojeda B, Criado-Garcia O, Fernandez-Sanchez E, Medraño-Fernandez L, Domínguez J, Garcia-Rocha M, Soriano E, Rodriguez de Cordoba S, Guinovart JJ. Mechanism suppressing glycogen synthesis in neurons and its demise in progressive myoclonus epilepsy. *Nat Neurosci* 10(11), 1407–1413, 2007.

Volman V, Bazhenov M, Sejnowski TJ. Computational models of neuron-astrocyte interaction in epilepsy. *Front Comput Neurosci* 6(58), 1–10, 2012.

Waagepetersen HS, Qu H, Sonnewald U, Shimamoto K, Schousboe A. Role of glutamine and neuronal glutamate uptake in glutamate homeostasis and synthesis during vesicular release in cultured glutamatergic neurons. *Neurochem Int* 47, 92–102, 2005.

Waagepetersen HS, Sonnewald U, Larsson OM, Schousboe A. A possible role of alanine for ammonia transfer between astrocytes and glutamatergic neurons. *J Neurochem* 75(2), 471–479, 2000.

Waagepetersen HS, Sonnewald U, Schousboe A. Glutamine, glutamate and GABA: Metabolic aspects. In: Lajtha A (Ed.), *Handbook of Neurochemistry and Molecular Neurobiology* (3rd Edition, Vol. 6). Springer-Verlag, Berlin Heidelberg, pp. 1–21, 2007.

Waldvogel D, van Gelderen P, Muellbacher W. The relative metabolic demand of inhibition and excitation. *Nature* 406, 995–998, 2000.

Wang H, Hitron IM, Iadecola C. Synaptic and vascular associations of neurons containing cyclooxygenase-2 and nitric oxide synthase in rat somatosensory cortex. *Cereb Cortex* 15, 1250–1260, 2005.

Yu ACH, Drejer J, Hertz L, Schousboe A. Pyruvate carboxylase activity in primary cultures of astrocytes and neurons. *J Neurochem* 41, 1484–1487, 1983.

Yu ACH, Hertz L. Metabolic sources of energy in astrocytes. In: Hertz L, Kvamme E, McGeer EG, Schousboe A (Eds.), *Glutamine, Glutamate and GABA in the Central Nervous System*. Alan R. Liss, Inc., New York, pp. 431–438, 1983.

Yu ACH, Schousboe A, Hertz L. Metabolic fate of 14C-labeled glutamate in astrocytes in primary cultures. *J Neurochem* 39, 954–960, 1982.

Yudkoff M. Brain metabolism of branched-chain amino acids. *Glia* 21, 92–98, 1997.

Zaganas I, Waagepetersen HS, Georgopoupolos P, Sonnewald U, Plaitakis A, Schousboe A. Differential expression of glutamate dehydrogenase in cultured neurons and astrocytes from mouse cerebellum and cerebral cortex. *J Neurosci Res* 66, 909–913, 2001.

Zheng Y, Martindale J, Johnston D, Jones M, Berwick J, Mayhew J. A Model of the Hemodynamic response and oxygen delivery to brain. *NeuroImage* 16, 617–637, 2002.

Zwingmann C, Richter-Landsberg C, Brand A, Leibfritz D. NMR spectroscopic study on the metabolic fate of [3-(13)C]alanine in astrocytes, neurons, and cocultures: Implications for glia-neuron interactions in neurotransmitter metabolism. *Glia* 3, 286–303, 2000.

10 The Vascular Response

10.1 OVERVIEW

The brain receives about 20 % of the resting cardiac output. Blood perfusion into the brain tissue takes place by means of a complex vascular network with multiple components with a hierarchical organization. At each hierarchical level, there are several mechanisms to dynamically regulate vessel diameters in a demand-based principle. Blood reaches the brain through two major arteries carrying oxygenated blood into the entire brain: **a)** the internal carotid and **b)** the vertebral. The internal carotid arteries in both hemispheres bifurcate into the anterior and the middle cerebral arteries. At the level of the pons, both right and left vertebral arteries fuse to form the midline basilar artery on the surface of the brainstem (ventral). At the base of the brain, near the hypothalamus and cerebral peduncles, blood from the basilar artery joins that coming from the internal carotids forming an arterial ring (i.e., *the circle of Willis*). These two blood supply systems irrigate blood to the cerebral cortex through pial branches running from the corpus-callosum and the external section of the cerebral cortex. The cerebellum is directly irrigated by cerebellar branches originated from the vertebral/basilar arteries (i.e., superior, and anterior/posterior-inferior). Therefore, blood is supplied to the brain according to well-defined non-overlapping vascular territories. These pial arteries communicate with the cerebral microvasculature by means of penetrating arterioles crossing a perivascular space, or the Virchow-Robin compartment. Inside the brain parenchyma, these penetrating arterioles branch into multiple precapillary arterioles, each of them forming a network of thin capillaries to densely infiltrate brain tissues. Multiple capillaries join and widen to become postcapillary venules. Finally, several postcapillary venules within a brain region converge into a single venule ascending to the pial surface. After passing an intricate network of veins (i.e., ophthalmic, anterior/posterior-intercavernous, cortical, thalamo-striate, Trolard, Labbe, internal, superficial-middle, and basal Rosenthal) and sinuses (i.e., spheno-parietal, cavernous, inferior/superior-sagittal, sigmoid, transverse, and straight), deoxygenated blood is finally drained out of the brain via the vein of Galen to the internal jugular veins. These veins and sinuses, which passively communicate with each other, drain not only the brain parenchyma, but also the three meninges (i.e., the dura mater, the arachnoid mater, and the pia mater). The subarachnoid space containing cerebrospinal fluid in specific cisterns is

DOI: 10.1201/9780367901509-10

located between the arachnoid mater and the pia mater. There is a set of blood vessels traversing the subarachnoid space. In this chapter, we will learn the most important structural properties of brain vasculature. We will also discuss mechanisms by means of which blood perfusion is not only kept stable but regionally modified according to local demand. We will describe the role of endothelial cells (EC), smooth muscle cells (SMC) and sphincters in blood flow control.

10.2 VASCULAR SEGMENTS: ARTERY/VEIN, ARTERIOLE/VENULE, AND CAPILLARY

Pial arteries are extracerebral blood vessels that distribute blood to the entire brain. Those pial arteries from the corpus-callosum cover six cortical territories (i.e., orbital, fronto-polar, calloso-marginal, peri-collosal, parieto-occipital, and calcarine), while those from the external section of the cerebral cortex cover the other nine cortical territories (i.e., anterior/middle-temporal, temporo-occipital, angular, postcentral, parietal central, pre-central, anterior-cerebral, and pre-frontal). Blood flow in these pial arteries is controlled by means of the direct release of large vasoactive molecules from extrinsic neuronal innervations (Hamel, 2006; Girouard and Iadecola, 2006). These extrinsic innervations originate from the autonomic (sympathetic and parasympathetic) nerve system, as well as from the trigeminal ganglion. Sympathetic innervations originate from the superior cervical ganglion, causing vasoconstriction by the release of norepinephrine (NE) and neuropeptide Y (NPY) direct into the pial arteries. In addition, NPY might also play a role in trophic effects, e.g., to shift toward higher pressures the upper limit of the autoregulation curve. Parasympathetic innervations originating from the sphenopalatine and otic ganglia can cause vasodilation by the release three major substances into extracerebral blood vessels: vasoactive intestinal peptide (VIP), acetylcholine (ACh), and nitric oxide (NO). However, this strong vasodilatory system appears to be neither involved in cerebrovascular responses nor in autoregulation. Innervations from the trigeminal ganglion can release calcitonin gene-related peptide (CGRP), substance P (SP), neurokinin A (NKA), and pituitary adenylate-cyclase activating polypeptide (PACAP). CGRP from trigemino-vascular pathway can help restore tone after vasocontractile stimuli. The trigemino-vascular system is involved in migraine headache.

Pial arteries branch into penetrating arteries that after crossing the Virchow-Robin compartment become the initial segments of the vasculature in the brain parenchyma. Extrinsic innervations of pial arteries vanish after crossing the Virchow-Robin compartment. Inside the brain parenchyma, arteries divide into arterioles forming specific non-overlapping vascular communities (Blinder et al., 2013). These communities are interconnected with noncolumnar patterns of blood flow. The weighted graph of edges and vertices of all vectorized vessels has been referred by these authors as the cortical angiome. Cortical arterioles are surrounded by a muscular wall comprising layers of smooth muscle cells (SMC). Dilation or constriction of arterioles take place due to the activity of SMCs; hence, most vascular resistance is determined at the level of arterioles. By changes in vessel diameter, arterioles are capable of causing local variations in blood pressure and blood flow velocity at the level of cerebral capillaries. Arterioles are intrinsically innervated by a subset of local GABAergic interneurons and subcortical vasoactive pathways. The local GABAergic interneurons can cause either dilation or constriction of brain vessels. For example, a small number of GABAergic interneurons innervating locally the brain vasculature shows immunoreactivity to neuronal nitric oxide synthase (nNOS, Type I); hence, they are able to create vasodilation when activated. These nitrergic GABAergic interneurons show regular spiking firing patterns and have multipolar morphology with dendrites in all directions. There is also a variety of non-nitrergic GABAergic interneurons neurons that are able to releases vasoactive substances directly into the brain microvasculature, e.g., *contraction:* NPY ($\mathbf{Y_1R}$) and somatostatin (SOM, **SSTR**); *dilation:* VIP ($\mathbf{VPAC_1}$), ACh ($\mathbf{M_5}$), cholecystokinin (CCK, $\mathbf{CCK_2}$), and GABA ($\mathbf{GABA_A}$). Some of these interneurons exhibit regular spiking firing patterns, while others reveal either irregular spiking patterns or have triangular-like shapes. They show bipolar/multipolar morphometries. The network of GABAergic interneurons involved in vascular regulation has shown a laminar profile in the neocortex.

On the other hand, the existence of subcortical vasoactive pathways helps coordinate local and global strategies to control both the vascular tone and blood flow in the neocortex. These pathways comprise neurons that innervate not only ECs and SMCs, but also astrocytes and sphincters in close contact with the cortical vasculature. They also target some of the previously mentioned GABAergic interneurons with role in vascular controllability. These intrinsic pathways originate from the nucleus basalis (ACh), raphe nucleus (5-HT), locus coeruleus (NE). ACh-linked perivascular afferents target muscarinic receptors (e.g., $\mathbf{M_5}$) mediating vasodilation. Microvessels in the cerebral cortex contain several 5-HT receptors underlying vessel contraction (e.g., $\mathbf{5\text{-}HT_1B}$). NE afferents target mostly perivascular astrocytes; however, their role is being strongly debated.

Each arteriole divides into a network of capillaries of diameters between $5-10\,\mu m$. Arterioles has a diameter ranging $38-55\,\mu m$. As a consequence of the Bernoulli's principle, blood pressure in the capillary increases while its flow velocity decreases. As the osmotic pressure is much lower outside the capillary, the movement of gases and nutrients from the blood plasma into the surrounding tissue is facilitated. The thin walls of the capillary facilitate the exchange of gases and metabolites, which occurs by passive diffusion. Capillary do neither dilate nor constrict; hence, increases in cerebral blood flow in the arteriole are translated into surges in capillary recruitment. Multiple capillaries from the same vascular community merge to give rise to a postcapillary venule. Local increases in cerebral blood flow in a precapillary arteriole cause inflation of the postcapillary venules as the results of their high compliance, which can lead to a balloon-type of deformation. During increases in blood flow rate, the diameter of a postcapillary venule can variate between $10-100\,\mu m$. Therefore, blood pressure drops at the entrance of the venules, leading to the diffusion of residual gases and metabolite toward the venule lumen due to the high osmotic pressure in the extracellular space. Multiple venules unite to form a vein. There are sphincters located between the capillaries and the arterioles. Precapillary sphincters help regulate local cerebral blood flow into the

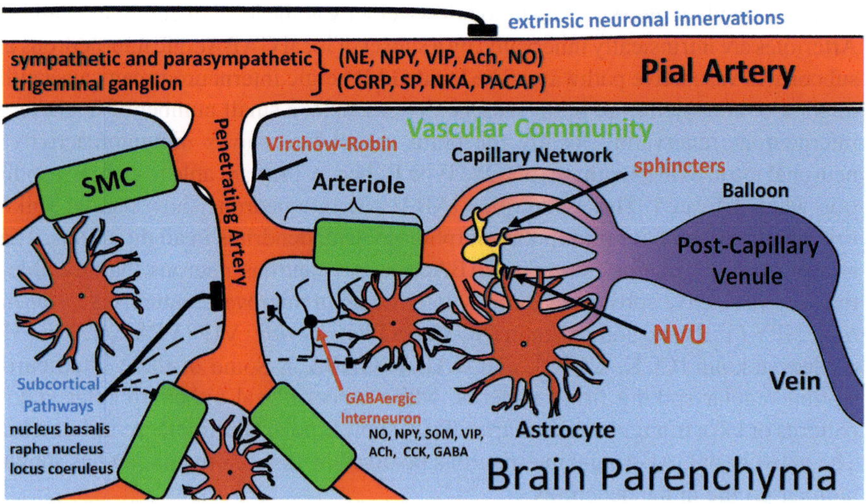

FIGURE 10.1 The organization of the brain vasculature comprising the NVU, extrinsic and intrinsic neuronal innervating pathways.

capillary bed with more special precision. These sphincters are circular mural cells that adjust blood flow by contractile forces (Grubb et al., 2020). Sphincters between capillaries and postcapillary venules have been also reported. Extrinsic and intrinsic regulatory pathways of the brain vascular network are illustrated in Figure 10.1.

10.3 THE NEUROVASCULAR UNITS

In 1885, Paul Ehrlich suggested the notion of a blood-brain barrier (BBB) separating the peripheral circulation and the central nervous system (CNS) mostly at the level of brain capillaries. He noticed that water-soluble dyes injected into the circulation did not stain the brain and spinal cord but all other organs. The concept was strongly debated until the 60s of the XX century. The BBB is formed by ECs strongly coupled by tight-junctions (TJ) (Reese and Karnovsky, 1967). A TJ is an agglomerate of channels connecting contiguous ECs for direct electrical and metabolic communication. ECs are also bound together by cell-cell adhesion points (CCAP). Adherens and TJ require the nectin–afadin complex. TJ channels are formed by a variety of isoforms of connexins (i.e., Cx37, Cx40, Cx43). ECs also created TJs with SMC and lymphocytes/monocytes. The electrical coupling between ECs underlies electrotonic propagation of electric signal along vessels. The propagation of a membrane hyperpolarization from capillaries to arterioles has been proposed as a crucial mechanism for CBF control. High-conductance electric coupling between ECs and SMCs via TJs is also crucial in CBF control. ECs and SMCs are also bound together by focal adhesion sites (FAS). In the

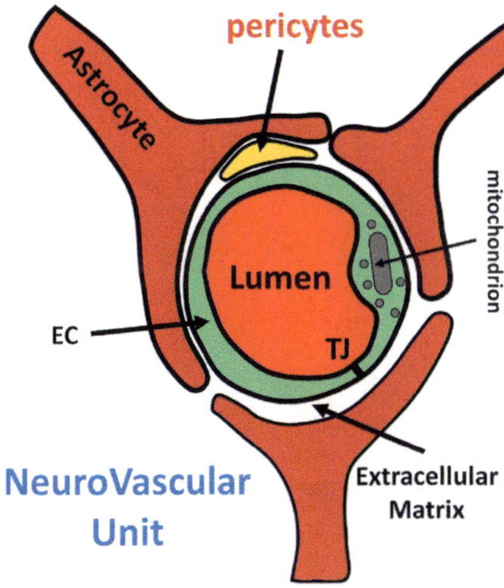

FIGURE 10.2 Microstructure of the neurovascular unit.

CNS, ECs have a high density of mitochondria and perform minimal pinocytosis. At the level of capillaries, ECs, together with astrocytes, pericytes, and the extracellular matrix, form the neurovascular unit (NVU, Figure 10.2, Hawkins and Davis, 2005). The NVU could also include sometimes neuronal terminals. The NVU is a key element in the exchange of metabolites (e.g., Glc, lactate) and gases (e.g., O_2) between the brain tissue and the microvasculature. It also helps maintain the extracellular concentration of Na^+, K^+, Cl^-, and Ca^{2+} in brain tissue within narrow ranges.

Previous experimental findings suggest a critical role of astrocytes in maintaining tissue vascularization. As discussed in Chapter 9, astrocytic end-feet surround brain vessels. Perivascular astrocytes have been also demonstrated in vivo (Figure 10.3). Astrocytic end-feet and perivascular astrocytes, essential components of the NVU, facilitate interactions between vascular and nervous tissues, which is critical for the development of the BBB. It has been demonstrated that brain tissues with astrocyte aggregations are rapidly vascularized. Similarly, pericytes surrounding ECs, another important part of the NVU, help maintaining BBB integrity. For example, occluding, an important constituent of TJs, is created from the pericyte-derived angiopoetin (Hori et al., 2004). Both ECs and pericytes are enclosed by the basal lamina, a $30-40\,nm$ membrane composed of (Farkas and Luiten, 2001): **a)** extracellular matrix proteins (e.g., collagen IV), **b)** glycoproteins (e.g., laminin, fibronectin), and **c)** macromolecules (e.g., heparin sulfate proteoglycans). ECs are anchored by interaction of glycoproteins from the extracellular matrix and endothelial integrin receptors. The basal lamina is bordered by the plasma membranes of astrocyte end-feet. As mentioned before, some neurons directly innervate brain vessels; hence, some neuronal terminals are also part of the NVU. These neurons regulate some BBB functions.

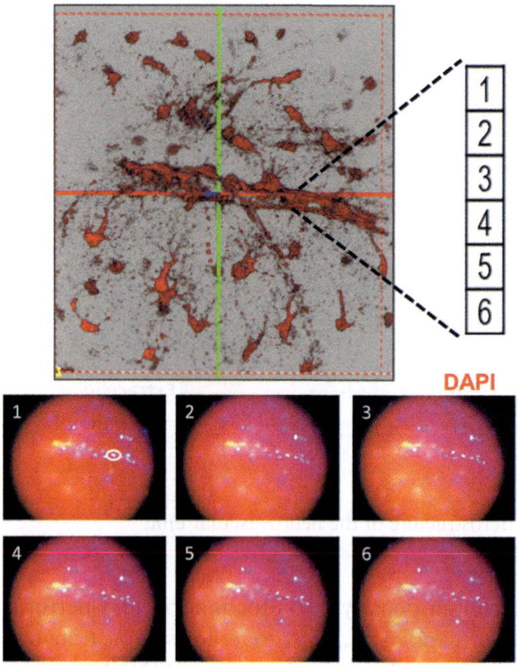

FIGURE 10.3 Preliminary data demonstrating the existence of perivascular astrocytes. Astrocytes are stained with SR101 and DAPI was used to label cellular nuclei. The astrocyte nucleus is indicated with a circle in slice #1.

10.4 ENDOTHELIAL CELLS – IONIC COMPONENTS

ECs transduce signals in both directions across the membranes in contact with the extracellular matrix (luminal) and the vascular lumen (abluminal). To that end, these cells use a large variety of ionic channels and sensors, which play a role in three basic processes: **a**) EC migration/proliferation and angiogenesis, **b**) intercellular permeability control, and **c**) blood flow regulation, including the vascular tone. Via endocytosis, transcytosis, exocytosis, and biosynthetic secretory, ECs also regulate macromolecule traffics (e.g., von Willebrand factor, tissue plasminogen activator, and tissue factor pathway inhibitor). Let us illustrate with few examples the role of the intracellular Ca^{2+} concentration $[Ca^{2+}]_i$ in most of these EC functions. First, changes in $[Ca^{2+}]_i$ have a direct impact on the regulation of gene expression in ECs by means of Ca^{2+}-dependent activation of certain protein kinases (e.g., c-Jun N-terminal kinase "JNK", extracellular signal-regulated kinase "ERK", focal adhesion kinases "FAK", c-Src). Second, Ca^{2+}-dependent c-Src activation, which can be induced by shear stress, plays a key role in the tyrosine-phosphorylation of the p130 Crk-associated substrate (Cas), a step crucial for actin filament assembly. This constitutes a basic mechanosensing mechanism underlying vascular morphogenesis. Third, the regulation of hydraulic permeability and EC contraction depend on Ca^{2+} influx, through Ca^{2+}/CaM-dependent action on a non-muscle myosin light-chain kinase (MLCK) isoform present in ECs. This mechanism is also key in trans-endothelial leukocyte migration. Fourth, 5,6-epoxyeicosatrienoic acids (5,6-EET), an endothelium-derived hyperpolarizing factor (EDHF) produced by cytochrome P-450 from AA, is also synthesized/released in a Ca^{2+}-dependent way. Lastly, EC membrane potential can change with variations in the cytosolic Ca^{2+} concentration $[Ca^{2+}]_i$ through Ca^{2+}-dependent K^+-channels and Ca^{2+}-dependent Cl^- channels (CaCC) both causing membrane hyperpolarization. However, the fact that Ca^{2+} signal in ECs promotes the production of vasoactive factors, e.g., NO and prostacyclin (PGI_2), is one of the most relevant examples for the topics discussed henceforth in connection with brain imaging. Although PGI_2 is a vasodilator, it has been mostly studied for its strong inhibition of platelet aggregation by rising cAMP levels. NO signal from ECs is a key component in the neurovascular coupling and in vascular tone regulation. In response to shear stress, endothelial NO synthase (eNOS, Type III) is upregulated, a process initiated by Ca^{2+}/CaM-dependent phosphorylation. Henceforward, we will discuss important channels regulating directly and indirectly cytosolic Ca^{2+} concentration in ECs. Channels in ECs can be classified into three major categories (Nilius and Droogmans, 2001): **1)** Ca^{2+}-entry channels, **2)** voltage-dependent channels, and **3)** mechano-sensitive channels. We will describe the most important channels in these categories and provide some of the biophysical models proposed in the literature.

10.4.1 The Ca^{2+}-Entry Channels

The concept of a Ca^{2+}-entry into ECs has been mostly associated with the existence of multiple types of transient receptor potential (TRP) channels. A TRP channel is a six transmembrane homo- or hetero-tetrameric ion channel that allows cations to flux across the cell membrane in response to internal and external stimuli. TRP can be divided into 6 mammalian subfamilies: TRPA (Ankyrin), TRPC (Canonical), TRPM (Melastatin), TRPML (Mucolipin), TRPP (Polycystin), and TRPV (Vanilloid). Each subfamily has multiple subtypes indicated by numbers. With only two exceptions (TRPM4&5), these channels are permeable to Ca^{2+} ions. In ECs, TRPM4 causes a decrease in NO in response to stretch, or intensification of PIP_2 and NO production. Participation of TRPM5 in EC functions has not been yet reported. Hence, most TRP channels in ECs are involved in the regulation of two important variables, i.e., $[Ca^{2+}]_i$ and V_m. Activation and/or modulation of several TRP isoforms in ECs take place in response to signaling molecules (e.g., 5,6-EET, diacylglycerol "DAG", ATP, bradykinin, and histamine). TRPV4 channels have been identified in ECs and associated with the genesis and propagation of electrotonic potentials from capillaries to arterioles. A specific biophysical model for the TRPV4 is discussed later in this chapter. Depending on the activation mode, TRP channels in ECs can be also classified into two major subtypes: **a)** nonselective cation channels (NSC) with permeability to Ca^{2+} ions, and **b)** store-operated Ca^{2+} (SOC) channels (or capacitative Ca^{2+} entry mechanism, CCE).

- There are two main TRP-types of Ca^{2+}-permeable NSCs in ECs (i.e., redox and receptor-activated). ECs produce different types of reactive oxygen species (ROS), such as superoxide anions, H_2O_2, peroxynitrite, and hydroxyl radicals. Redox NSCs are activated during oxidative stress by the effect of these ROSs, mostly in response to superoxide anions. In ECs, two redox TRP channels (TRPC3&4) have been identified. Apparently, TRPC3&4 form redox-sensitive heteromeric channels that open to cations in response to oxidative stress. TRPM2&7 have been reported in ECs, which are perhaps associated with EC response to damages caused by ROSs. There are multiple families of receptor-activated cation channels (RACC). This subtype of NSC is activated/modulated by the action of certain agonists on EC membrane receptors, involving phospholipase C (PLC) signaling cascades. A subfamily of RACC can be activated by the action of vasoactive agonists. Activation of another RACC family depends also on the intracellular Ca^{2+} concentration. RACCs activated by either ATP (TRPC5), ACh (TRPC4), DAG (TRPC3,6&7), or bradykinin (TRPC6) have also been reported in ECs. An amiloride-sensitive NSCs have been also found in the brain microvasculature. Amiloride produced a rapid flickering of NSC activity at first, which is then followed by its complete blockade. The TRPP3 subtype, expressed in retina, brain, heart, testis, and kidney, has been proposed as a good candidate for this amiloride-sensitive NSCs, but functions are yet to be clarified. Theoretical characterizations of these Ca^{2+}-permeable NSCs are very scarce in the literature. However, a phenomenological model has been proposed to describe Ca^{2+} flux (j_{NSC}, $mM\,s^{-1}$) through Ca^{2+}-permeable NSCs, which directly depend on the EC membrane potential V_m and intracellular Ca^{2+} levels (Koenigsberger et al., 2006):

$$j_{NSC} = G_{NSC}\frac{\left(E_{Ca} - V_m\right)}{2}\left(1 + tanh\left(\frac{log_{10}\left[Ca^{2+}\right] - m_3}{m_4}\right)\right) \tag{10.1}$$

$G_{NSC} = 0.66\ nM\ mV^{-1}s^{-1}$ is the whole-cell cation channel conductivity and $E_{Ca} = 50\ mV$ is the Ca^{2+} equilibrium potential in ECs. This model includes two parameters $m_3 = -0.18\ \mu M$ and $m_4 = 0.37\ \mu M$.

- SOCs cause a Ca^{2+} influx that is mostly dependent on the filling degree of Ca^{2+} intracellular stores, and perhaps on other mechanisms for sequestration/buffering inside the EC cytosol. In ECs, the CCE effect is less affected by mitochondria buffering and it is modulated by the activity of the plasma membrane Ca^{2+} pump. One of the most relevant mechanisms underlying CCE is the Ca^{2+}-release-activated Ca^{2+} channel (CRAC). In ECs, CRACs have high single channel conductance. In Chapter 8, we discussed two mechanisms underlying CRAC pore-formation via conformational coupling and presented two theoretical models for CRAC current j_{CCE} in astrocytes (i.e., exhaustive and phenomenological). In a more recent study (Sriram et al., 2016), a cGMP-dependent mechanism was proposed for ECs: $j_{CCE} = \theta_{CCE}\psi\left([cGMP]\right)\left([Ca^{2+}]_{ER,0} - [Ca^{2+}]_{ER}\right)\left([Ca^{2+}]_o - [Ca^{2+}]_i\right)$. In this model, the dependency with cGMP was characterized by $\psi([cGMP]) = 1 - \xi[cGMP]$, with $\xi = 0.0075\mu M^{-1}$. The CCE rate was defined as $\theta_{CCE} = 8\times10^{-7}\mu M^{-1}s^{-1}$. The basal Ca^{2+} in the ER was $[Ca^{2+}]_{ER,0} = 2828\mu M$ and the extracellular Ca^{2+} concentration was $[Ca^{2+}]_o = 1500\mu M$. CRAC currents have been linked to the existence and interplay of three important components: TRPC1, Orai1, and STIM1. In EC, the Ca^{2+}-dependent modulation of store-related Ca^{2+} entry seems to be mediated by *"short"* TRPC1 proteins containing several calmodulin-binding sites. The recruitment of TRPC1 into the plasma membrane is triggered by Orai1-mediated Ca^{2+} entry. In this stage, TRPC1 is activated by stim1. The orai1/stim1 complex lies behind other mechanisms for conformational CRAC pore-formation in ECs. The orai1/stim1 complex has been proposed to help endothelial tube formation and angiogenesis via the activation of vascular endothelial growth factor (VEGF) receptors (VEGFRs). Apparently, orai1 and VEGFR are partially co-localized. It has been also suggested that vessel sprouting is driven by tissue hypoxic stages via VEGFR activation. Hence, Ca^{2+} influx through orai1 due to the depletion of intracellular Ca^{2+} store is evoked by VEGF. ECs respond to different types of tissue dysfunctions, e.g., coagulation and vascular inflammatory reactions, and fibribolysis. The VEGF-evoked capacitative Ca^{2+} entry effect may be also involved in the vascular response to these dysfunctions, and hence, requires further investigation. Together with the IP_3 production that leads to SOC, PLC also produces DAG, which could activate some types of TRPC (3,6&7) channels in ECs causing additional Ca^{2+} influx through RACCs. A GPCR-dependent mechanism for production of IP_3, linked to deformation of EC membrane due to shear stress, has also been proposed in the past (Sriram et al., 2016). In this case, the rate of the G-protein activation $[G]$ follows: $d[G]/dt = k_a[R^*]([G_t]-[G]) - k_d[G]$, with $k_a = 0.017s^{-1}$ and $k_d = 0.15s^{-1}$ being the respective forward and backward rate constants for G-protein activation. $[G_t] = 0.33\mu M$ represents the total concentration of G-proteins. The fraction of G-protein activated by shear stress τ $(dynes/cm^2)$ is defined as $[R^*] = tanh(\pi\tau/\Lambda)$, with $\Lambda = 15 dynes/cm^2$. Formation of IP_3 from the hydrolysis of PIP_2 (phosphatidylinositol 4,5-bisphosphate) is caused by activation of the G-protein in accordance with the rate equation: $d[IP_3]/dt = r_f[PIP_2] - \mu_1[IP_3]$, with $r_f = \alpha K_{cp} M_{ATP}[G]$. Parameters in this equation are defined as $\mu_1 = 0.2s^{-1}$, $\alpha = 8.37s^{-1}$, $K_{cp} = 0.002\mu M^{-1}$, $M_{ATP} = 0.7937$. The cycling of PIP_2 to IP_3 and back to PIP_2 is characterized by a system of two equations: $d[PIP_2]/dt = -(r_f + r_r)[PIP_2] - r_r[IP_3] + r_r[PIP_{2_t}] - K^+_{PIP_2}[PIP_2] + K^-_{PIP_2}[PIP_3]$ and $d[PIP_3]/dt = K^+_{PIP_2}[PIP_2] - K^-_{PIP_2}[PIP_3]$, with parameters defined as $[PIP_{2_t}] = 10\mu M$, $r_r = 10s^{-1}$, and $K^-_{PIP_2} = 0.7024s^{-1}$. In contrast, parameter $K^+_{PIP_2}$ is not constant. It depends on the activation level of the phosphatidylinositide 3-kinases (PI3K) enzymes. PI3K activation is linked to mechanical forces on the integrins that anchor ECs to the extracellular matrix. These mechanical forces caused the tyrosine-phosphorylation of focal adhesion kinases (FAKs), resulting in the stimulation of PI3K. Once activated, PI3K returns to its basal level as an exponential function of time, $exp(-\eta t)$ with decaying factor $\eta = 0.003s^{-1}$. Experimental data have been used to obtain a parametric model for $K^+_{PIP_2} = k_{1p}(1 + \alpha_{KI3K} exp(-\eta t) tanh(\pi\tau/\delta))/(1 + \alpha_{KI3K}) + k_{2p}$. Parameters were fitted to experimental data from several studies: $\alpha_{KI3K} = 2.5$, $\delta = 24\,dynes / cm^2$, $k_{1p} = 0.021\,s^{-1}$, and $k_{2p} = 0.022\,s^{-1}$.

There are two other major mechanisms for Ca^{2+} entry in ECs: **1)** the cyclic nucleotide-activated NSC channels and **2)** the Na-Ca exchanger (NCX). The cyclic nucleotide-activated NSC channels comprise two families: cyclic nucleotide-gated channels (CNG) and hyperpolarization-activated cyclic nucleotide–modulated (HCN) channels. Both families are expressed in ECs and are activated by the direct binding of cAMP and/or cGMP on the C-terminal region of the enzyme. These channels are composed of four identical subunits. Each subunit has six transmembrane segments (S1–S6). The pore region is formed by the segments S6. The C-terminal contains the cyclic nucleotide-binding domains and the C-linkers. The C-linkers connect the binding domains to the pore. It has been suggested that some CNG and HCN channels are more permeable to monovalent than to divalent cations. Therefore, although Ca^{2+} ions might flow through some of these channel subtypes in ECs (e.g., CNG1&2), their impact on EC function is caused by a substantial influx of Na^+ ions. Increases of intracellular Na^+ concentration in ECs through these channel subtypes lead to a subsequent Ca^{2+} influx via the activation of the NCX. CNG channels might play a crucial role in NO and cGMP cross-talking in ECs. Ca^{2+}-dependent NO production causes cGMP increases, hence, the activation of multiple subtypes of CNG channels. Activation of these CNG channels leads to both direct (minor) and indirect (major) Ca^{2+} influxes in ECs, in what constitutes a positive closed-loop mechanism. The role of redox- and CNG- NSC channels in the regulation of the vascular tone and regional blood flow is still elusive. Activation of these two NSC subtypes in ECs might lead also to complex negative feedback impacting vascular resistance. For example, redox NSC activation by superoxide anions will stimulate the production of NO via Ca^{2+}/CaM-dependent activation of the eNOS. However, superoxide anions are strong NO scavengers, and they also inhibit eNOS expression. Increases in NO production will cause cGMP signaling that can inhibit several TRP isoforms in ECs, such as TRPM4 and TRPC3&6. Finally, increases in Ca^{2+} activity in ECs could cause SMC contractions via ion-diffusion and electrotonic effect through myoendothelial gap junctions. Interestingly, CNG channels seem to be inhibited by DAG. HCN channels are mainly modulated by cAMP, and they are involved in cellular pacemaker functions. The CNG and HCN channels also comprise a voltage sensor (S4). The channels open more under hyperpolarization.

Three mathematical models have been proposed to describe CNG and HCN channel kinetics: the sequential model, the coupled dimer model, and the cyclic allosteric model. The latter constitutes the most widely used in literature; hence, it will be discussed in detail. In 1965, Jacques Monod, Jeffries Wyman and Jean-Pierre Changeux (MWC) proposed a comprehensive mathematical model to describe allosteric enzymes (Monod et al., 1965). Here, we will use the MWC model to describe the dynamics of CNG and HCN channels. Because of their tetrametric structure, CNG and HCN channels have four binding sites for cyclic nucleotides. These channels are oligomers made of 4 protomers (identical subunits) occupying equal positions

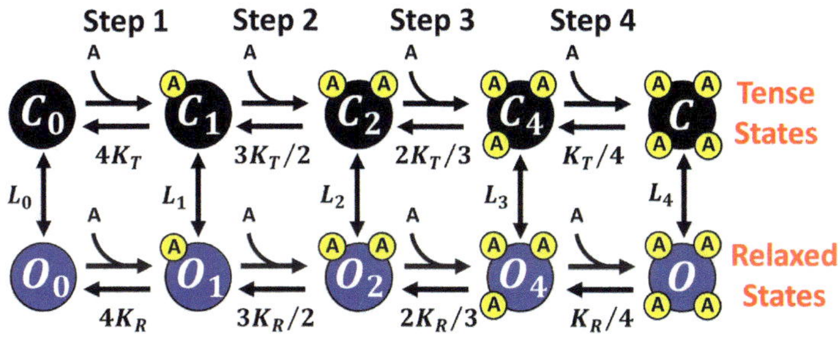

FIGURE 10.4 The cyclic allosteric model.

along a symmetry axis; hence, they are considered allosteric systems (Figure 10.4). Depending on their potential energy, these channels can accept two conformation states: tense (T) and relaxed (R). The channel opens after the sequential binding of four ligand A molecules (i.e., cGMP, cAMP) to the enzyme in any of its conformation states. The ligand forms a stereospecific complex with the enzyme. In the absence of a ligand A, the transition between the T and R states occurs spontaneously with an equilibrium constant $L_0 = [C_0]/[O_0]$. In this condition, the equilibrium is in favor of the low affinity state T ($L_0 > 1$). At any time, all four protomers in the enzyme must be in either T or R state. The molecular symmetry is conserved during inter-state transitions. According to the general binding theory for n binding sites, microscopic (intrinsic) binding constants (K_s) for individual sites of a macromolecule in state $s = \{T, R\}$ at the i-th biding step can be represented by macroscopic dissociation constants (K_s') (Eq. 10.2).

$$K_s' = K_s \frac{i}{n-i+1} \tag{10.2}$$

The constant K_s represents the microscopic dissociation constant of ligand A bound to each of the four stereospecific sites in the conformation state s. For these allosteric systems, the macromolecule has higher affinity for the ligand in the R state than in the T state, i.e., $f = K_R / K_T < 1$. Spontaneous transitions between the T and R states at the i-th biding step occur with an equilibrium constant $L_i = f^i L_0$, with $i = \{1,2,3,4\}$.

According to the MWC model, the fraction $F_R([A])$ of the enzyme in an R state for a ligand concentration $[A]$ is defined by the following equation:

$$F_R([A]) = \frac{\left(1+[A]/K_R\right)^4}{\left(1+[A]/K_R\right)^4 + L_0\left(1+[A]/K_T\right)^4} = \frac{1}{1+L_0\left(\dfrac{1+[A]/K_T}{1+[A]/K_R}\right)^4} \tag{10.3}$$

A channel opens when all subunits are in the "relaxed" state which will be characterized by variable P. Under this formalism, the free-energy variations associated with spontaneous T-R transitions is $\Delta G = -RTln(1/L_0)$, with R the gas constant and T is the absolute temperature. Equation (10.3) does not depend on the membrane potential V_m. This dependency is stronger in HCN channels, a mechanism proposed to explain pacemaker properties of different heart/brain cells with f/h channels. Therefore, the channel open probability must be a function of V_m and $[A]$, i.e., $P(V_m,[A])$. There exist two proposals to include a dependency with V_m under the MWC formalism by: **1)** using Hodgkin & Huxley (HH) scheme and **2)** introducing S4-based allosteric structural modifications (S4A).

MWC-HH formalism: In this case, an o-order HH kinetics is proposed to explain the electric current through cyclic nucleotide-activated NSC channels in response to applications of ligand A: $I_c = g^c_{max} P(V_m,[A])^o (V_m - V_c)$. V_c and g^c_{max} are the reversal potential and maximal conductance for the channel. Under this formalism, the activation curve of the channel just shifts after the ligand binds to a COOH-terminal CNBD. Hence, the probability for the gate of the channel to be open satisfies at the equilibrium a Boltzmann relation (Eq. 10.4) with an additional shift in the half-activation voltage $s([A])$ that depends on the concentration of the ligand A.

$$P(V_m,[A]) = 1/\left(1+exp\left(\left(V_m - V_{1/2} - s([A])\right)/v\right)\right) \tag{10.4}$$

Wang et al. (2001) used cell-free inside-out patches on Xenopus Oocytes with mutant HCN1 and HCN2 channels, which are the most abundant in the brain, to evaluate the response of these channels to cAMP applications. For these two HCN isoforms, the authors used a 1-order HH model ($o = 1$). According to the study, cAMP can cause a shift of up to $4\,mV$ and $17\,mV$ toward more positive values for HCN1 and HCN2, respectively. The Hill equation was used to fit the cAMP dose–response data: $\Delta V_{1/2} = V_{1/2} + s([A]) = \Delta V^{max}_{1/2} / (1 + (K_{1/2} / [cAMP])^h))$. The parameter $\Delta V^{max}_{1/2}$ indicates the shift produced by a saturating cAMP. $K_{1/2}$ and h are the cAMP concentration producing half of the maximal effect and the Hill coefficient, respectively. The fitted model parameters for HCN1 and HCN2 are:

HCN1: $V_{1/2} = -115.3\ mV$, $v = 6.2\ mV$, $\Delta V^{max}_{1/2} = 4.1\ mV$, $K_{1/2} = 0.06\ \mu M$, and $h = 1.0$.

HCN2: $V_{1/2} = -135.7\ mV$, $v = 5.1\ mV$, $\Delta V^{max}_{1/2} = 17.4\ mV$, $K_{1/2} = 0.10\ \mu M$, and $h = 1.1$.

Note that a theoretical model of $P(V_m,0)$ for the HC1 isoform (h-type, absence of a ligand) can be found in Chapter 6 (Table 6.1).

DiFrancesco (1999) proposed a 2-order HH model ($o = 2$) and the following equation for $s([A]) = vnln\left(\frac{1+[A]/K_R}{1+[A]/K_T}\right)$ to explain HCN4 kinetics. Parameters in this model ($V_{1/2} = -113.9\,mV$, $v = 7.103 \pm 0.231\,mV$, $K_R = 0.0732\,\mu M$, and $K_T = 0.4192\,\mu M$) were fitted to experimental data obtained using macropatches on SA node (pacemaker) cells from New Zealand white rabbits. The authors measured the shifts after perfusing cAMP on the intracellular side macropatches. Note that it was assumed that the f-type channels (HCN4) in these cells have a single binding site for cAMP (i.e., $n = 1$) despite their tetrametric structure. Although f-type channels are primary expressed in the heart, h-type channels are found in several types of excitatory neurons in the brain.

S4-based allosteric structural modifications: In this case, channel activation/deactivation kinetics cannot be explained with a HH formalism (Altomare et al., 2001). Instead, the allosteric MWC model was generalized to include a voltage sensor in each of the four identical channel subunits. The voltage sensor can have two configurations: reluctant (hidden) and willing (protruding). The model assumes that the K_T rate constant depends on the membrane potential, which is described by function $K_T = K(V_m)$. The model presumes that the voltage dependency favors left-to-right and top-to-bottom transitions (Figure 10.4) upon hyperpolarization, with the following relation holding $K_R = aK(V_m)$. Based on a previously developed theoretical construct (DiFrancesco, 1984), Altomare et al. (2001) proposed that the probability for the channel to open satisfies the following generalized MWC model:

$$P\left(V_m,[A]\right) = \frac{1}{1+L\left(V_m\right)\left(\dfrac{1+[A]/K\left(V_m\right)}{1+[A]/aK\left(V_m\right)}\right)^4} \tag{10.5}$$

With rate constants $L(V_m) = exp((2z_\beta)F(V_m - V_L)/RT)$ and $K(V_m) = exp((2z_\delta)F(V_m - V_K)/RT)$ depending explicitly on the membrane potential V_m in a non-HH way. Model parameters were fitted to electrophysiological data (whole-cell configuration) obtained from modified HEK 293 cells previously transfected with either mHCN1 (mouse), hHCN2 (human) or hHCN4 (human). Results from the fitting are shown in Table 10.1.

Reductions of Na^+ transmembrane gradients have been associated with NCX-mediated Ca^{2+} entry in ECs. Hence, the interplay between Na^+-selective CNG/HCN channels and NCX constitutes one important Ca^{2+}-entry mechanism in ECs. A NCX isoform has been reported in caveolin-rich EC membranes and associated with Ca^{2+} -dependent eNOS activation (Teubl et al., 1999). A theoretical model for the NCX can be found in Chapter 6 (Eqs. 6.21 and 6.22).

It has also been reported Ca^{2+} influx into the EC cytosol through mechanosensing ion channels (MSICs), which is assumed linearly proportional to the fraction of ion channels open at a given wall shear stress τ (Sriram et al., 2016): $j_{MSIC} = q_{max}/(1+Nexp(-W(\tau)))$, where $q_{max} = 17.6\,\mu Ms^{-1}$ is the maximum Ca^{2+}

TABLE 10.1
Model parameters for the generalized MWC model with a=0.2

Subtype	z_β	z_δ	V_L(mV)	V_K(mV)
HCN1	0.9369	1.018	−150.04	−44.19
HCN2	1.066	1.021	−145.24	−96.57
HCN4	0.8042	0.8934	−170.64	−62.15

influx rate due to shear stress, $N = 2$ is the no-load channel constant, and the function $W(\tau) = W_0(\tau + \sqrt{16\chi^2 + \tau^2} - 4\chi)^2 / (\tau + \sqrt{16\chi^2 + \tau^2})$. Further model parameters are defined as $W_0 = 1.4874$ and $\chi = 28.6\, dynes\, cm^{-1}$. Mechanistic basis for the shear stress τ are discussed later in this chapter.

10.4.2 Channels Regulating the Membrane Potential in ECs

Changes in the EC membrane potential (V_m) are small and slow; hence, EC are considered in general non-excitable cells. However, slight variations in EC membrane potential could lie behind electrotonic signal propagation along ECs and toward SMCs via gap junctions. Therefore, it is important to list all channels regulating V_m in ECs. The presence of classical Na^+ and Ca^{2+} voltage-gated channels has been recognized in the past in both large vessel ECs and microvasculature ECs: **1)** Na^+ channels with sensitivity to tetrodotoxin (low) and saxitoxin (high); and **2)** L-, T- and R- types of Ca^{2+} channels. The R-type Ca^{2+} channel in ECs seems to be activated by platelet activating factor (PAF). Kinetic models for some of these channel isoforms can be found in Chapters 6 and 7. K^+ channels are the most important in setting the membrane potential in ECs. There are multiple types of K^+ channels in ECs with different functions: **1)** delayed-rectifier K^+ channels (Kv1.5), **2)** ATP-sensitive K^+ channels, and **3)** 4-aminopyridine-sensitive, voltage-dependent, transient outward channels (A-type K^+). However, the two most relevant K^+ channels in the context of brain imaging are: **a)** Ca^{2+}-activated K^+ channels and **b)** inwardly rectifying K^+ channels (K_{IR}).

Ca^{2+} influx causes EC depolarization and increase in the cytosolic Ca^{2+} concentration $[Ca^{2+}]_i$. This depolarization is balanced by the activation of Ca^{2+}-activated K^+ channels, leading to EC membrane hyperpolarization. There are three subtypes of Ca^{2+}-activated K^+ channels in brain ECs, i.e., large-conductance (BK, $165-240\, pS$), intermediate-conductance (IK, $15\, pS$) and small-conductance (SK, $\leq 10\, pS$) channels. Let us define I_K as the total charge current (pA) via large (BK) and small (SK) conductance K^+ channels. A theoretical model (Eqs. 10.6 and 10.7) to represent total ionic currents through BK and SK channels was proposed by Dormanns et al. (2015), with parameters defined in Table 10.2.

$$I_K = G_K^{tot}\left(V_m - E_K\right)\left(n_{BK} + n_{SK}\right)$$

$$n_{BK} = 0.2\left(1 + tanh\left(\frac{\left(\log_{10}\left([Ca^{2+}]_i\right) - c\right)\left(V_m - b\right) - a_1}{m_{3b}\left(V_m + a_2\left(\log_{10}\left([Ca^{2+}]_i\right) - c\right) - b\right)^2 + m_{4b}}\right)\right) \quad (10.6)$$

$$n_{SK} = 0.3\left(1 + tanh\left(\frac{\log_{10}\left([Ca^{2+}]_i\right) - m_{3s}}{m_{4s}}\right)\right) \quad (10.7)$$

TABLE 10.2
Model parameters for the BK and SK channels in ECs

Parameter	Value	Description
G_K^{tot}	$6.927\,nS$	Total potassium channel conductivity
E_K	$-80\,mV$	K^+ equilibrium potential
c	$-0.4\,\mu M$	constants
b	$-80\,mV$	
a_1	$53.3\,\mu M\,mV$	
a_2	$53.3\,mV\,\mu M^{-1}$	
m_{3b}	$1.32\times 10^{-3}\,\mu M\,mV^{-1}$	
m_{4b}	$0.3\,\mu M\,mV$	
m_{3s}	$-0.28\,\mu M$	
m_{4s}	$0.389\,\mu M$	

Note that BK channels depend on the membrane potential V_m (blocked by $[Mg^{2+}]_o$) and the cytosolic Ca^{2+} concentration $[Ca^{2+}]_i$; however, SK channels only depend on $[Ca^{2+}]_i$. Functions n_{BK} and n_{SK} are dimensionless.

The BK channels in ECs are blocked by charybdotoxin ($50\,nM$), and TEA ($1\,mM$). Other blockers have also been identified (iberiotoxin, D-tubocurarine, and quinine). The SK channels in ECs are blocked by apamin ($10\,nM$) and D-tubocurarine ($100\,\mu m$). IK channels in ECs are activated by IP_3-sensitive Ca^{2+} release from the action of ATP, ACh, and bradykinin.

The inwardly rectifying K^+ channels (IRK or K_{IR}) have been suggested as a crucial sensor and amplifier of neuronal signals by the brain microvasculature. At typical extracellular K^+ concentration (~6 mM), K_{IR} channels conduct inwardly at V_m below the K^+ equilibrium (E_K). Small increases in extracellular K^+ shift the reversal potential E_K toward a more positive value and increase the inward conductance of K_{IR} for $V_m < E_K$ (Fig. 8 in Nilius and Droogmans, 2001). Increases in extracellular K^+ concentration might be caused by neuronal-glial-vascular interactions via large-conductance BK channels abundant in astrocytic perivascular end-feet (i.e., K^+ siphoning, Chapter 9). Moshkforoush et al. (2020) proposed that vasodilatory signals originate in ECs at the level of capillaries and ascend in the form of a hyperpolarization wave ($0.8-2.4\,mm\,/\,s$) to dilate penetrating arterioles and surface pial arteries, a phenomenon mediated by K_{IR}. In addition to the K_{IR} ($K_{IR}2.1$) current in ECs, they incorporated a background current summarizing all ionic currents in EC described above as well as leaking currents throughout gap junctions between ECs and with SMCs. K_{IR} channel gating probability depends on both EC membrane potential V_m and extracellular K^+ concentration $[K^+]_o$, as represented by the following equation:

$$I_{K_{IR}} = \frac{G_{K_{IR}}\sqrt{[K^+]_o}\left(V_m - E_K\right)}{1+exp\left(\left(V_m - V_{0.5}\right)/k\right)} \tag{10.8}$$

The maximum conductance of the channels at $[K^+]_o = 1\,mM$ is defined as $G_{K_{IR}} = 0.18\,nS\,mM^{-1/2}$. Parameters $V_{0.5} = E_K + 25\,mV$ (half-maximal inactivation) and $k = 7\,mV$ (steepness factor) describe a sigmoidal inhibition of K_{IR} with V_m. According to this model, K_{IR} current underlies a sustained hyperpolarization in response to increases in $[K^+]_o$. Moshkforoush et al. (2020) proposed that the transient hyperpolarizing V_m signaling is also mediated by the TRPV4 channel. The TRPV4 channels was also required for the returning of ECs to equilibrium after the $[K^+]_o$ increases are terminated. A stochastic TRPV4 model with three states (blocked – B, open – O, and shut – S) was proposed $\left(B \underset{k_1}{\overset{k_2}{\rightleftarrows}} O \underset{k_4}{\overset{k_3}{\rightleftarrows}} S \right)$ with transition rates $\{k_1 \cdots, k_4\}$. Experimental values of the mean open time $(k_3 + k_1)^{-1} = 90\,ms$, the mean blocked time $k_2^{-1} = 33\,ms$, the mean shunt time $k_4^{-1} = 4\,min$, and the mean bursting time $(1 + k_1 / k_2) / k_3 = 20\,s$ were used to estimate the transition rates in this model. Simulations of TRPV4 gating probability were performed using a three-state discrete time Markov chain model. The state transition probability matrix used in the Markov chain model to generate temporal traces of TRPV4 openings $N(t)$ in each EC was defined as $T = exp\left(\begin{bmatrix} -(k_3 + k_1) & k_1 & k_3 \\ k_2 & -k_2 & 0 \\ k_4 & 0 & -k_4 \end{bmatrix} \right)$.

The total current through the TRPV4 channels was defined as:

$$I_{TRPV4}(V_m, t) = N(t) \sum_{k=\{Ca,Na,K\}} I_{TRPV4}(V_m, [S_k]_i) \tag{10.9}$$

With contribution from each ionic species calculated by the Goldman–Hodgkin–Katz equation (Eq. 10.10). The ionic permeabilities for each specie satisfy the following ratios $P_{TRPV4,Ca} : P_{TRPV4,K} : P_{TRPV4,Na} = 7.1 : 1.4 : 1.0$, with $P_{TRPV4,Ca} = 1 \times 10^{-7}\,cm\,s^{-1}$.

$$I_{TRPV4}\left(V_m, \left[S_k\right]_i\right) = P_{TRPV4,S_k} \frac{z_{S_k}^2 F^2}{RT} V_m \frac{\left[S_k\right]_i - \left[S_k\right]_o exp\left(-z_{S_k} V_m F / RT\right)}{1 - exp\left(-z_{S_k} V_m F / RT\right)} \tag{10.10}$$

It was additionally proposed that the depletion of PIP_2 by the hydrolytic action of PGE_2 and Gq-protein-coupled receptors (Gq-PCR) inhibits K_{IR} and increases TRPV4 activity.

There are several other ionic channels reported in ECs, which are listed below. However, these channels are not discussed in detail because their relevance to brain imaging is still unclear. There is an amiloride-sensitive Na^+ channel in ECs that is involved in fluid transport. Different types of Cl^- channels are also expressed in ECs. Among them, the volume-regulated anion channels (VRAC) play a crucial role in housekeeping and cell-swelling/shrinkage. The channel is partially activated at rest, so it contributes to the overall resting membrane potential in ECs. VRAC is activated by shear stress. ECs express also Ca^{2+}-dependent Cl^- channels (CaCC), high-conductance Cl^- channels (BCl) and cAMP-stimulated cystic fibrosis transmembrane conductance regulator (CFTR) channels. Previous biophysical models of EC membrane potentials (e.g., Koenigsberger et al., 2006; Moshkforoush et al., 2020) have grouped all non-relevant ionic channels (Na^+, Ca^{2+}, K^+ and Cl^-) in a background current $I_B = G_B(V_m - E_B)$, with $G_B = 0.06\,nS$ and $E_B = -30\,mV$. However, it has been suggested that the conductance of some of these channels can increase significantly in response to shear stress, which has been phenomenologically modeled (Eq. 10.12). Electrotonic potential is transmitted along ECs via TJs. EC-EC electrical coupling can be modeled as $v^{EC}_{coupling} = -G^{EC}_{coupling}\sum_l (V_m - V_m^{EC(l)})$, with the homo-cellular electric coupling coefficient $G^{EC}_{coupling} = 1000\,s^{-1}$ and $V_m^{EC(l)}$ representing the membrane potential of the l-th neighbor EC. There is also an electrotonic coupling between ECs and SMCs, which has been modeled previously as $v^{EC-SMC}_{coupling} = -G^{EC-SMC}_{coupling}\sum_l (V_m - V_m^{SMC(l)})$, with an hetero-cellular electrical coupling coefficient $G^{EC-SMC}_{coupling} = 50\,s^{-1}$ and $V_m^{SMC(l)}$ representing the membrane potential of the l-th neighbor SMC. Given the factor $\alpha = 10^3 A_{SMC}$ (A_{SMC} is the SMC surface area in cm^2), EC membrane potential is described by the following equation:

$$\frac{dV_m}{dt} = \frac{1}{C_m}\left(I_K + I_{K_{IR}} + \alpha I_{TRPV4} + I_B\right) + v^{EC}_{coupling} + v^{EC-SMC}_{coupling} \tag{10.11}$$

10.4.3 The Mechano-Sensitive Channels

Blood vessels can sense two types of mechanical forces on their internal wall: **i)** perpendicular forces caused by biaxial tensile stress (e.g., stretch and pressure) and **ii)** tangential forces caused by blood flow (i.e., shear stress). These forces are transduced into intracellular signaling by means of a variety of stretch-activated channels (SACs) in ECs and SMCs. The open probability of SACs is represented by the Boltzmann equation.

$$j_{stretchy} = \frac{G_{stretch}\left(V_m - E_{SAC}\right)}{\left(1 + exp\left(-\alpha_{stretch}\left(\sigma_{\theta\theta} - \sigma_0\right)\right)\right)} = \frac{G_{stretch}\left(V_m - E_{SAC}\right)}{\left(1 + exp\left(-\alpha_{stretch}\left(R\Delta p / h - \sigma_0\right)\right)\right)} \tag{10.12}$$

In this model, $\sigma_{\theta 6}$ is the average circumferential stress in the artery/arteriole, or also known as the cellular membrane tensile stress. A relation between changes in arterial radius R and the circumferential stress $\sigma_{\theta\theta}$ can be obtained using a generalized form of the Laplace's law, named the Lamé and Clapeyron's law $\sigma_{\theta\theta} = R\Delta p / h$. In this case, it is assumed that changes in $\sigma_{\theta 6}$ produce negligible acceleration of the arterial wall. Δp and h are the transmural pressure and the arterial wall thickness. Experimental observations suggest that $\alpha_{stretch} = 0.0074\,mmHg^{-1}$ and $\sigma_0 = 500\,mmHg$. The reversal potential and whole-cell conductance for SACs are $E_{SAC} = -18\,mV$ and $G_{stretch} = 0.0061\,\mu M\,mV^{-1}s^{-1}$, respectively.

Henceforth, we use a theoretical model for viscoelasticity to describe changes in the vessel radius, which result from multiple forces causing dynamic changes in the circumferential stress $\sigma_{\theta 6}$. First, the cytoskeleton of ECs is formed by a network of cross-linked F-actin microfilaments permeating the cytoplasm (density – $\rho_s = 732\,mg\,/\,ml$). Together with actin-binding proteins (e.g., myosin), microfilaments form bundles of ~350 nm in diameter, referred to as stress fibers. These reversible elastic deformations can be characterized by the Young's module E associated with mechanical deformation of the network, a phenomenon well-characterized by Hookean elastic **spring**, i.e., $\sigma^s_{\theta\theta} = E\varepsilon^s_{\theta 6}$. The strain on the spring associated to the vessel wall satisfies the following equation: $\varepsilon^s_{\theta\theta} = \Delta\omega\,/\,l$, where $\Delta\omega$ is the angle displacement created by the stress on the F-actin elements in the network with length l. Different experimental settings have been used to predict a Young's module in ECs associated with elastic deformations on the order of $10^5\ dynes\,/\,cm^2$ (Satcher and Dewey, 1996). Second, blood is a viscous liquid with dynamic viscosity coefficient μ. In response to the movement of blood, ECs suffer a shear stress due to changes in blood flow velocity $v(r)$ as a function of the radial distance r from the vessel wall. For example, a longitudinal gradient of pressure will create a steady-state flow in a cylindrical vessel of radius R, with a radial r dependency of the blood velocity satisfying the following equation: $v(r) = -\dfrac{(R^2 - r^2)}{4\mu}\dfrac{dP}{dx}$. This expression can be used to calculate the volumetric flow rate $Q = Av$ in a vessel of length L subject to a pressure difference ΔP. Considering the transversal area of the vessel $A = \pi R^2$ and the average velocity of the blood flow $\langle v \rangle = \int_0^R 2\pi r v(r) dr\,/\,A$, the Hagen–Poiseuille equation can be obtained:

$$\Delta P = \frac{8\mu L Q}{\pi R^4} \tag{10.13}$$

Theoretically, the viscous shear stress is proportional to the radial gradient of the velocity $\tau = \mu\dfrac{dv(r)}{dr}$. However, the viscous shear stress for linear material can be modeled as a dashpot (**damper**). In this case, the circumferential shear stress is proportional to temporal variations in circumferential strain rates $\dot{\varepsilon}^d_{\theta\theta} = d\varepsilon^d_{\theta\theta}\,/\,dt$, with the dynamic viscosity as the linear coefficient, i.e., $\sigma^d_{\theta\theta} = \mu\dot{\varepsilon}^d_{\theta 6}$. If μ does not depend on the amount of shear at a given temperature, the fluid is called Newtonian. Although blood has been considered a Newtonian fluid, recent findings suggest a non-Newtonian behavior (Brust et al., 2013). By definition, circumferential shear stress causes a change in the radius of a cylindrical vessel from an initial value R_0 to a final value R; i.e., $\varepsilon^d_{\theta\theta} = 2\pi(R - R_0)\,/\,2\pi R_0$. Therefore, the following equation holds $\sigma^d_{\theta\theta} = \mu\dot{R}\,/\,R_0$. Finally, SMCs also exert an **active** circumferential stress on vessels, which is proportional to the normalized degree of activation (A_{AMC}) of the actin-myosin contractile system. The degree of activation is determined by the amount of myosin attached state (see below), as both phosphorylated (AM_p) and dephosphorylated (AM) states can generate force. If the vessel can be exposed to a maximal active stress $\sigma^{a,max}_{\theta\theta}(R)$, which depend on the radius R, then we can define an active stress by SMCs on the vessel wall as:

$$\sigma_{\theta\theta}^{a}\left(A_{AMC},R\right)=\sigma_{\theta\theta}^{a,max}\left(R\right)A_{AMC} \tag{10.14}$$

$$A_{AMC}=\frac{AM_{p}+AM}{\left(AM_{p}+AM\right)_{max}} \tag{10.15}$$

Therefore, the total circumferential stress comprises three components under the equilibrium of tangential forces: **a)** an elastic spring stress $\sigma_{\theta\theta}^{s}$, **b)** a viscous damper stress $\sigma_{\theta\theta}^{d}$, and **c)** an active SMC stress $\sigma_{\theta\theta}^{a}(A_{AMC},R)$, the latter depending on the actin-myosin activation and radius. Based on certain assumptions and the arrangement of the spring and damper components, linear models for viscoelasticity can be classified into: Kelvin–Voigt model (parallel system, equal strain: $\varepsilon_{\theta\theta}=\varepsilon_{\theta\theta}^{s}=\varepsilon_{\theta\theta}^{d}$) and Maxwell model (series system, equal stress: $\sigma_{\theta\theta}=\sigma_{\theta\theta}^{s}=\sigma_{\theta\theta}^{d}$). More complex linear models (e.g., Burgers model, the standard linear solid model, and the generalized Maxwell model) add additional components. Among all these models, the Kelvin–Voigt model has been the most used in the literature to phenomenologically represent mechanical deformations in arterial walls. The model comprises the Hookean elastic spring and a Newtonian viscous damper in parallel. If we generalize the model to include the active stress from SMCs, also in parallel and subject to same strain, the total circumferential stress $\sigma_{\theta\theta}$ can be described by the following equation.

$$\sigma_{\theta\theta}=\sigma_{\theta\theta}^{s}+\sigma_{\theta\theta}^{d}+\sigma_{\theta\theta}^{a}(A_{AMC},R)=E\varepsilon_{\theta\theta}+\frac{\mu}{R_{0}}\frac{dR}{dt}+\sigma_{\theta\theta}^{a}(A_{AMC},R) \tag{10.16}$$

Considering the Lamé and Clapeyron's law, the equation takes the following final form:

$$\frac{dR}{dt}=\frac{R_{0}}{\mu}\left(\frac{R\Delta p}{h}-E\varepsilon_{\theta\theta}-\sigma_{a}\left(A_{AMC},R\right)\right) \tag{10.17}$$

There have been several proposals to parametrically represent the passive and active components in the viscoelastic model. Here, we describe only the work by Koenigsberger et al. (2006), which proposed:

$$E\varepsilon_{\theta\theta}=\begin{cases}\sigma_{p,0}\left(exp\left(k_{p}\left(R-R_{0}\right)\right)\right) & R\geq R_{0}\\ \sigma_{p,0}k_{p}\left(1-\left(\frac{R^{2}}{R_{0}^{2}}\right)^{-3/2}\right) & R\leq R_{0}\end{cases} \tag{10.18}$$

$$\sigma_{\theta\theta}^{a,max}\left(R\right)=\sigma_{a,0}exp\left(-k_{a}\left(R+h-r_{a}\right)^{2}\right) \tag{10.19}$$

The model comprises the following parameters: the elastic stress $\sigma_{p,0} = 0.0191\,mmHg$, the elastic coefficient $k_p = 0.15\,\mu m^{-1}$, the maximal active stress $\sigma_{a,0} = 1.8\times10^5\,N/m^2$, the muscular coefficient $k_a = 0.0006\,\mu m$, and the optimal radius for active stress $r_a = 95\,\mu m$. The unstressed radius was assumed $R_0 = 50\,\mu m$. It was also assumed that the maximal fraction of attached cross-bridges $(AM_p + AM)_{max} \approx 0.8$. Assuming that the wall is incompressible, the wall volume should be constant; hence, $h = -R + \sqrt{R^2 + 2R_b h_b + h_b^2}$. In this formula, $R_b = 56.3\,\mu m$ and $h_b = 15\,\mu m$ are the basal values for the luminal radius and wall thickness, respectively. Farr and David (2011) proposed an alternative approach by introducing the concept of radius perturbation.

In addition to electrotonic potentials, ECs transmit ion/molecule signals along the endothelium network via TJs and toward surrounding SMCs via myoendothelial gap junctions. Ca^{2+} ions and IP_3 molecules are the two most important intercellular signaling in brain vessels. An IP_3-dependent release of Ca^{2+} ions from ER have been proposed in ECs, with production rate described by the following equation $j_{IP_3} = F[IP_3]_i^2/(K_r^2 + [IP_3]_i^2)$. Parameters $F = 0.23\,\mu M\,s^{-1}$ and $K_r = 1\,\mu M$ represent the maximum rate of activation-dependent Ca^{2+} influx and half-saturation constant for agonist-dependent Ca^{2+} entry. Ca^{2+} ions are pumped back into the ER by a pump with rate $j_{ER-Pump} = B[Ca^{2+}]_i^2/(c_b^2 + [Ca^{2+}]_i^2)$, with ER uptake rate constant $B = 0.5\,\mu M\,s^{-1}$ and half-point of the sigmoidal function for the pump activation $c_b = 1.0\,\mu M$. Ca^{2+} ions are also extracted out to the extracellular space by another pump $j_{Extrusion} = D[Ca^{2+}]_i$ with rate $D = 0.24\,s^{-1}$. Ionic (Ca^{2+}) and molecular (IP_3) couplings between ECs and SMCs have been modeled as: $j^{EC-SMC}_{Ca^{2+}-coupling} = -P_{Ca^{2+}}([Ca^{2+}]_i^{SMC} - [Ca^{2+}]_i^{EC})$ and $j^{EC-SMC}_{IP_3-coupling} = -P_{IP_3}([IP_3]_i^{SMC} - [IP_3]_i^{EC})$, respectively, with hetero-cellular coupling coefficients $P_{Ca^{2+}} = P_{IP_3} = 0.05\,s^{-1}$.

In addition to K^+, changes in the extracellular concentrations of H^+ and Ca^{2+} in the perivascular space have an impact on blood flow regulation (Heuser, 1978). By the opening of K^+ channels, increased concentrations of H^+ cause vasodilation. Reduction in extracellular Ca^{2+} has been also linked to vasodilation. Sustained brain activation leads to ATP depletion; hence, causing the opening of ATP-sensitive K^+ channels on vessels.

Finally, we discuss transport phenomena through EC membranes. The Starling's law provides a description of the flow through capillary wall.

$$J = K(\Delta p - \sigma\Delta\pi) \tag{10.20}$$

Δp represents the different in hydrostatic pressure between the capillary p_c to the tissue p_t; i.e., $\Delta p = p_c - p_t$. $\Delta\pi$ is the different in oncotic pressure (or also referred to as colloid osmotic pressure) between the capillary π_c to the tissue π_t; i.e., $\Delta\pi = (\pi_c - \pi_t)$. J is the rate of volume flow per unit area (e.g., water, solutes) from capillary to tissue. Parameter K is the hydraulic conductivity. The reflection coefficient σ represents the degree of permeability of the capillary wall to the specific solute. $\sigma = 1$ if the wall is impermeable to the solute. The smaller σ for a solute is, the higher the permeability of the wall to it. For $J > 0$, the solute tends to leave the capillary (filtration). Absorption of the solute takes place for $J < 0$. A mathematical model to describe gas exchange (e.g., oxygen) across the capillary-tissue interface must have two components: **a)** diffusive and **b)** convective. Although the tissue does not move, the convective component originates from the movement of blood within the capillary. Considering a circular cylindrical tube (Krogh, 1919) and axisymmetric gas distribution, the concentration c of the gas must satisfy the following partial differential equations in cylindrical coordinates: inside (Eq. 10.21) and outside (Eq. 10.22) the cylindrical capillary.

$$\frac{\partial c}{\partial t}+v_z\frac{\partial c}{\partial z}=D_b\left(\frac{\partial^2 c}{\partial r^2}+\frac{1}{r}\frac{\partial c}{\partial r}+\frac{\partial^2 c}{\partial z^2}\right)+d(c) \tag{10.21}$$

$$\frac{\partial c}{\partial t}=D_t\left(\frac{\partial^2 c}{\partial r^2}+\frac{1}{r}\frac{\partial c}{\partial r}+\frac{\partial^2 c}{\partial z^2}\right)-g(c) \tag{10.22}$$

Parameters D_b and D_t are the diffusion coefficient of the gas in the blood and tissue, respectively. v_z represents the velocity of blood along the cylindrical capillary (z-direction). The production of the gas inside the capillary is represented by function $d(c)$. For oxygen, this function models the dissociation of oxyhemoglobin in the blood red cells. Function $g(c)$ accounts for gas consumption by the tissue. These two rate functions are in moles/unit-volume/unit-time. The following boundary conditions must be satisfied: 1) $D_b\left.\frac{\partial c}{\partial r}\right|_{r=r_c}=D_t\left.\frac{\partial c}{\partial r}\right|_{r=r_c}$, 2) $\left.\frac{\partial c}{\partial r}\right|_{r=r_t}=0$, and 3) $\left.\frac{\partial c}{\partial r}\right|_{r=0}=0$.

The partial pressure of oxygen PO_2 (mmHg) decreases quickly from brain arteries to capillaries. In hamsters, the hemoglobin saturation decreases along the vessel network (Duling and Berne, 1970): arteries (70 % – 92 %), arterioles (41 % – 70 %), and capillaries (14 % – 41 %).

10.5 SMOOTH MUSCLE CELLS – THE ARTERIOLE CONTRACTION SYSTEM

The force exerted by SMCs on the cylindrical vessel wall is determined by the cross-bridges between the actin and myosin filaments in this cell type. At rest, SMCs are in a semi-contracted stage, which requires constant energy consumption. A relaxation of the contractile force implies vasodilation, while vasoconstriction results from an increase in the contractile force. Actin is a protein in SMCs producing thin and short contractile filaments. Actin filaments have a smooth surface and are free at one end. They do not form cross-bridges. In contrast, myosin forms thick and long filaments, which does form cross-bridges. There are multiple classes of myosin proteins. However, myosin II is the class associated with SMC contraction/dilation. Myosin II is a type of motor protein containing two heavy chains and four light chains. The two heavy chains are held together by their C-terminal tails forming a coiled-coil structure. Each heavy chain has a N-terminal head domain at the other end. There are two myosin light chains (MLC) per head. The SMC contractile complex is stabilized by interactions of α-helical neck region of the heavy chains with the MLCs. MLCs contain two Ca^{2+}-binding EF-hand motifs. By binding the filamentous actin, the heads walk along the filament in the direction of the positive end, causing contraction. The force required to cause the displacement is fueled by ATP hydrolysis. There are two important enzymes regulating the mechanisms for the SMC contraction/dilation system: **1)** the myosin light-chain kinase (MLCK) and **2)** the myosin light-chain phosphatase (MLCP). The MLCK enzyme phosphorylates the regulatory 20-kDa light chain of myosin II. Increases in the cytosolic Ca^{2+} concentration and, hence, in the probability of formation of the Ca^{2+}-calmodulin ($CaCaM$) complex, cause a conformational change in MLCK. Then, the enzyme is activated, resulting in the phosphorylation of MLC (serine residue 19), which initiates contraction by enabling the myosin cross-bridge to bind to the actin filament. Therefore, the relative level of intracellular Ca^{2+} ions with respect to the basal value Ca_0^{2+} in SMC determine muscle stage, with contraction occurring for $[Ca^{2+}]_i > Ca_0^{2+}$ and relaxation for $[Ca^{2+}]_i < Ca_0^{2+}$. Relaxation can also be initiated through the dephosphorylation of the MLCs by the action of the MLCP. It has been proposed that cGMP signaling activates a family of serine/threonine protein kinase G (PKG) causing a MLCP-dependent SMC relaxation. This cGMP-dependent PKG phosphorylates the M110 regulatory subunit of MLCP, decreasing MCL phosphorylation levels. As a result, the response of contractile system is desensitized to Ca^{2+}, causing inhibition of MLC phosphorylation without changes in $[Ca^{2+}]_i$ (Lee et al., 1997). By the inactivation of MLCK, cAMP signaling also plays a role in the SMC contractive system. Apparently, phosphorylation of MLCK by cAMP-dependent PKA causes a reduction in its affinity to Ca^{2+} -CaM by increasing the concentration of CaM required for half-maximal activation. Therefore, the fate of the SMC contractile system is determined by two competing mechanisms: **a)** the Ca^{2+}-dependent MLCK activation; hence, MLC phosphorylation, and **b)** the PKG- and PKA- dependent reduction in MLC phosphorylation.

Hai and Murphy (1988) proposed an initial latch-state model (Eqs. 10.23–10.26) to describe actin-myosin contractile system in SMCs (Figure 10.5). According to this model, there are four potential states for the formation of myosin: **1)** free non-phosphorylated $[M]$, **2)** free phosphorylated $[M_p]$, **3)** attached phosphorylated $[AM_p]$, and **4)** attached dephosphorylated $[AM]$.

$$\frac{d[M]}{dt} = -K_1[M] + K_2[M_p] + K_7[AM] \tag{10.23}$$

$$\frac{d[M_p]}{dt} = K_4[AM_p] + K_1[M] - (K_2 + K_3)[M_p] \tag{10.24}$$

$$\frac{d[AM_p]}{dt} = K_3[M_p] + K_6[AM] - (K_4 + K_5)[AM_p] \tag{10.25}$$

$$\frac{d[AM]}{dt} = K_5[AM_p] - (K_7 + K_6)[AM] \tag{10.26}$$

The myosin is allowed to be only in these states; hence, $[M]+[M_p]+[AM]+[AM_p]=1$. The rate constants K_i, with $i=\{1,\cdots,7\}$, determine both phosphorylation stage and

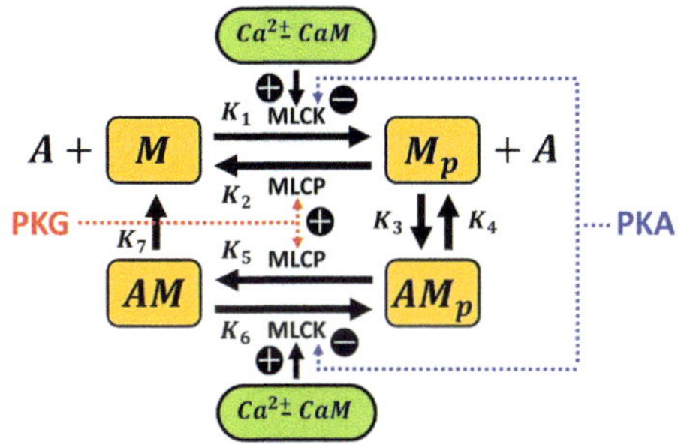

FIGURE 10.5 The actin-myosin contractile model in SMCs.

bridge formation. As mentioned above, SMC stress (A_{AMC}) is proportional to the total fraction of attached cross-bridge ($[AM]+[AM_p]$). The Ca^{2+} and cAMP dependency in the model was introduced by the following relationship: $K_1 = K_6 = \frac{\delta}{\lambda cAMP+1}(Ca^{2+}_{smc})^3$, with $\delta = 9.21\,\mu M^{-3}s^{-1}$ and $\lambda = 2.7\,\mu M^{-1}s^{-1}$ defining the scaling effect of free Ca^{+2} ions on MLCK activation and the desensitization effect of the enzyme by cAMP (Suarez et al., 2025). The model was inspired in that proposed by Hai and Murphy (1989) for the Ca^{2+} dependency. To my knowledge, this constitutes the first mathematical model characterizing MLCK inhibition by PKA phosphorylation. In the model, it is considered that the duty cycle of phosphorylated cross-bridges was 4:1 (i.e., $K_3 = 0.4\,s^{-1}$ and $K_4 = 0.1\,s^{-1}$). Experimental data indicate that $K_7 = 0.1s^{-1}$. Yang et al. (2005) proposed a theoretical framework to account for the cGMP-dependent Ca^{2+} desensitization of MLC phosphorylation. According to this purely empirical model, the dependency of cGMP on the effect that PKG has on a physical quantity (Q) can be explained by using the following general equation: $Q = Q_b + Q_c R_{cGMP}(K_m, n_H)$, with Q_b and Q_c representing the basal level and maximal change of the physical magnitude Q due to the cGMP signaling, respectively. $R_{cGMP}(K_m, n_H) = [cGMP]^{n_H} / (K_m^{n_H} + [cGMP]^{n_H})$ is a normalized Hill equation with operating point K_m and Hill coefficient n_H (steepness). In principle, Q could be either (**a**) the rate constant of an enzymatic reaction or (**b**) of electrophysiological origin (e.g., the maximal conductance or the half-activation voltage of an ionic channel). Based on this general model, the authors could represent the cGMP-dependent Ca^{2+} desensitization of MLC phosphorylation via PKG (Type I) by imposing the following relationship $K_2 = K_5 = K_{KLCP,b} + K_{KLCP,c} R_{cGMP}(K_{m,KLCP}, n_{H,KLCP})$, with parameters fitted to experimental data from SMCs, which were dissected from rabbit femoral arteries: $K_{KLCP,b} = 0.0573\,s^{-1}$, $K_{KLCP,c} = 0.005\,s^{-1}$, $K_{m,KLCP} = 1.83\,\mu M$ and $n_{H,KLCP} = 2$ (assumed).

10.6 cGMP AND cAMP SIGNALING IN SMCS

Henceforth, we will focus on existing mechanisms underlying the formation of cGMP and cAMP in SMCs.

10.6.1 NO/cGMP-Dependent SMC Relaxation

Sriram et al. (2016) proposed a very comprehensive model for the shear-dependent production of NO in ECs due to increases in cytosolic Ca^{2+}, which can be caused by IP_3 formation and several other mechanisms previously discussed in this chapter. NO is produced in ECs by the eNOS. Similar to nNOS, eNOS is a constitutive isoform; hence, it is regulated by Ca^{2+} and calmodulin. There are multiple isoforms of the $CaCaM$ complex. However, the Ca_4CaM isoform constitutes the one recruiting eNOS into the eNOS-CaM complex because of its high cytosolic concentration and affinity for eNOS. The concentration of the Ca_4CaM complex has been modeled with a Hill function, with Hill coefficient $\beta = 2.7$ and dissociation constant $k_{dCaM} = 1\mu M$.

$$\frac{d\left[Ca_4CaM\right]}{dt} = k_{Ca_4CaM}\left(\theta\frac{\left[Ca^{2+}\right]_i^{\beta}}{k_{dCaM} + \left[Ca^{2+}\right]_i^{\beta}}\left[CaM\right]_{tot} - \left[Ca_4CaM\right]\right) \tag{10.27}$$

The other parameters were defined as: $\theta = 0.0045$, $k_{Ca_4CaM} = 100\,s^{-1}$, and $[CaM]_{tot} = 30\,\mu M$.

A three-step process was proposed to model activation of eNOS, which takes place by the phosphorylation of the Ser-1197.

Step-1: Caveolin-bound eNOS ($eNOS_{cav}$) forms a complex with Ca_4CaM, which is represented by $eNOSCaM$.

$$\frac{d[eNOSCaM]}{dt} = \frac{k_{CaM}^{+}\left[Ca_4CaM\right]}{k_{0.5CaM} + \left[Ca_4CaM\right]}\left[eNOS_{cav}\right] - k_{CaM}^{-}\left[eNOSCaM\right] - \frac{d\left[eNOSCaM^{*}\right]}{dt} \tag{10.28}$$

Experimental data have been used to estimate the rate constants (forward $k_{CaM}^{+} = 7.5\,s^{-1}$ and backward $k_{CaM}^{-} = 0.01\,s^{-1}$) and the Michaelis-Menten constant $k_{0.5CaM} = 3\,\mu M$. In equation (10.28), the state variable $eNOSCaM^{*}$ represents the $eNOSCaM$ complex after being stabilized by phosphorylation due to the action of protein kinase B (AKT) at Ser-1197.

Step-2: Dynamic variations of $eNOSCaM^{*}$ follows first-order kinetics.

$$\frac{d\left[eNOSCaM^{*}\right]}{dt} = k_{eAKT}^{max}\frac{\left[AKT^{*}\right]}{\left[AKT\right]_{tot}}\left[eNOSCaM\right] - k_{eAKT}^{-}\left[eNOSCaM^{*}\right] \tag{10.29}$$

Parameters $k_{eAKT}^{max} = 0.004\,s^{-1}$ and $k_{eAKT}^{-} = 2.22\times10^{-4}\,s^{-1}$ were also estimated experimentally. In comparison to the $eNOSCaM$ complex, the stabilized $eNOSCaM^{*}$ is more active to produce NO from L-Arg.

Step-3: $eNOS_{cav}$ phosphorylation by protein kinase C (PKC) at Thr-495 blocks the formation of $eNOSCaM$. The state variable $eNOS_{cav}^{0}$ represents the $eNOS_{cav}$ in the inactivated state (i.e., phosphorylated at Thr-495), which follows also first-order kinetics.

$$\frac{d\left[eNOS_{cav}^{0}\right]}{dt}=k_{Thr}^{+}\frac{\left[PKC^{*}\right]}{\left[PKC\right]_{tot}}\left[eNOS_{cav}\right]-k_{Thr}^{-}\left[eNOS_{cav}^{0}\right] \tag{10.30}$$

With $k_{Thr}^{+}=0.002\,s^{-1}$ and $k_{Thr}^{-}=2.22\times10^{-4}\,s^{-1}$ determined experimentally.

The total concentration of AKT and PKC is conserved, i.e., $[AKT]_{tot}=[AKT]+[AKT^{*}]$ and $[PKC]_{tot}=[PKC]+[PKC^{*}]$, with $[X]$ the respective inactivated and $[X^{*}]$ the activated (phosphorylated) version of the kinase $X=\{AKT,PKC\}$. The activation (phosphorylation) of these two kinases is mediated by PIP_3, a process modeled by rate laws, with constants $k_{AKT}=k_{PKC}=0.1155\,s^{-1}$.

$$\frac{d\left[AKT^{*}\right]}{dt}=0.1k_{AKT}\left[AKT\right]\frac{\left[PIP_3\right]-\left[PIP_3\right]_{min}}{\left[PIP_3\right]_{max}-\left[PIP_3\right]_{min}}-k_{AKT}\left[AKT^{*}\right] \tag{10.31}$$

$$\frac{d\left[PKC^{*}\right]}{dt}=0.1k_{PKC}\left[PKC\right]\frac{\left[PIP_3\right]-\left[PIP_3\right]_{min}}{\left[PIP_3\right]_{max}-\left[PIP_3\right]_{min}}-k_{PKC}\left[PKC^{*}\right] \tag{10.32}$$

Sriram et al. (2016) proposed $[PIP_3]_{min}=0.0031[PIP_2]_{tot}$ and $[PIP_3]_{max}=0.031[PIP_2]_{tot}$. The equations describing the $[PIP_3]$ dynamics were discussed earlier in this chapter. Note that total amount of eNOS must be conserved; i.e., $[eNOS]_{tot}=[eNOS_{cav}]+[eNOS_{cav}^{0}]+[eNOSCaM]+[eNOSCaM^{*}]$.

Production of NO due to the oxidation of L-Arg is catalyzed by the *eNOSCaM* and *eNOSCaM** complexes. Under constant O_2 supply, the production rate of NO (Q_{eNOS}) has been assumed to follow a linear equation with parameter $\rho_{eNOS}=300\,s^{-1}$ and the AKT-related scaling factor $\vartheta=9$ adjusted from experimental data.

$$Q_{eNOS}=\rho_{eNOS}\left(\left[eNOSCaM\right]+\vartheta\left[eNOSCaM^{*}\right]\right) \tag{10.33}$$

There are multiple sinks for NO: **a)** autooxidation ($NO+O_2\rightarrow NO_2^-$), **b)** reaction with superoxide ($NO+O_2^-\rightarrow ONOO^-$), and **c)** heme-protein binding (e.g., hemoglobin “Hb”, soluble guanylate cyclase “sGC”, myoglobin, and red blood cells). Theoretical models for these scavenging mechanisms can be found in Buerk (2001). For the purpose of this book, we will focus on NO scavenging by the hemoglobin and the sGC. NO association with Hb inside blood vessels is almost an irreversible process that can be modeled with an exponential decay ($S_{Hb}=-\lambda_{NO}[NO]$), with $\lambda_{NO}=382\,s^{-1}$. Sriram et al. (2016) proposed a theoretical for the NO scavenging by sGC:

$$S_{sGC}=R_{NO}^{max}\frac{\left[NO\right]^2+b_1\left[NO\right]}{a_0+a_1\left[NO\right]+\left[NO\right]^2}\left[sGC\right] \tag{10.34}$$

The values and dimensions for these parameters can be found in table 1 of Sriram et al., (2016).

Parameters in this equation are: Rmax_NO=0.022 s^{-1}, b1=15.15 nM, a0=1200.16 nM^2, and a1=37.33 nM. Therefore, NO production by ECs and its degradation can be modeled as:

$$\frac{d[NO]}{dt} = Q_{eNOS} - S_{sGC} - S_{Hb} \tag{10.35}$$

A phenomenological model has been also proposed in the past to represent the NO production due to the shear stress τ (Yamazaki and Kamiyama, 2014): $d[NO]/dt = R_{NO}(\tau) - \alpha_{NO}[NO]$, with $R_{NO}(\tau) = R_{NO,max}/(1+exp(-P_1\tau + P_2)) + R_{NO,min}$. In this model, a sigmoidal relationship between the eNOS concentration and the wall shear stress was assumed. Parameters $R_{NO,min} = -110.84\,nM/s$ and $R_{NO,max} = 350.99\,nM/s$ represent the minimal and maximal production rate of NO. Previously in this chapter, τ was modeled for linear material as a viscous circumferential "damper" stress $\sigma^d_{\theta\theta}$. Parameters P_1 and P_2 were fitted from experimental data (i.e., the mean value of two subjects, Yamazaki and Kamiyama, 2014). However, Suarez et al. (2025) scaled them up to compensate for a phenomenological equation to adequately represent the impact of shear stress on NO production, $P_1 = 1.5\,cm^2/dyn$ and $P_2 = 7.5$. $\alpha_{NO} = 1$ characterizes the diffusion of NO ($1/s$).

NO produced by ECs (described above), and also by neurons (Chapter 9), can diffuse to nearby SMCs where it produces cGMP by the activation of the sGC (Francis et al., 2010). The sGC is a heterodimeric hemoprotein with α- and β- subunits that catalyzes the conversion of guanosine triphosphate (GTP) into cGMP, which has been modeled by a two-state process (Figure 10.6). By binding to basal sGC (E_B), NO forms an intermediate six-coordinate (6c) ferrous nitrosyl heme complex (E_{6c}) in the first step (Eq. 10.36) with rate constants $k_1 = 2.0\,nM^{-1}\,s^{-1}$ and $k_{-1} = 100\,s^{-1}$. The second step describes E_{6c}-to-E_{5c} transitions by means of both NO-dependent and NO-independent pathways (Eqs. 10.37 and 10.38). Through His105 residue dissociation, E_{6c} naturally decays to E_{5c} following a first-order irreversible kinetics with a reaction rate $k_2 = 0.1\,s^{-1}$. A Michaelis-Menten kinetics was introduced to describe the NO-dependent pathway, which requires NO binding to sGC in a second site at the heme to facilitate the Fe-His105 cleavage. The reaction rate for this second decay is $k_3 = 0.003\,nM^{-1}\,s^{-1}$.

The activated 5c complex (E_{5c}) decays spontaneously to the basal state with a rate proportional to cGMP $k_4 = K_4[cGMP]^m$. Note that the total sGC concentration E_0 is conserved: $[E_0] = [E_B] + [E_{6c}] + [E_{5c}]$.

$$\frac{d[E_B]}{dt} = -k_1[NO][E_B] + k_{-1}[E_{6c}] + k_4[E_{5c}] \tag{10.36}$$

$$\frac{d[E_{6c}]}{dt} = k_1[NO][E_b] - (k_{-1} + k_2 + k_3[NO])[E_{6c}] \tag{10.37}$$

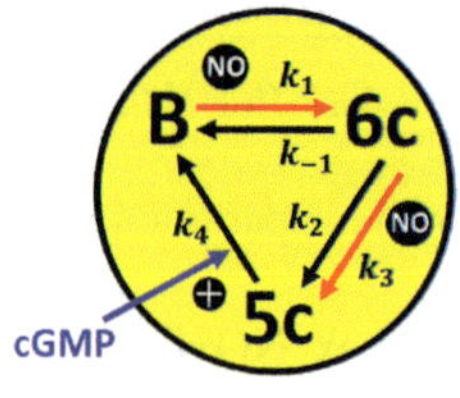

FIGURE 10.6 The sGC kinetic model.

$$\frac{d[E_{5c}]}{dt} = (k_3[NO] + k_2)[E_{6c}] - k_4[E_{5c}] \tag{10.38}$$

The synthesis of cGMP is associated with the active sGC state E_5. Multiple isoforms of cyclic nucleotide phosphodiesterases (PDEs) regulate the lifetime of cGMP by hydrolyzing it to the inactive 5'-GMP form. Therefore, cGMP production and degradation in SMCs is modeled by a balance equation.

$$\frac{d[cGMP]}{dt} = V_{sGC}[E_5] - \frac{\gamma_{PDE}[cGMP]^2}{K_{PDE} + [cGMP]} \tag{10.39}$$

10.6.2 PGE2/cAMP-Dependent SMC Relaxation

Parameters in the sGC model have been fitted from experimental data (Table 10.3).

In Chapter 9, we discussed the production of PGE_2 from AA by the constitutive COX-1 enzyme in perivascular astrocytic end-feet. PGE_2 is then secreted to the extracellular space through MRPs, which can diffuse and bind to EP4 prostaglandin receptors in nearby SMCs. This mechanism has been modeled in Chapter 9 (Eqs. 9.14 and 9.16). Activated EP4 PGE_2 receptor causes increases in the intracellular concentration of cAMP in SMCs, which has been recently modeled by a set of two differential equations (Tesler et al., 2023; Suarez et al., 2025).

$$\frac{dR_{PGE_2}}{dt} = -\frac{R_{PGE_2}}{\tau_R} + O_R[PGE_2](1 - R_{PGE_2}) \tag{10.40}$$

$$\frac{d[cAMP]}{dt} = -\frac{[cAMP]}{\tau_{cAMP}} + O_{cAMP}R_{PGE_2} \tag{10.41}$$

In this model, R_{PGE_2} is the portion of activated EP4 PGE_2 receptors. The relaxation time and binding rate for the PGE_2 receptor are $\tau_R = 0.8\,s$ and $O_R = 0.2 \times 10^4\ mM^{-1}s^{-1}$, respectively. The relaxation time and maximum rate of cAMP production for cAMP are defined as $\tau_{cAMP} = 2\,s$ and $O_{cAMP} = 2 \times 10^{-3}\ mM\,s^{-1}$, respectively. Note that as for cGMP, PDE participates in the abolition of the cAMP signaling by converting this second messenger into its inactivated form 5'-AMP. cAMP activates BK channels via

TABLE 10.3
Model parameters for the sGC model in SMCs

m	$V_{sGC}\ (\mu M / s)$	$\gamma_{PDE}\ (s^{-1})$	$K_4\ (\mu M^{-m}\ s^{-1})$
0	1.26	0.0695	0.4
1	1.09	0.032	0.098
2	0.8520	0.0195	0.011

Sources: Yang et al. (2005) and data from Bellamy et al. (2000).
Note that $K_{PDE} = 2\,\mu M$.

the phosphorylation of the serine-869 residue by PKA, causing hyperpolarization and hence vasodilation due to L-type Ca^{2+} channel deactivation in SMCs.

10.6.3 AA/20-HETE-Dependent SMC Contraction

Groups II and V of $sPLA_2$ can be found in astrocytic end-feet (sPLA2-IIA, Sun et al., 2005). After synthesis intracellularly, the $sPLA_2$ enzymes are secreted from astrocytes. Secreted $sPLA_2$ can also be internalized and degraded. Secreted $sPLA_2$ acts on different types of extracellular phospholipids (e.g., lipoproteins, microvesicles/exosomes, and surfactants). A crosstalk between $cPLA_2$ and $sPLA_2$ has been previously reported in different cell types. The sensitivity of $sPLA_2$ to extracellular Ca^{2+} varies for different groups. Kamitani et al. (2012) reported 75 % – 100 % activation of groups X and V of $sPLA_2$ at physiological levels of extracellular Ca^{2+} ions (2 mM). Therefore, it is expected that exogenous $sPLA_2$ continuously produces larges amount of extracellular AA. It can be assumed that AA produced by $cPLA_2$ in either astrocytic end-feet or other cells in the neurovascular units is negligible compared to that created by $sPLA_2$. Perivascular AA can freely diffuse to nearby SMCs, serving as a substrate for a particular subtype (CYP4A) of the cytochrome P450. CYP4A produces 20-HETE acid (Chapter 9). Suarez et al. (2025) proposed a theoretical model to characterize the impact on SMCs of perivascular AA produced by $sPLA_2$.

$$\frac{d[AA]_p}{dt} = -\frac{[AA]_p}{\tau_{AA_p}} + O_{AA_{max}} \tag{10.42}$$

The maximum production rate $O_{AA_{max}} = 2\,mMs^{-1}$ and the relaxation time $\tau_{AA_p} = 1\,s$ of $[AA]_p$ were adjusted from experimental data. The production of 20-HETE was modeled by Suarez et al. (2025, Eq 10.43), with parameters $\tau_{HETE} = 2\,s$ and $K_{HETE} = 1\,mM$.

$$\frac{d\left[{}^{20}HETE\right]}{dt} = -\frac{\left[{}^{20}HETE\right]}{\tau_{HETE}} + \frac{O_{\mathrm{O\_HNO(NO)}}[AA]_p}{K_{HETE} + [AA]_p} \tag{10.43}$$

NO constitutes an inhibitory agent for CYP4A (Metea and Newman, 2006; Liu et al., 2008). Local increases in local NO might cause additional vasodilation by inhibiting the constricting effect of 20-HETE on SMCs. In Suarez et al. (2025), the inhibitory effect of NO on the production of 20-HETE was modeled by a parametric function $O_{\mathrm{O\_HNO(NO)}} = O_{HETE}(1 + tanh((NO - NO_0)/O_{NO}))$, with $O_{HETE} = 0.2\,mMs^{-1}$, $NO_0 = 0.075\,\mu M$, and $O_{NO} = 0.534\,\mu M$. In SMCs, 20-HETE direct blocks BK channels (phosphorylation) and activates the L-type calcium channels in a PKC-dependent process. Blocking BK channels depolarizes SMCs, which also causes the indirect activation of L-type calcium channels. Therefore, the contractile effect of 20-HETE was modeled by Suarez et al. (2025) only due to its effect on BK. In SMCs, the BK channels are activated also by intracellular cGMP (Eq. 10.39) and cAMP (Eq. 10.41) produced inside SMCs. Released EET, a lipophilic acid produced in astrocytic end-feet by CYP2C (Eq. 9.15), can also diffuse to nearby SMCs causing

SMC hyperpolarization by the activation of BK channels $I_{BK}^{(SMC)}$. Suarez et al. (2025) introduced a BK kinetic model like the one proposed for astrocytic end-feet (Eqs. 9.19–9.23), but with conductance $g_{BK} = 94\,pS$, equilibrium potential $V_{BK} = -80\,mV$, and an instantaneous open-gating probability $n_{\infty,BK}$ (Eq. 10.44) affected by the four intracellular signals (EET, 20-HETE, cAMP and cGMP).

$$n_{\infty,BK} = 0.5\left(1 + \tanh\left(\frac{\begin{array}{l} V_m + k_{EET}[EET] - k_{HETE}[{}^{20}HETE] \\ +k_{cAMP}[cAMP] + k_{cGMP}[cGMP] - v_{3,BK} \end{array}}{v_{4,BK}}\right)\right) \tag{10.44}$$

The following model parameters were assumed to be like those used for the BK channels in astrocytic end-feet: ψ_n, $v_{4,BK}$, $v_{5,BK}$, and $v_{6,BK}$. Experimental data were used to estimate the remaining parameters: $Ca_{3,BK} = 0.749\,\mu M$, $Ca_{4,BK} = 0.209\,\mu M$, $k_{EET} = 98.1\,mV / \mu M$, $k_{HETE} = 1210\,mV / \mu M$, $k_{cAMP} = 187.1\,mV / \mu M$, $k_{cGMP} = 33.34\,mV / \mu M$, and $g_{BK} = 13\,nS$.

As for the MLCP, the transfer of γ-phosphate from ATP to several proteins [i.e., ion channels (e.g., BK, L-type Ca^{2+}), pumps (Ca^{2+} ATPases), exchanges (NCX), receptors (IP_3R), and regulatory proteins] is activated by PKG. It has been assumed that both the PKG activation and the associated phosphorylation of these target proteins occur very fast compared to cGMP synthesis. Therefore, cGMP has been considered the mediating agent for most of these cases. Because of their complexity and/or lack of experimental validation, we are not including in this book biophysical models for PKG-based activation of these target proteins.

10.7 CA^{2+} SIGNALING IN SMCS

As mentioned above, the Ca^{2+}-dependent MLCK activation is a key mechanism in SMC contractibility. Therefore, mechanisms for Ca^{2+} signaling in SMCs are described henceforth, and biophysical models are provided for those more relevant. According to Hill-Eubanks et al. (2011), graded Ca^{2+} changes in the SMC cytoplasm take place either in the form of global (cell-wide) variations, or as highly localized small-scale increases (e.g., flashes, sparklets). Ca^{2+} flashes have been associated with non-vascular SMCs (e.g., ureter, urinary, bladder); hence, they are not discussed in this chapter. Contractability of vascular SMCs has been mostly associated with global Ca^{2+} variations. There are two general pathways underlying changes in SMC cytosolic Ca^{2+} concentrations: **1)** entry from the extracellular space and **2)** release from intracellular storages.

10.7.1 Mechanisms for Ca^{2+} Entry from the Extracellular Space

The influx of Ca^{2+} ions from the extracellular space occurs via SOC, NSC, ionotropic purinergic (PX2) receptors, and VGCCs. Apparently, a STIM1-Orai mechanism underlies Ca^{2+} influx through SOC channels in SMCs, a mechanism described as follows. Ca^{2+} depletion in the SR causes Orai activation by the aggregation of STIM1 proteins, which leads to the formation of discrete plasmalemmal-proximate clusters that tether Orai proteins. Orai is a highly selective Ca^{2+} channel on the SMC membrane promoting a large influx of this ion if activated. The SR Ca^{2+} depletion is sensed by the low-affinity Ca^{2+}-binding protein STIM1. STIM1 protein can also negatively regulate the VGCCs. On the other hand, an important role of NSC channels in the regulation of SMC contractability has been suggested in recent years. Multiple TRP subtypes have been found in vascular SMCs, with regulatory properties for Ca^{2+} and Na^+ influxes (CRAC-type) via a SOC-like mechanism (e.g., TRPC1, TRPP2, and TRPC4/5). TRP channels in SMCs are also associated with other functions (e.g., myogenic tone, cell proliferation, and responses to noradrenaline, diacylglycerol, store-depletion, and hypo-osmotic stress). To my knowledge, theoretical models for SOC and TRP channel kinetics in SMCs are yet to be developed. The role and impact of PX2 receptors in vascular SMCs need to be further studied. We recommend seeing Manninen et al. (2018) for models that include a P2X component in the Ca^{2+} signal genesis.

By far, VGCCs are the major source of Ca^{2+} influx in SMCs. The L-type high-voltage activated VGCC (Cav1.2 and perhaps Cav1.3) is the most abundant in SMCs, and it has been associated with global changes in cytosolic Ca^{2+} concentration in these cells. However, expression of T-type low-voltage activated VGCC (Cav3 pore-forming α subunit) has also been reported, but to a lesser extent and not directly associated to vascular SMCs. A kinetic model for Ca^{2+} influx ($\mu M\, s^{-1}$) through the L-type Ca^{2+} channels (j_{CaL}) was proposed by Koenigsberger et al. (2006), with whole-cell conductance defined as $G_{CaL} = 0.00129\,\mu M\, mV^{-1} s^{-1}$.

$$j_{CaL} = G_{CaL} \frac{\left(V_m - E_{Ca}\right)}{1 + exp\left(-\left(V_m - v_{Ca}\right)/R_{Ca}\right)} \tag{10.45}$$

The half-point and slope for the sigmoidal activation of the L-type Ca^{2+} channel are defined as $v_{Ca} = -24\,mV$ and $R_{Ca} = 8.5\,mV$, respectively. The reversal potential for Ca^{2+} is set to $E_{Ca} = 100\,mV$. Influx through L-type VGCC has been suggested to also depend on the transmural pressure ΔP ($mmHg$). Gonzalez-Fernandez and Ermentrout (1994) proposed the following simplified model to include such dependence in the L-type Ca^{2+} current (fA): $I_{CaL} = g_{CaL} m_\infty (V_m - E_{Ca})$, with $m_\infty = 0.5(1 + tanh(V_m - v_1 / v_2))$. For this model, $g_{CaL} = 157\,pS$, $v_1 = -17.4 - 12\Delta P / 200$, and $v_2 = 25\,mV$.

Highly localized ($\sim 0.8\,\mu m^2$ – membrane surface) and dihydropyridine-sensitive Ca^{2+} sparklets in vascular SMCs are caused by the opening of a single VGCC channel. These Ca^{2+} quantal events elevate Ca^{2+} concentration locally in about $35\,nM$. Persistent and synchronized sparklets create global increases in Ca^{2+} concentrations in vascular SMCs.

Depolarization of the SMC membrane by $\sim 15\,mV$ from its resting membrane potential ($-45\,mV$) causes global increases in Ca^{2+} concentration up to $400\,mM$. Hyperpolarization of SMCs by $\sim 15\,mV$ reduces Ca^{2+} levels in about $100\,mM$. Changes in SMC membrane potential occur by the action of several other ion channels. In addition to the BK channel described above, there are two other important ionic channels in SMCs contributing the membrane potential V_m: **1)** Cl^- channels (j_{Cl}) and **2**). K_{IR} channels ($j_{K_{IR}}$). A simple kinetic model for the Cl^- ion flux ($\mu M\, s^{-1}$) through the

Cl⁻ channels was proposed by Koenigsberger et al. (2006), i.e., $j_{Cl} = G_{Cl}(V_m - E_{Cl})$, with whole-cell conductance defined as $G_{Cl} = 0.00134\,\mu M\,mV^{-1}s^{-1}$. The reversal potential for Cl⁻ was set to $E_{Cl} = -25\,mV$.

Hyperpolarization in SMCs is also caused by elevations ($10 - 20\,mM$) of extracellular K^+ (i.e., the perivascular space) by the activation of the inward rectifier K_{IR} ($K_{IR}2.1$ and possible $K_{IR}2.2$) channels. K_{IR}-dependent reduction in intracellular Ca^{2+} and consequent vasodilation have been reported in cortical brain slices. The K_{IR} channels are tetramers with pore-forming α- subunits, each comprising two transmembrane domains. It has been suggested that PIP_2 activates K_{IR} channels by interactions with amino acids and cholesterol in the plasma membrane (Jackson, 2017). A theoretical model for the ion flux through K_{IR} channels in SMCs has been proposed in the past.

$$j_{K_{IR}} = \frac{F_{K_{IR}} g_{K_{IR}} \left(V_m - E_{K_{IR}}\right)}{\gamma} \tag{10.46}$$

Parameter $F_{K_{IR}} = 750\,mV\,\mu M^{-1}$ represents a scaling factor of K^+ efflux via K_{IR} in SMCs (González-Fernandez and Ermentrout, 1994). The Nernst potential and conductance for the K_{IR} channel in SMCs can be estimated using the following formulas $E_{K_{IR}} = z_1[K^+]_p - z_2$ in (mV), and $g_{K_{IR}} = exp(z_5 V_m + z_3[K^+]_p - z_4)$ in ($\mu M\,mV^{-1}s^{-1}$), respectively. Dormanns et al. (2015) fitted this theoretical model to data obtained by Filosa et al. (2006), with the following results: $z_1 = 4.5\times 10^3\,mV\,\mu M^{-1}$, $z_2 = 112\,mV$, $z_3 = 4.2\times 10^2\,mV^{-1}\,s^{-1}$, $z_4 = 12.6\,\mu M\,mV^{-1}s^{-1}$, and $z_5 = -7.4\times 10^{-2}\,\mu M\,mV^{-2}s^{-1}$. The conversation parameter $\gamma = 1970\,mV\,\mu M^{-1}$ was proposed in a previous study to relate ion fluxes to the membrane potential (Koenigsberger et al., 2006). Farr and David (2011) proposed an alternative model for $J_{K_{IR}}$. Increases in the concentration of K^+ in the perivascular space can also occur due to K^+ efflux through BK channels in ECs and astrocytes, causing both vasodilation due to SMC hyperpolarization. Therefore, changes in K^+ concentration in the perivascular space is determined by Eq. (10.47).

$$\frac{[K^+]_p}{dt} = \frac{1}{FV_p}\left(A_{ast,p} I_{BK}^{(ast)} + A_{smc,p} I_{BK}^{(SMC)}\right) + \frac{j_{K_{IR}}}{V_{R_{ps}}} - R_{decay}\left([K^+]_p - K^+_{o,0}\right) \tag{10.47}$$

V_p and $V_{R_{ps}}$ are the volume of the perivascular space and the volume ratio of perivascular space to SMC, respectively. $A_{ast,p}$ and $A_{smc,p}$ are surface cell membrane in direct contact with the perivascular space for astrocyte and SMC, respectively. The K^+ ion decay due to multiple processes is linearly modeled, with $R_{decay} = 0.02\,s^{-1}$ (Suarez et al., 2025). $K^+_{o,0} = 3\,mM$ is the extracellular basal levels of K^+ ions.

The role of other types of K^+ channels (K_v and K_{ATP}) in SMCs are discussed in Ko et al. (2008). The compound 4-Aminopyridine (4-AP, $0.3 - 1.1\,mM$) and glibenclamide ($10\,\mu M$) constitute specific blockers for K_v and K_{ATP} channels in vascular SMCs, respectively. As the role of these two channels in CBF control has not been demonstrated, their kinetics models will not be discussed in this book. In general, Eq. 10.48 describes the dynamics of the SMC membrane potential (Dormanns et al., 2015), which I have modified to include a $I_{BK}^{(SMC)}$ current with the kinetics proposed by Farr and David (2011).

$$\frac{dV_m}{dt} = -\frac{1}{C_m} I_{BK}^{(SMC)} - \gamma(j_{Cl} + 2j_{CaL} + j_{K_{IR}} + j_{Na/K} + j_{Na/Ca} + j_{stretchy}) + v_{coupling}^{SMC} + v_{coupling}^{SMC-EC} \tag{10.48}$$

In Dormanns et al. (2015), the flux through the BK channel was modeled as: $j_{BK}^{(SMC)} = G_{BK}\omega(V_m - E_K)$, with $G_{BK} = 4.46 \times 10^{-3}\,\mu MmV^{-1}s^{-1}$, $E_K = -94\,mV$, and ω satisfying the differential equation $d\omega / dt = K\_act([CA^{2+}]i, V_m)$. Function $K\_act([CA^{2+}]i, V_m)$ characterizes the dependency with $[Ca^{2+}]_i$ and V_m.

$$K\_act([CA^{2+}]i, V_m) = \frac{\left([Ca^{2+}]_i + c_\omega\right)^n}{\left([Ca^{2+}]_i + c_\omega\right)^n + \beta exp\left(-\left(V_m - V_{Ca}\right) / R_{Ca}\right)}$$

Interestingly, the model proposes a Hill's function for the Ca^{2+} dependency with Hill's coefficient $n = 2$, and voltage-dependent apparent dissociation constant. If needed, a translation factor c_ω could be introduced for sigmoidal Ca^{2+} dependence. The dependence with V_m was modeled as sigmoidal with half-point $V_{Ca} = -27\,mV$, maximum slope $R_{Ca} = 12\,mV$, and translation factor $\beta = 0.13\,\mu M^2$, respectively. I_{BK} is defined by Eqs. 9.19–9.23, but with $n_{\infty,BK}$ as described by Eq. 10.44 for SMCs. The coupling currents between SMCs and with ECs are defined as: $v_{coupling}^{SMC} = -G_{coupling}^{SMC} \sum_l (V_m - V_m^{SMC(l)})$ and $v_{coupling}^{SMC-EC} = -G_{coupling}^{SMC-EC} \sum_l ((V_m - V_m^{EC(l)}))$, with the homo-cellular $G_{coupling}^{SMC} = 1000\,s^{-1}$ and hetero-cellular $G_{coupling}^{SMC-EC} = 50\,s^{-1}$ electric coupling coefficients. Dormanns et al. (2015) provided a different estimate for $G_{coupling}^{SMC-EC} = 0.5\,s^{-1}$. Note that we defined early in this chapter the electric coupling between ECs and SMCs through TJ channels as $v_{coupling}^{EC-SMC}$. $j_{stretchy}$ is modeled according to Eq. 10.12, with R satisfying Eq. 10.17 and the arterial wall thickness h as for an incompressible wall. $J_{Na/K} = 0.2\,\mu M\,s^{-1}$ represents the net whole-cell flux via the Na-K ATPase. Ion fluxes in the Na-Ca exchanger (NCX) were described by Eqs. 6.21 and 6.22. However, Koenigsberger et al. (2006) proposed a simplified model $j_{Na/Ca} = G_{Na/Ca} \frac{[Ca^{2+}]_i}{([Ca^{2+}]_i + C_{Na/Ca})}(V_m - v_{Na/Ca})$, with $G_{Na/Ca} = 0.007\,\mu MmV^{-1}s^{-1}$ the whole-cell conductance for the exchanger, $C_{Na/Ca} = 0.5\,\mu M$ the half-point for the activation of the exchanger by Ca^{2+}, and $v_{Na/Ca} = -30.0\,mV$ the reversal potential for the exchanger.

10.7.2 Mechanisms for Ca^{2+} Release from the Intracellular Storages

The sarcoplasmic reticulum (SR) is the most important Ca^{2+} intracellular store in SMCs. Ca^{2+} release from the SR occurs via RyR and IP_3R. Cytosolic Ca^{2+} is transported to the SR by the action of SR/ER Ca^{2+} ATPase (SERCA). PKA and PKG activities regulate SERCA through the protein phospholamban. The inhibitory effect of phospholamban on SERCA is lost on phosphorylation, causing increases in Ca^{2+} uptakes by the SR. Localized Ca^{2+} release from SR through RyR channels, mostly subtype RyR2, are called sparks. Typically, 4-6 RyR channels must concurrently open to create a Ca^{2+} spark, which causes a localized increase of Ca^{2+} of $10-30\,\mu M$. The rise time of a spark is approximately $\sim 30\,ms$ and they last (half-time) for $\sim 55\,ms$. In SMCs, RyR channels are in close proximity to cluster of BK channels, ensuring a functional coupling. Activation of BK channels by Ca^{2+} sparks produced spontaneous transient outward currents (STOC). Ca^{2+} sparks also cause spontaneous transient inward currents (STIC) through the activation of Cl^- channels. STOC and STIC area associated with hyperpolarization and depolarization of SMC membrane and hence, more global changes in Ca^{2+} concentrations by means of VGCCs. The Ca^{2+} flux (μMs^{-1}) through the RyR channel is modeled by Eq. 10.49 (Dormanns et al. 2015).

$$j_{RyR} = C_{RyR} \frac{\left([Ca^{2+}]_{SR}\right)^2}{\left(\left([Ca^{2+}]_{SR}\right)^2 + C_{SR}^{\,2}\right)} \frac{\left([Ca^{2+}]_i\right)^4}{\left(\left([Ca^{2+}]_i\right)^4 + C_i^{\,4}\right)} \tag{10.49}$$

The channel rate is defined as $C_{RyR} = 55\,\mu Ms^{-1}$. Parameters $C_{iSR} = 2.0\,\mu M$ and $C_{iSR} = 0.9\,\mu M$ represent the half-point of the RyR sigmoidal function for the efflux and activation components, respectively. $[Ca^{2+}]_{SR}$ is the concentration in the SR (μM). A RyR kinetic model for neuronal synapses was also proposed earlier in this book (Eq. 7.3).

Ca^{2+} ions are also released from the SR via IP_3R. The most abundant subtypes in SMCs are IP_3R1-3. Ca^{2+} release by IP_3R is regulated by two second messengers: IP_3 and Ca^{2+}. In SMCs, increases of IP_3 molecules take place due to the activity of GPCR ($G_{q/11}$ subtype), resulting in the hydrolysis of PIP_3 by the stimulation of a PLC. Ca^{2+} release into the SMC cytosol through the IP_3R channels can be modeled in a same way as for ECs; i.e., see equation for j_{IP_3} early in this chapter. A kinetic model for IP_3R in astrocytic ER was also presented in Chapter 8. Small and localized Ca^{2+} releases from SR are called puffs, which are not prominent in vascular SMCs. However, Ca^{2+} release by IP_3R can also induce large Ca^{2+} waves, a phenomenon associated with the CICR regenerative effect of RyR in cerebral SMCs.

Mathematical models have been proposed to describe changes in the cytosolic Ca^{2+} concentration in SMCs. In general, these models propose a differential equation to describe Ca^{2+} variations in SMCs due to the contributions of combinations of the following fluxes: **1**) VGCC, specifically the L-type j_{CaL} (Eq. 10.45), 2) RyR release j_{RyR} (Eq. 10.49), 3) IP_3R release j_{IP_3}, **4**) NCX $j_{Na/Ca}$, **5**) Ca^{2+} factor in $j_{stretchy}$(Eq. 10.12), and **5**) TJ between EC-SMC $j^{EC-SMC}_{Ca^{2+}-coupling}$, **6**) SR Ca^{2+} pumping $j_{SR-Pump}$, and 7) plasma membrane extrusion $j_{Extrusion}$. The j_{IP_3} and $j_{stretchy}$ fluxes were defined in the EC subsection. $j_{SR-Pump}$ is modeled like $j_{ER-Pump}$, but with $B = 2.025\ \mu M\ s^{-1}$. Koenigsberger et al. (2006) proposed a voltage dependent Ca^{2+} extrusion by the Ca^{2+} ATPase pump (PMCA), $j_{Extrusion} = D[Ca^{2+}]_i\left(1 + \frac{V_m - v_d}{R_d}\right)$, with intercept $v_d = 100.0\ mV$ and slope

$R_d = 250.0\ mV$ for the voltage dependence. Taking all these fluxes into account, variations in SMC cytosolic $[Ca^{2+}]_i$ are described as:

$$\frac{d[Ca^{2+}]_i}{dt} = j_{CaL} + j_{RyR} + j_{IP_3} + 0.1 j_{stretchy} + j^{EC-SMC}_{Ca^{2+}-coupling} - j_{Na/Ca} - j_{SR-Pump} - j_{Extrusion} \tag{10.50}$$

Note that although not included in Eq. 10.50, contributions from NSC can also be considered j_{NSC}. Changes in Ca^{2+} concentration in the SR (i.e., $[Ca^{2+}]_{SR}$) is modeled as:

$$\frac{d\left[Ca^{2+}\right]_{SR}}{dt} = -j_{RyR} - j_{IP_3} + j_{SR-Pump} \tag{10.51}$$

10.8 THE PRECAPILLARY SPHINCTERS

As mentioned in this chapter, CBF is regulated by SMCs circumscribing arteries and arterioles, as well as by pericytes that enclose capillaries close to the penetrating arterioles. It has been proposed, the existence of brain precapillary sphincters that help homogeneously redistribute increases in blood perfusion toward the capillary network, thus supporting active tone regulation and protecting them from high blood pressure (Grubb et al., 2020). These precapillary sphincters are contractile mural cells, which have been identified as either SMCs or pericytes. The sphincters are positive to neuron-glial antigen 2 (NG2) and contain α-smooth muscle actin (α-SMA). To my knowledge, their contractility mechanism of sphincters has neither been understood nor modeled yet.

10.9 PROBLEMS AND EXERCISES

Problem 10.1 – Calculate the pressure drop along a mouse cerebral vessel of radius $r = 10.5\ \mu m$ and length $200\ \mu m$. The mean speed of blood is $u = 9.2\ mm/s$. The viscosity of the blood is $\mu = 3.5\ cP$.

Problem 10.2 – The characteristics of a flow in a pipe is determined by the Reynolds number $R_e = 2\rho ur/\mu$, with ρ, u, and μ defining density (kg/m^3), mean speed (m/s), and viscosity (Ns/m^2) of the fluid, respectively. r is the pipe diameter (m). A laminar flow occurs if $R_e < 2300$. For $R_e > 2900$ the flow will be turbulent. Given that the blood density is $1056.5\ Kg/m^3$, determine whether the blood flow regime in Problem II.6.1 is laminar or turbulent.

REFERENCES

Altomare C, Bucchi A, Camatini E, Baruscotti M, Viscomi C, Moroni A, DiFrancesco D. Integrated allosteric model of voltage gating of HCN channels. *J Gen Physiol* 117, 519–532, 2001.

Bellamy TC, Wood J, Goodwin DA, Garthwaite J. Rapid desensitization of the nitric oxide receptor, soluble guanylyl cyclase, underlies diversity of cellular cGMP responses. *Proc Natl Acad Sci USA* 97, 2928–2933, 2000.

Blinder P, Tsai PS, Kaufhold JP, Knutsen PM, Suhl H, Kleinfeld D. The cortical angiome: An interconnected vascular network with noncolumnar patterns of blood flow. *Nat Neurosci* 16(7), 889–897, 2013.

Brust M, Schaefer C, Doerr R, Pan L, Garcia M, Arratia PE, Wagner C. Rheology of human blood plasma: Viscoelastic versus Newtonian behavior. *Phys Rev Lett* 110(7), 078305, 2013.

Buerk DG. Can we model nitric oxide biotransport? A survey of mathematical models for a simple diatomic molecule with surprisingly complex biological activities. *Annu Rev Biomed Eng* 3, 109–143, 2001.

DiFrancesco D. Characterization of the pacemaker current kinetics in calf Purkinje fibres. *J Physiol* 348, 341–367, 1984.

DiFrancesco D. Dual allosteric modulation of pacemaker (f) channels by cAMP and voltage in rabbit SA node. *J Physiol* 512, 367–376, 1999.

Dormanns K, van Disseldorf EMJ, Brown RG, David T. Neurovascular coupling and the influence of luminal agonists via endothelium. *J Theoretical Biol* 364, 49–70, 2015.

Duling BR, Berne RM. Longitudinal gradients in periarteriolar oxygen tension. *Circ Res* 27, 669–678, 1970.

Farkas E, Luiten PG. Cerebral microvascular pathology in aging and Alzheimer's disease. *Prog Neurobiol* 64, 575–611, 2001.

Farr H, David T. Models of neurovascular coupling via potassium and EET signalling. *J Theor Biol* 286, 13–23, 2011.

Filosa JA, Bonev AD, Straub SV, Meredith AL, Wilkerson MK, Aldrich RW, Nelson MT. Local potassium signaling couples neuronal activity to vasodilation in the brain. *Nat Neurosci* 9(11), 1397–1403, 2006.

Francis SH, Busch JL, Corbin JD. cGMP-Dependent protein kinases and cGMP phosphodiesterases in nitric oxide and cGMP action. *Pharmacol Rev* 62, 525–563, 2010.

Girouard H, Iadecola C. Neurovascular coupling in the normal brain and in hypertension, stroke, and Alzheimer disease. *J Appl Physiol* 100, 328–335, 2006.

Gonzalez-Fernandez B, Ermentrout JM. On the origin and dynamics of the vasomotion of small arteries. *Math Biosci* 167(2), 127–167, 1994.

Grubb S, Cai C, Hald BO, Khennouf L, Murmu RP, Jensen AGK, Fordsmann J, Zambach S, Lauritzen M. Precapillary sphincters maintain perfusion in the cerebral cortex. *Nat Commun* 11, 395, 2020.

Hai C-M, Murphy RA. Cross-bridge phosphorylation and regulation of latch state in smooth muscle. *AM J Physiol Cell Physiol* 254, C99–C106, 1988.

Hai C-M, Murph RA. Ca2+, Crossbridge phosphorylation, and contraction. *Annu Rev Physiol* 51, 285–298, 1989.

Hamel E. Perivascular nerves and the regulation of cerebrovascular tone. *J Appl Physiol* 100, 1059–1064, 2006.

Hawkins BT, Davis TP. The blood-brain barrier/neurovascular unit in health and disease. *Pharmacolo Rev* 57(2), 173–185, 2005.

Heuser D. The significance of cortical extracellular H+, K+ and Ca2+ activities for regulation of local cerebral blood flow under conditions of enhanced neuronal activity. *Ciba Found Symp* 56, 339–353, 1978.

Hill-Eubanks DC, Werner ME, Heppner TJ, Nelson MT. Calcium signaling in smooth muscle. *Cold Spring Harb Perspect Biol* 3, a004549, 2011.

Hori S, Ohtsuki S, Hosoya K, Nakashima E, and Terasaki T. A pericyte derived angiopoietin-1 multimeric complex induces occludin gene expression in brain capillary endothelial cells through Tie-2 activation in vitro. *J Neurochem* 89, 503–513, 2004.

Jackson WF. Boosting the signal: Endothelial inward rectifier K+ channels. *Microcirculation* 24, e12319, 2017.

Kamitani S, Yamada K, Yamamoto S, Ishimoto Y, Ono T, Saiga A, Hanasaki K. Differences between group X and group V secretory phospholipase A2 in lipolytic modification of lipoproteins. *Cell Mol Biol Lett* 17, 459–478, 2012.

Ko EA, Han J, Jung ID, Park WS. Physiological roles of K+ channels in vascular smooth muscle cells (Hirosi Kuriyama Award 2007 Memorial Review). *J Smooth Muscle Res* 44(2), 65–81, 2008.

Koenigsberger M, Sauser R, Bény J-L, Meister J-J. Effects of arterial wall stress on vasomotion. *Biophys J* 91, 1663–1674, 2006.

Krogh A. The number and distribution of capillaries in muscle with calculation of the oxygen pressure head necessary for supplying the tissue. *J Physiol* 52, 409–415, 1919.

Lee MR, Li L, Kitazawa T. Cyclic GMP causes Ca2+ desensitization in vascular smooth muscle by activating the myosin light chain phosphatase. *J Biol Chem* 272, 5063–5068, 1997.

Liu X, Li C, Falck JR, Roman RJ, Harder DR, Koehler RC. Interaction of nitric oxide, 20-HETE, and EETs during functional hyperemia in whisker barrel cortex. *Am J Physiol Heart Circ Physiol* 295, H619–H631, 2008.

Manninen T, Havela R, Linne M-L. Computational models for calcium-mediated astrocyte functions. *Front Comput Neurosci* 12(article 14), 1–27, 2018.

Metea MR, Newman EA. Glial cells dilate and constrict blood vessels: A mechanism of neurovascular coupling. *J Neurosci* 26, 2862–2870, 2006.

Monod J, Wyman J, Changeux JP. On the nature of allosteric transitions: A plausible model. *J Mol Biol* 12, 88–118, 1965.

Moshkforoush A, Ashenagar B, Harraz OF, Dabertrand F, Longden TA, Nelson MT, Tsoukias NM. The capillary Kir channel as sensor and amplifier of neuronal signals: Modeling insights on K+-mediated neurovascular communication. *Proc Natl Acad Sci USA* 117(28), 16626–16637, 2020.

Nilius B, Droogmans G. Ion channels and their functional role in vascular endothelium. *Physiol Rev* 81(4), 1415–1459, 2001.

Reese TS, Karnovsky MJ. Fine structural localization of a blood-brain barrier to exogenous peroxidase. *J Cell Biol* 34, 207–217, 1967.

Satcher RL, Dewey CF. Theoretical estimates of mechanical properties of the endothelial cell cytoskeleton. *Biophys J* 71, 109–118, 1996.

Sriram K, Laughlin JG, Rangamani P, Tartakovsky DM. Shear-induced nitric oxide production by endothelial cells. *Biophys J* 111, 208–221, 2016.

Suarez A, Fernandez L, Riera J. Characterizing astrocyte-mediated neurovascular coupling by combining optogenetics and biophysical modeling. *J Cereb Blood Flow Metab*, 2025. https://doi.org/10.1177/0271678X241311010

Sun GY, Xu J, Jensen MD, Yu S, Wood WG, González FA, Simonyi A, Sun AY, Weisman GA. Phospholipase A_2 in astrocytes – Responses to oxidative stress, inflammation, and G protein-coupled receptor agonists. *Mol Neurobiol* 31, 27–41, 2005.

Tesler F, Linne ML, Destexhe A. Modeling the relationship between neuronal activity and the BOLD signal: Contributions from astrocyte calcium dynamics. *Sci Rep* 13, 6451, 2023.

Teubl M, Groschner K, Kohlwein SD, Mayer B, Schmidt K. Na+/Ca2+ exchange facilitates Ca2+-dependent activation of endothelial nitric-oxide synthase. *J Biol Chem* 274, 29529–29535, 1999.

Wang J, Chen S, Siegelbaum SA. Regulation of hyperpolarization-activated HCN channel gating and cAMP modulation due to interactions of COOH terminus and core transmembrane regions. *J Gen Physiol* 118, 237–250, 2001.

Yamazaki Y, Kamiyama Y. Mathematical model of wall shear stress-dependent vasomotor response based on physiological mechanisms. *Comput Biol Med* 45, 126–135, 2014.

Yang J, Clark JW, Bryan RM, Robertson CS. Mathematical modeling of the nitric oxide/cGMP pathway in the vascular smooth muscle cell. *Am J Physiol Heart Circ Physiol* 289, H886–H897, 2005.

Appendix A
Introduction to Vector Calculus

In this appendix, operators, and identities relevant to vector calculus are listed. Vector calculus is defined as differentiation and integration of vector $\vec{F}(\vec{r})$ and scalar $f(\vec{r})$ fields. For this book, a vector (scalar) field consists of an assignment of a 3D vector (1D scalar) to each point in a subset of the physical space. The space on which a vector (scalar) field is defined could, in general, be either on the Euclidian 3D space or on an arbitrary 2D manifold (surface).

Def 1: Assuming an infinitesimal volume $V_{\vec{r}}$ enclosing a point $\vec{r}$ in the space through its limiting surface S, the divergence operator ($\nabla\cdot$, or div) of a vector field $\vec{F}(\vec{r})$ at $\vec{r}$ is defined as the limit of the ratio of the flux of $\vec{F}(\vec{r})$ across a surface S and $V_{\vec{r}}$, as the latter reduces to zero:

$$\nabla\cdot\vec{F}(\vec{r}) = \lim_{V_{\vec{r}}\to 0}\frac{1}{V_{\vec{r}}}\oint_S \vec{F}(\vec{r}')\cdot d\vec{S}'$$

$$The\,Gauss's\ Theorem: \int_V \nabla\cdot\vec{F}(\vec{r})dV = \oint_S \vec{F}(\vec{r})\cdot d\vec{S} \tag{A.1}$$

Def 2: Assuming that an infinitesimal area $A_{\vec{r}}$ with boundary C contains a point $\vec{r}$ in the space, the curl operator ($\nabla\times$, or rot) of a vector field $\vec{F}(\vec{r})$ at $\vec{r}$ is defined as the limit of the ratio of the line integral of $\vec{F}(\vec{r})$ along C and $A_{\vec{r}}$, as the latter reduces to zero:

$$\nabla\times\vec{F}(\vec{r}) = \lim_{A_{\vec{r}}\to 0}\frac{1}{A_{\vec{r}}}\oint_C \vec{F}(\vec{r}')\cdot d\vec{l}'$$

$$The\,Stoke's\,Theorem: \int_S (\nabla\times\vec{F}(\vec{r}))\cdot d\vec{S} = \oint_C \vec{F}(\vec{r})\cdot d\vec{l} \tag{A.2}$$

Def 3: The gradient of a scalar field $f(\vec{r})$ is a vector field defined as:

$$\nabla f(\vec{r}) = \frac{\partial f(\vec{r})}{\partial x}\hat{x} + \frac{\partial f(\vec{r})}{\partial y}\hat{y} + \frac{\partial f(\vec{r})}{\partial z}\hat{z} \tag{A.3}$$

∇ is named the Del or nabla operator. The gradient can be applied to each component $k=\{x,y,z\}$ of the vector field $\vec{F}(\vec{r})$.

$$\nabla F_k(\vec{r})=\frac{\partial F_k(\vec{r})}{\partial x}\hat{x}+\frac{\partial F_k(\vec{r})}{\partial y}\hat{y}+\frac{\partial F_k(\vec{r})}{\partial z}\hat{z}=\sum_{l=\{x,y,z\}}\frac{\partial F_k(\vec{r})}{\partial l}\hat{l}$$

The gradient of a vector field $\vec{F}(\vec{r})$ is a 2nd order tensor defined as:

$$\nabla\vec{F}(\vec{r})=\left\{\frac{\partial F_k(\vec{r})}{\partial l}\right\},\text{ with } k=\{x,y,z\}\text{ and } l=\{x,y,z\} \tag{A.4}$$

A 2nd order tensor is defined as $\vec{F}(\vec{r})=\{\vec{F}^x,\vec{F}^y,\vec{F}^z\}=\{F_l^k\}$. In this case, $\vec{F}^l=\nabla F_l(\vec{r})$, with $l=\{x,y,z\}$.

Def 3: The Laplacian operators on scalar and vector fields are defined as $\nabla^2 f(\vec{r})=\nabla\cdot\nabla f(\vec{r})$ and $\nabla^2\vec{F}(\vec{r})=\nabla\cdot\nabla\vec{F}(\vec{r})$, respectively. Note that the definition of divergence operator was generalized to a 2nd order tensor, consisting of applying separately the divergence to each vector component *k-th* of the tensor $\nabla\cdot\vec{F}^k(\vec{r})$. Such operation, which belongs to tensor calculus, is only used in this book to define the Laplacian of magnetic vector potential $\vec{A}$.

FUNDAMENTAL IDENTITIES

Given the vector $\vec{F}(\vec{r})$ and scalar $f(\vec{r})$ fields, the following identities are demonstrated:

$$\nabla\cdot\nabla\times\vec{F}(\vec{r})=0 \tag{A.5}$$

$$\nabla\times\nabla f(\vec{r})=0 \tag{A.6}$$

$$\nabla\times\nabla\times\vec{F}(\vec{r})=\nabla\left(\nabla\cdot\vec{F}(\vec{r})\right)-\nabla^2\vec{F}(\vec{r}) \tag{A.7}$$

$$\nabla\cdot\left(f(\vec{r})\vec{G}(\vec{r})\right)=\left(\nabla f(\vec{r})\right)\cdot\vec{G}(\vec{r})+f(\vec{r})\left(\nabla\cdot\vec{G}(\vec{r})\right) \tag{A.8}$$

$$\nabla\left(f(\vec{r})g(\vec{r})\right)=\left(\nabla f(\vec{r})\right)g(\vec{r})+f(\vec{r})\left(\nabla g(\vec{r})\right) \tag{A.9}$$

With $g(\vec{r})$ and $\vec{G}(\vec{r})$ another arbitrary scalar and vector fields, respectively.

$$\nabla\times\left(f(\vec{r})\vec{G}(\vec{r})\right)=\left(\nabla f(\vec{r})\right)\times\vec{G}(\vec{r})+f(\vec{r})\left(\nabla\times\vec{G}(\vec{r})\right) \tag{A.10}$$

Appendix B
The Green Identities

A Green's function $G(\vec{r},\vec{r}')$ can be defined for any linear differential operator $LG(\vec{r},\vec{r}') = -4\pi\delta(\vec{r}-\vec{r}')$. The Dirac delta function $\delta(\vec{r}-\vec{r}')$ is ∞ at $\vec{r}=\vec{r}'$ and zero elsewhere. However, the one of interest for this book is defined from the Laplacian linear operator, i.e., $L=\nabla^2$. As ∇^2 is a self-adjunct operator, this Green's function is symmetric $G(\vec{r},\vec{r}') = G(\vec{r}',\vec{r})$, a property called "reciprocity". The Green's function for the Laplacian operator is linked to the Green's identities.

Applying (A.8) to a vector function $u(\vec{r})\nabla v(\vec{r})$, taking the integral over a volume V (limited by surface S) and using the Gauss's theorem (A.1), it can be obtained the Green's first identity.

$$\int_V \left(u(\vec{r})\nabla^2 v(\vec{r}) + \nabla u(\vec{r})\cdot\nabla v(\vec{r})\right)dV = \oint_S u(\vec{r})\nabla v(\vec{r})\cdot d\vec{S} \tag{B.1}$$

Similarly, applying (A.8) to vector functions $u(\vec{r})\nabla v(\vec{r})$ and $v(\vec{r})\nabla u(\vec{r})$, and subtracting them $\nabla\cdot(u(\vec{r})\nabla v(\vec{r})) - \nabla\cdot(v(\vec{r})\nabla u(\vec{r}))$, it can be proven $\nabla\cdot(u(\vec{r})\nabla v(\vec{r}) - v(\vec{r})\nabla u(\vec{r})) = u(\vec{r})\nabla^2 v(\vec{r}) - v(\vec{r})\nabla^2 u(\vec{r})$. The Green's second identity is obtained by taking the integral over a volume V (limited by surface S) and using the Gauss's theorem (A.1).

$$\int_V \left(u(\vec{r})\nabla^2 v(\vec{r}) - v(\vec{r})\nabla^2 u(\vec{r})\right)dV = \oint_S \left(u(\vec{r})\nabla v(\vec{r}) - v(\vec{r})\nabla u(\vec{r})\right)\cdot d\vec{S} \tag{B.2}$$

These identities are very useful to solve boundary problems. By defining $v(\vec{r}) = G(\vec{r},\vec{r}')$, equations (B.1) and (B.2) transform into:

$$u(\vec{r}) = \oint_S u(\vec{r}')\nabla' G(\vec{r},\vec{r}')\cdot d\vec{S}' - \int_V \left(\nabla u(\vec{r}')\cdot\nabla' G(\vec{r},\vec{r}')\right)dV' \tag{B.3}$$

$$u(\vec{r}) = \int_V \left(G(\vec{r},\vec{r}')\nabla^2 u(\vec{r}')\right)dV' + \oint_S \left(u(\vec{r}')\nabla G(\vec{r},\vec{r}') - G(\vec{r},\vec{r}')\nabla u(\vec{r}')\right)\cdot d\vec{S}' \tag{B.4}$$

There are several methods to obtain the Green's function for a particular boundary value problem (e.g., Laplace transforms, the method of images, and separation of variables). However, one of the most useful methods is based on the complete set of eigenfunctions $\psi_n(\vec{r})$ of the Laplacian operator: $\nabla^2\psi_n(\vec{r}) = \lambda_n\psi_n(\vec{r})$.

$$G(\vec{r},\vec{r}\,') = \sum_{n=1}^{\infty} \frac{\psi_n^*(\vec{r})\,\psi_n(\vec{r}\,')}{\lambda_n} \tag{B.5}$$

The symbol * represents the complex conjugation.

The Green function for the Laplacian in the 3D Euclidian linear space $\Re$ is defined as:

$$G(\vec{r},\vec{r}\,') = \frac{1}{|\vec{r}-\vec{r}\,'|} + F(\vec{r}) \tag{B.6}$$

With $F(\vec{r})$ being any Harmonic function in $\Re$. By definition, a scalar (or vector) harmonic function has zero Laplacian.

Appendix C Magnetic LF – Isotropic Piecewise Homogeneous Conductor

In this appendix, the eddy current density $\vec{J}_M(\vec{r},\omega)$ inside a volume conductor V induced by energizing a single coil with an ac current $I_c(\omega)$ will be computed using the magnetic reciprocity theorem (Figure 2.12).

REALISTIC ISOTROPIC PIECEWISE HOMOGENEOUS VOLUME CONDUCTOR

As in Chapter 2, let us consider the head as an isotropic piecewise homogenous volume conductor model, i.e., a set of embedded compartments V_k $(k=1,\cdots,N)$ of isotropic conductivity σ_k $(k=0,\cdots,N)$, with the brain the innermost region V_0. The surface S_k separates compartments V_k and V_{k+1}. Energizing the single coil with the ac current induces an eddy current density inside each compartment V_k, which satisfies: **a)** $\vec{J}_k(\vec{r},\omega)$ has no flux sources inside V_k, and consequently, its divergence is equal to zero (i.e., charge continuity law), and **b)** the $\vec{J}_k(\vec{r},\omega)$ vortex sources inside V_k will be due to a time varying magnetic induction vector. Thus, the eddy current density inside compartment V_k is described by the following equations in the frequency domain:

$$\nabla\cdot\vec{J}_k(\vec{r},\omega)=0 \tag{C.1}$$

$$\nabla\times\vec{J}_k(\vec{r},\omega)=-i\omega\sigma_k\vec{B}_k(\vec{r},\omega) \tag{C.2}$$

The magnetic induction vector in the compartment V_k $(k=1,\cdots,N+1)$ has three vortex sources: (**1**) the primary coil current $\vec{J}_c(\vec{r},\omega)$, and the secondary eddy $\vec{J}_k(\vec{r},\omega)$ (**2**) and capacitive $i\omega\varepsilon(\omega)\vec{E}_k(\vec{r},\omega)$ (**3**) currents.

$$\nabla\times\vec{B}_k(\vec{r},\omega)=\mu_o\left(\vec{J}_c(\vec{r},\omega)+\vec{J}_k(\vec{r},\omega)+i\omega\varepsilon_k(\omega)\vec{E}_k(\vec{r},\omega)\right)$$

Therefore, $\vec{J}_k(\vec{r},\omega)$ satisfies the following homogeneous vector Helmholtz equation:

$$\nabla^2\vec{J}_k(\vec{r},\omega)+K^2(\omega)\vec{J}_k(\vec{r},\omega)=0$$

where $K^2(\omega) = \left[\left(\frac{\omega}{c_k(\omega)}\right)^2 - i\omega\mu_o\sigma_k\right]$. The velocity of light in the compartment depends on the frequency ω, i.e., $c_k(\omega) = \frac{1}{\sqrt{\mu_o \varepsilon_k(\omega)}}$. Therefore, the dispersion relationship can be written as:

$$K(\omega) = \frac{1}{\sqrt{2}}\left[\left(\left(\frac{\omega}{c_k(\omega)}\right)^4 + (\omega\mu_o\sigma_k)^2\right)^{\frac{1}{2}} + \left(\frac{\omega}{c_k(\omega)}\right)^2\right]^{\frac{1}{2}} + \frac{i}{\sqrt{2}}\left[\left(\left(\frac{\omega}{c_k(\omega)}\right)^4 + (\omega\mu_o\sigma_k)^2\right)^{\frac{1}{2}} - \left(\frac{\omega}{c_k(\omega)}\right)^2\right]^{\frac{1}{2}}$$

However, brain activity appears at a very low frequency range $(0 \le \omega \le 500)Hz$. Taking into consideration, the velocity of light in living tissues, the magnitude $\left(\frac{\omega}{c_k(\omega)}\right)^2 << 1$. In this case, the wavelength is around several meter larger than the dimension of the head, hence the dispersion relationship transforms to: $k(\omega) = \frac{(1+i)}{\delta(\omega)}$, where the depth $\delta(\omega) = \sqrt{\frac{2}{\omega\mu_o\sigma_j}}$ is defined as the distance inside the compartment in which the magnetic induction vector decrease e^{-1} times its amplitude. The depth is too large compared to the dimension of the head for this frequency range and, therefore $k(\omega) = 0$, for which the quasi-static approach of living tissues is satisfied. Under the quasi-static approach, the magnetic induction vector inside each compartment will be only that produced by the single coil in an infinite and homogeneous volume conductor $\vec{B}_c(\vec{r}, \omega)$.

$$\nabla \cdot \vec{B}_c(\vec{r}, \omega) = 0$$

$$\nabla \times \vec{B}_c(\vec{r}, \omega) = \mu_o \vec{J}_c(\vec{r}, \omega)$$

$$\lim_{\vec{r} \to \infty} \vec{B}_c(\vec{r}, \omega) = 0$$

Under the Coulomb gauge $\nabla \cdot \vec{A}_c(\vec{r}, \omega) = 0$, the solution for the magnetic potential satisfies:

$$\vec{A}_c(\vec{r}, \omega) = \frac{\mu_o I_c(\omega)}{4\pi} \int_{\Delta S_c} d\vec{r}'^2 \cdot \left(\nabla' \times \vec{G}(\vec{r}, \vec{r}')\right)$$

The dyadic Green function $\overleftarrow{G}(\vec{r},\vec{r}')$ was used in a previous study to compute the electric LF (Riera and Fuentes, 1998). Since the single coil area ΔS_c is small enough as to consider $\overleftarrow{G}(\vec{r},\vec{r}')$ a constant function, it follows:

$$\vec{A}_c(\vec{r},\omega)=\frac{\mu_o I_c(\omega)\Delta S_c}{4\pi}\nabla\times\left(\overleftarrow{G}(\vec{r},\vec{r}_c)\cdot\vec{o}\right)$$

Finally, the magnetic induction vector produced by a single coil in an infinite and homogeneous volume conductor is:

$$\vec{B}_c(\vec{r},\omega)=\frac{\mu_o I_c(\omega)\Delta S_c}{4\pi}\nabla\times\nabla\times\left(\overleftarrow{G}(\vec{r},\vec{r}_c)\cdot\vec{o}\right) \tag{C.3}$$

Under the quasi-static approach, $\vec{J}_k(\vec{r},\omega)$ in the compartment V_k satisfies:

$$\nabla\times\vec{J}_k(\vec{r},\omega)=-i\omega\sigma_k\vec{B}_c(\vec{r},\omega) \tag{C.4}$$

$$\nabla^2\vec{J}_k(\vec{r},\omega)=0 \tag{C.5}$$

The normal component of $\vec{J}_k(\vec{r},\omega)$ has no discontinuity on any surface:

$$\left(\vec{J}_k(\vec{r},\omega)-\vec{J}_{k+1}(\vec{r},\omega)\right)\cdot\vec{n}_k(\vec{r})\Big|_{S_k}=0 \quad (k=1,\cdots,N)$$

$\vec{n}_k(\vec{r})$ is the normal to the surface S_k at position $\vec{r}$. The tangential component of the electric field has no discontinuity on any surface, hence $\vec{J}_k(\vec{r},\omega)$ will have a surface vortex source on every surface:

$$\left(\frac{\vec{J}_k(\vec{r},\omega)}{\sigma_k}-\frac{\vec{J}_{k+1}(\vec{r},\omega)}{\sigma_{k+1}}\right)\times\vec{n}_k(\vec{r})\Bigg|_{S_k}=0 \quad (k=1,\cdots,N)$$

Considering that the normal and tangential components of $\vec{B}_c(\vec{r},\omega)$ are continuos on any surface, $\vec{J}_l(\vec{r},\omega)$ in the compartment V_l can be computed from the vector Green second identity:

$$4\pi\vec{J}_l(\vec{r},\omega)=4\pi\vec{J}_\infty(\vec{r},\omega)+\sum_{k=1}^{N}\frac{(\sigma_k-\sigma_{k+1})}{\sigma_k}\oint_{S_k}\left(\nabla'\times\overleftarrow{G}(\vec{r},\vec{r}')\right)\cdot\left(\vec{n}_k(\vec{r}')\times\vec{J}_k(\vec{r}',\omega)\right)d\vec{r}'^2 \tag{C.6}$$

$$\text{where: } \vec{J}_\infty(\vec{r},\omega)=\frac{i\omega}{4\pi}\sum_{k=1}^{N}(\sigma_k-\sigma_{k+1})\oint_{S_k}\overleftarrow{G}(\vec{r},\vec{r}')\cdot\left(\vec{n}_k(\vec{r}')\times\vec{B}_c(\vec{r}',\omega)\right)d\vec{r}'^2 \tag{C.7}$$

In the Cartesian coordinate system, the expression (C.6) can be transformed in:

$$4\pi\vec{J}_l(\vec{r},\omega)=4\pi\vec{J}_\infty(\vec{r},\omega)-\sum_{k=1}^{N}\frac{(\sigma_k-\sigma_{k+1})}{\sigma_k}\oint_{S_k}\nabla'\left(\frac{1}{|\vec{r}-\vec{r}'|}\right)\times\left(\vec{n}_k(\vec{r}')\times\vec{J}_k(\vec{r}',\omega)\right)d\vec{r}'^2 \tag{C.8}$$

Note that equation (C.8) is similar to that proposed in Riera and Fuentes (1998) for the computation of the electric LF. Therefore, to compute the magnetic LF for a realistic isotropic piecewise homogeneous volume conductor, the vector boundary element method (BME) presented in Riera and Fuentes (1998) must be solved with the specification of a new inhomogeneous term.

$$\vec{J}_\infty(\vec{r},\omega_c)=\frac{i\omega}{4\pi}\sum_{k=1}^{N}(\sigma_k-\sigma_{k+1})\oint_{S_k}\frac{\vec{n}_k(\vec{r}')\times\vec{B}_c(\vec{r}',\omega_c)}{|\vec{r}-\vec{r}'|}d\vec{r}'^2 \tag{C.9}$$

If the magnetic induction vector is assumed constant in each triangle and equal to its value in the center of mass $\vec{B}_{a}^{\,k}(\omega_c)$, the equation (C.9) transforms to:

$$\vec{J}_\infty(\vec{r},\omega_c)=\frac{i\omega_c}{4\pi}\sum_{k=1}^{N}(\sigma_k-\sigma_{k+1})\sum_{n=1}^{N_k}\vec{n}_n^k\times\vec{B}_{cn}^{\,k}(\omega_c)\,\eta_n^k(\vec{r}) \tag{C.10}$$

where $\eta_n^k(\vec{r})=\int_{\Delta_n^k}\frac{d\vec{r}'^2}{|\vec{r}-\vec{r}'|}$. N_k represents the number of triangles in the surface S_k.

An analytical expression to compute $\eta_n^k(\vec{r})$ has been proposed in de Munck (1992, equation 25). Note that it is necessary to specify the transversal component of the magnetic induction vector $\vec{B}_{cn}^{\,k}(\omega_c)$ in the triangles of every surface. This component is computed from the evaluation of expression (C.3) in the center $\vec{r}_n^{\,k}$ of each triangle Δ_n^k for all surfaces S_k in Cartesian coordinate system.

$$\vec{B}_{cn}^{\,k}(\omega_c)=\frac{\mu_o I_c(\omega_c)\Delta S_c}{4\pi}\left(\frac{3\vec{o}\cdot(\vec{r}_n^{\,k}-\vec{r}_c)\circ(\vec{r}_n^{\,k}-\vec{r}_c)}{|\vec{r}_n^{\,k}-\vec{r}_c|^5}-\frac{\vec{o}}{|\vec{r}_n^{\,k}-\vec{r}_c|^3}\right)$$

THREE CONCENTRIC SPHERES

Let us consider the case of three concentric spheres with radii $a<b<c$, and conductivities σ_a, σ_b and σ_c, respectively. The ohmic current density inside the inner, middle, and outer spherical compartments must satisfy equation (C.6), which leads to a *"system of equations"*. The eigenfunctions of the vector Laplacian operator comprise: a longitudinal ($\vec{N}_j(\vec{r})$) and two transversal ($\vec{M}_j(\vec{r})$ and $\vec{G}_j(\vec{r})$) vector fields. These vector fields satisfy the following relationship in the whole space: **1)**

$\nabla \cdot \vec{N}_j(\vec{r}) = 0$, $\nabla \times \vec{N}_j(\vec{r}) = \vec{0}$; **2)** $\nabla \cdot \vec{M}_j(\vec{r}) = 0$, $\nabla \times \vec{M}_j(\vec{r}) \neq \vec{0}$; and **3)** $\nabla \cdot \vec{G}_j(\vec{r}) \neq 0$, $\nabla \times \vec{G}_j(\vec{r}) \neq \vec{0}$. In the spherical symmetry ($j = \{n,m,s\}$), each of these eigenfuctions has two components. These components satisfy that: $\vec{N}^1_{nms}(\vec{r})$, $\vec{M}^1_{nms}(\vec{r})$ and $\vec{G}^1_{nms}(\vec{r})$ neglect for $|\vec{r}| \to 0$ and $\vec{N}^2_{nms}(\vec{r})$, $\vec{M}^2_{nms}(\vec{r})$ and $\vec{G}^2_{nms}(\vec{r})$ neglect for $|\vec{r}| \to \infty$. The transversal fields $\vec{M}^1_{nms}(\vec{r})$ and $\vec{M}^2_{nms}(\vec{r})$ satisfies the conditions (C.1 and C.2) for $\vec{J}_k(\vec{r},\omega)$ in each compartment V_k ($k = a,b,c$). Therefore, neither the longitudinal components $\vec{N}^1_{nms}(\vec{r})$ and $\vec{N}^2_{nms}(\vec{r})$ nor the transversal components $\vec{G}^1_{nms}(\vec{r})$ and $\vec{G}^2_{nms}(\vec{r})$ contribute to $\vec{J}_k(\vec{r},\omega)$ in any compartment.

$$\vec{J}_a(\vec{r},\omega) = \sum_{nms} A_{nms}(\omega)\vec{M}^1_{nms}(\vec{r}) \quad r < a$$

$$\vec{J}_b(\vec{r},\omega) = \sum_{nms} B_{nms}(\omega)\vec{M}^1_{nms}(\vec{r}) + \sum_{nms} C_{nms}(\omega)\vec{M}^2_{nms}(\vec{r}) a < r < b \tag{C.11}$$

$$\vec{J}_c(\vec{r},\omega) = \sum_{nms} D_{nms}(\omega)\vec{M}^1_{nms}(\vec{r}) + \sum_{nms} E_{nms}(\omega)\vec{M}^2_{nms}(\vec{r}) b < r < c$$

For the computation of $\vec{J}_k(\vec{r},\omega)$ in each compartment, it is necessary to obtain the coefficients $A_{nms}(\omega)$, $B_{nms}(\omega)$, $C_{nms}(\omega)$, $D_{nms}(\omega)$, and $E_{nms}(\omega)$. To compute the inhomogeneous term of the equation (C.6), the magnetic induction vector $\vec{B}_c(\vec{r},\omega)$ produced by an energized single coil positioned on $\vec{r}_c$ and oriented as $\vec{o}$ must be obtained. The expression (C.3) provides us with an analytical formula to compute the magnetic induction vector $\vec{B}_c(\vec{r},\omega)$ being generated by the application of an ac current $I_c(\omega)$ to a single coil. Using the expression (19) in Riera and Fuentes (1998), the following result can be justified:

$$\vec{B}_c(\vec{r},\omega) = -\frac{\mu_o I_c(\omega)\Delta S_c}{4\pi}\sum_{nms}\varepsilon_m \frac{(n-m)!}{(n+m)!}\left(\vec{N}^2_{n+1,ms}(\vec{r}_c)\cdot\vec{o}\right)\vec{N}^1_{n-1,ms}(\vec{r})$$

Therefore, the following relationship is valid for all surfaces:

$$\vec{n} \times \vec{B}_c(\vec{r}',\omega) = \frac{\mu_o I_c(\omega)\Delta S_c}{4\pi}\sum_{nms}\varepsilon_m \frac{(n-m)!}{(n+m)!}\left(\vec{N}^2_{n+1,ms}(\vec{r}_c)\cdot\vec{o}\right)\sqrt{n(n+1)}\,r'^{n-1}\vec{C}^s_{nm}(\vartheta',\phi')$$

Finally, the inhomogeneous term in each compartment can be specified as:

$$\vec{J}_\infty(\vec{r},\omega) = \sum_{nms}\beta_{nms}(\omega)\sigma_a\vec{M}^1_{nms}(\vec{r}) \quad \vec{r} \in R_a$$

$$\vec{J}_\infty(\vec{r},\omega) = \sum_{nms}\beta_{nms}(\omega)\left(\sigma_b\vec{M}^1_{nms}(\vec{r}) + (\sigma_a - \sigma_b)a^{2n+1}\vec{M}^2_{nms}(\vec{r})\right) \quad \vec{r} \in R_b$$

$$\vec{J}_\infty(\vec{r},\omega) = \sum_{nms}\beta_{nms}(\omega)\left(\sigma_c\vec{M}^1_{nms}(\vec{r}) + \left((\sigma_a - \sigma_b)a^{2n+1} + (\sigma_b - \sigma_c)b^{2n+1}\right)\vec{M}^2_{nms}(\vec{r})\right)\vec{r} \in R_c$$

where: $\beta_{nms}(\omega)=\dfrac{i\omega\mu_o I_c(\omega)\Delta S_c}{4\pi}\varepsilon_m\dfrac{(n-m)!}{(n+m)!}\dfrac{\left(\vec{N}^2_{n+1,ms}(\vec{r}_c)\cdot\vec{o}\right)}{(2n+1)}$

Substituting these expressions for $\vec{J}_\infty(\vec{r},\omega)$ and (C.11) in the main system of equations (C.6), the following algebraic system for the coefficients is obtained:

$$A_{nms}(\omega)=\sigma_a\beta_{nms}(\omega)+\frac{n}{(2n+1)}\left[\frac{(\sigma_a-\sigma_b)}{\sigma_a}A_{nms}(\omega)+\frac{(\sigma_b-\sigma_c)}{\sigma_b}\left(B_{nms}(\omega)+C_{nms}(\omega)b^{-2n-1}\right)+\left(D_{nms}(\omega)+E_{nms}(\omega)c^{-2n-1}\right)\right]$$

$$B_{nms}(\omega)=\sigma_b\beta_{nms}(\omega)+\frac{n}{(2n+1)}\left[\frac{(\sigma_b-\sigma_c)}{\sigma_b}\left(B_{nms}(\omega)+C_{nms}(\omega)b^{-2n-1}\right)+\left(D_{nms}(\omega)+E_{nms}(\omega)c^{-2n1}\right)\right]$$

$$C_{nms}(\omega)=(\sigma_a-\sigma_b)a^{2n+1}\beta_{nms}(\omega)-\frac{(n+1)}{(2n+1)}\left[\frac{(\sigma_a-\sigma_b)}{\sigma_a}a^{2n+1}A_{nms}(\omega)\right]$$

$$D_{nms}(\omega)=\sigma_c\beta_{nms}(\omega)+\frac{n}{(2n+1)}\left(D_{nms}(\omega)+E_{nms}(\omega)c^{-2n-1}\right)$$

$$E_{nms}(\omega)=\left[(\sigma_a-\sigma_b)a^{2n+1}+(\sigma_b-\sigma_c)b^{2n+1}\right]\beta_{nms}(\omega)-\frac{(n+1)}{(2n+1)}\left[\frac{(\sigma_a-\sigma_b)}{\sigma_a}a^{2n+1}A_{nms}(\omega)+\frac{(\sigma_b-\sigma_c)}{\sigma_b}b^{2n+1}\left(B_{nms}(\omega)+C_{nms}(\omega)\right)b^{-2n-1}\right]$$

The coefficients resulting of the solution of this system are:

$$A_{nms}(\omega)=\frac{(2n+1)}{(n+1)}\sigma_a\beta_{nms}(\omega)$$

$$B_{nms}(\omega)=\frac{(2n+1)}{(n+1)}\sigma_b\beta_{nms}(\omega)$$

$$C_{nms}(\omega)=0$$

$$D_{nms}(\omega)=\frac{(2n+1)}{(n+1)}\sigma_c\beta_{nms}(\omega)$$

$$E_{nms}(\omega)=0$$

Note that the eigen functions $\vec{N}^1_{nms}(\vec{r})$, $\vec{N}^2_{nms}(\vec{r})$, $\vec{G}^1_{nms}(\vec{r})$ and $\vec{G}^2_{nms}(\vec{r})$ are specified by the two first vector spherical harmonics $\vec{P}^s_{nm}(\vartheta,\phi)$ and $\vec{B}^s_{nm}(\vartheta,\phi)$ (Riera and Fuentes,

1998). The eigenfunctions $\vec{M}^1_{nms}(\vec{r})$ and $\vec{M}^2_{nms}(\vec{r})$ are specified by the third vector spherical harmonic $\vec{C}^s_{nm}(\vartheta,\phi)$ (Riera and Fuentes, 1998). The dyadic Green function $\bar{G}(\vec{r},\vec{r}')$ can be expressed in terms of the vector fields $\vec{N}^1_{nms}(\vec{r})$, $\vec{N}^2_{nms}(\vec{r})$, $\vec{M}^1_{nms}(\vec{r})$, $\vec{M}^2_{nms}(\vec{r})$, $\vec{G}^1_{nms}(\vec{r})$, and $\vec{G}^2_{nms}(\vec{r})$. Additionally, $\bar{G}(\vec{r},\vec{r}')$ satisfies the relationships (18)-(21) in Riera and Fuentes (1998). The following relationships were also used:

$$\vec{n}\times\vec{M}^1_{nms}(\vec{r}')=\sqrt{n(n+1)}\,r'^{n}\,\vec{B}^s_{nm}(\vartheta',\phi')$$

$$\vec{n}\times\vec{M}^2_{nms}(\vec{r}')=\sqrt{n(n+1)}\,r'^{-n-1}\,\vec{B}^s_{nm}(\vartheta',\phi')$$

The functions $Y^s_{nm}(\vartheta,\phi)$ are the scalar spherical harmonics.

Finally, $\vec{J}_k(\vec{r},\omega)$ in the compartment V_k is calculated as:

$$\vec{J}_k(\vec{r},\omega)=\frac{i\omega\mu_o I_c(\omega)\Delta S_c\sigma_k}{4\pi}\sum_{nms}\varepsilon_m\frac{(n-m)!}{(n+m)!}\frac{\left(\vec{N}^2_{n+1,ms}(\vec{r}_c)\cdot\vec{o}\right)}{(n+1)}\vec{M}^1_{nms}(\vec{r})(k=a,b,c) \quad \text{(C.12)}$$

Each spherical compartment has the same magnetic LF, and it can be written as:

$$\vec{K}_M(\vec{r})=-\frac{\mu_o}{4\pi}\sum_{nms}\varepsilon_m\frac{(n-m)!}{(n+m)!}\frac{\left(\vec{N}^2_{n+1,ms}(\vec{r}_c)\cdot\vec{o}\right)}{(n+1)}\vec{M}^1_{nms}(\vec{r}) \quad \text{(C.13)}$$

Therefore, the magnetic LF $\vec{K}_M(\vec{r})$ inside the spherical compartments is represented by a transversal field $\vec{M}^1_{smn}(\vec{r})$ and it is also independent of radius and conductivity values.

Appendix D
The Spherical Harmonics

The spherical harmonics are the eigenfunction of the Laplace-Beltrani operator ∇^{*2} defined over the unit sphere.

$$\nabla^{*2} Y_{l,m}(\theta,\varphi) = l(l+1) Y_{l,m}(\theta,\varphi) \tag{D.1}$$

In spherical coordinates, ∇^{*2} is defined as:

$$\nabla^{*2} = -\frac{1}{(sin(\theta))^2}\left(\frac{\partial}{\partial\theta}\left(sin(\theta)\frac{\partial}{\partial\theta}\right)+\frac{\partial^2}{\partial\varphi^2}\right)$$

Hence, the spherical harmonics are represented in terms of the spherical coordinates θ and φ (A4.1.2). The angle θ is defined with respect to polar axis. The angle φ describes a rotation from the meridian plane.

$$Y_{l,m}(\theta,\varphi) = (-1)^m \sqrt{\left(\frac{2l+1}{4\pi}\right)\frac{(l-m)!}{(l+m)!}} P_l^m(cos(\theta)) e^{-im\varphi} \tag{D.2}$$

P_l^m are the associated Legendre polynomials.

FUNDAMENTAL PROPERTIES

$$Y_{l,-m}(\theta,\varphi) = (-1)^m Y_{l,m}^*(\theta,\varphi) \tag{D.3}$$

$Y_{l,m}^*$ is the complex conjugation of $Y_{l,m}$.

$$P_l(cos(\alpha)) = \frac{4\pi}{2l+1}\sum_{m=-l}^{l} Y_{l,m}(\theta,\varphi) Y_{l,m}^*(\theta',\varphi') \tag{D.4}$$

With $cos(\alpha) = cos(\theta)cos(\theta') - sin(\theta)sin(\theta')cos(\varphi - \varphi')$. P_l represents the Legendre polynomials of degree l. This property is known as the "addition theorem".

ORTHONORMALITY

$$\int_S Y_{l,m}(\theta,\varphi)Y_{l',m'}(\theta,\varphi)dS = \delta_{l,l'}\delta_{m,m'} \qquad dS = sin(\theta)d\theta d\varphi \tag{D.5}$$

RECURRENT RELATIONSHIPS

$$cos(\theta)Y_{l,m}(\theta,\varphi) = \sqrt{\frac{(l+m)(l-m)}{(2l+1)(2l-1)}}Y_{l-1,m}(\theta,\varphi) + \sqrt{\frac{(l+m+1)(l-m+1)}{(2l+1)(2l+3)}}Y_{l+1,m}(\theta,\varphi) \tag{D.6}$$

$$\begin{aligned} sin(\theta)e^{i\varphi}Y_{l,m}(\theta,\varphi) &= \sqrt{\frac{(l-m)(l-m-1)}{(2l+1)(2l-1)}}Y_{l-1,m+1}(\theta,\varphi) \\ &- \sqrt{\frac{(l+m+1)(l+m+2)}{(2l+1)(2l+3)}}Y_{l+1,m+1}(\theta,\varphi) \end{aligned} \tag{D.7}$$

$$\begin{aligned} sin(\theta)e^{-i\varphi}Y_{l,m}(\theta,\varphi) &= -\sqrt{\frac{(l+m)(l+m-1)}{(2l+1)(2l-1)}}Y_{l-1,m-1}(\theta,\varphi) \\ &+ \sqrt{\frac{(l-m+1)(l-m+2)}{(2l+1)(2l+3)}}Y_{l+1,m-1}(\theta,\varphi) \end{aligned} \tag{D.8}$$

Index

Note: Figures are indicated by *italics*. Tables are indicated by **bold**.

9780367901493